Physics Laboratory Manual

Third Edition

David H. Loyd
Angelo State University

Australia • Brazil • Japan • Korea • Mexico • Singapore • Spain • United Kingdom • United States

Physics Laboratory Manual, Third Edition
David H. Loyd

Publisher: David Harris
Acquisitions Editor: Chris Hall
Development Editor: Rebecca Heider
Editorial Assistant: Shawn Vasquez
Marketing Manager: Mark Santee
Project Manager, Editorial Production: Belinda Krohmer
Creative Director: Rob Hugel
Art Director: John Walker
Print Buyer: Rebecca Cross
Permissions Editor: Roberta Broyer
Production Service: ICC Macmillan Inc.
Copy Editor: Ivan Weiss
Cover Designer: Dare Porter
Cover Image: © Visuals Unlimited/Corbis
Compositor: ICC Macmillan Inc.

Library of Congress Control Number: 2007925773

ISBN-13: 978-0-495-11452-9

ISBN-10: 0-495-11452-9

Brooks/Cole
10 Davis Drive
Belmont, CA 94002-3098
USA

Cengage Learning is a leading provider of customized learning solutions with office locations around the globe, including Singapore, the United Kingdom, Australia, Mexico, Brazil, and Japan. Locate your local office at:
international.cengage.com/region

Cengage Learning products are represented in Canada by Nelson Education, Ltd.

For your course and learning solutions, visit **academic.cengage.com**

Purchase any of our products at your local college store or at our preferred online store **www.ichapters.com**

Printed in Canada
2 3 4 5 6 7 11 10 09 08

Contents

For each laboratory listed below the symbol ◆ preceding the laboratory means that lab requires a calculation of the mean and standard deviation of some repeated measurement. The symbol ◙ preceding the laboratory means that the laboratory requires a linear least squares fit to two variables that are presumed to be linear. The symbol www preceding the laboratory indicates a computer-assisted laboratory available to purchasers of this manual at academic.cengage.com/physics/loyd

Preface xi

Acknowledgements xiii

General Laboratory Information 1

Purpose of laboratory, measurement process, significant figures, accuracy and precision, systematic and random errors, mean and standard error, propagation of errors, linear least squares fits, percentage error and percentage difference, graphing

◆ LABORATORY 1

Measurement of Length 13

Measurement of the dimensions of a laboratory table to illustrate experimental uncertainty, mean and standard error, propagation of errors

◆ LABORATORY 2

Measurement of Density 23

Measurement of the density of several metal cylinders, use of vernier calipers, propagation of errors

LABORATORY 3

Force Table and Vector Addition of Forces 33

Experimental determination of forces using a force table, graphical and analytical theoretical solutions to the addition of forces

◆ ◙ LABORATORY 4

Uniformly Accelerated Motion 43

Analysis of displacement versus $(\text{time})^2$ to determine acceleration, experimental value for acceleration due to gravity g

www LABORATORY 4A

Uniformly Accelerated Motion Using a Photogate

Measurement of velocity versus time using a photogate to determine acceleration for a cart on an inclined plane

LABORATORY 5

Uniformly Accelerated Motion on the Air Table 53

Analysis to determine the average velocity, instantaneous velocity, acceleration of a puck on an air table, determination of acceleration due to gravity g

LABORATORY 6

Kinematics in Two Dimensions on the Air Table 63

Analysis of x and y motion to determine acceleration in y direction, with motion in the x direction essentially at constant velocity

LABORATORY 7

Coefficient of Friction 73

Determination of static and kinetic coefficients of friction, independence of the normal force, verification that $\mu_s > \mu_k$

www LABORATORY 7A

Coefficient of Friction Using a Force Sensor and a Motion Sensor

Measurement of coefficients of static and kinetic friction using a force sensor and a motion sensor

LABORATORY 8

Newton's Second Law on the Air Table 85

Demonstration that $F = ma$ for a puck on an air table and determination of the frictional force on the puck from linear analysis

LABORATORY 9

Newton's Second Law on the Atwood Machine 95

Demonstration that $F = ma$ for the masses on the Atwood machine and determination of the frictional force on the pulley from linear analysis

LABORATORY 10

Torques and Rotational Equilibrium of a Rigid Body 105

Determination of center of gravity, investigation of conditions for complete equilibrium, determination of an unknown mass by torques

LABORATORY 11

Conservation of Energy on the Air Table 117

Spring constant, spring potential energy, kinetic energy, conservation of total mechanical energy (kinetic + spring potential)

LABORATORY 12

Conservation of Spring and Gravitational Potential Energy 127

Determination of spring potential energy, determination of gravitational potential energy, conservation of spring and gravitational potential energy

www LABORATORY 12A

Energy Variations of a Mass on a Spring Using a Motion Sensor

Determination of the kinetic, spring potential, and gravitational potential energies of a mass oscillating on a spring using a motion sensor

◆ LABORATORY 13
The Ballistic Pendulum and Projectile Motion 137
Conservation of momentum in a collision, conservation of energy after the collision, projectile initial velocity by free fall measurements

LABORATORY 14
Conservation of Momentum on the Air Track 149
One-dimensional conservation of momentum in collisions on a linear air track

www LABORATORY 14A
Conservation of Momentum Using Motion Sensors
Investigation of change in momentum of two carts colliding on a linear track

LABORATORY 15
Conservation of Momentum on the Air Table 159
Vector conservation of momentum in two-dimensional collisions on an air table

LABORATORY 16
Centripetal Acceleration of an Object in Circular Motion 169
Relationship between the period T, mass M, speed v, and radius R of an object in circular motion at constant speed

◆ ◉ LABORATORY 17
Moment of Inertia and Rotational Motion 179
Determination of the moment of inertia of a wheel from linear relationship between the applied torque and the resulting angular acceleration

LABORATORY 18
Archimedes' Principle 189
Determination of the specific gravity for objects that sink and float in water, determination of the specific gravity of a liquid

◆ ◉ LABORATORY 19
The Pendulum—Approximate Simple Harmonic Motion 197
Dependence of the period T upon the mass M, length L, and angle θ of the pendulum, determination of the acceleration due to gravity g

◉ LABORATORY 20
Simple Harmonic Motion—Mass on a Spring 207
Determination of the spring constant k directly, indirect determination of k by the analysis of the dependence of the period T on the mass M, demonstration that the period is independent of the amplitude A

www LABORATORY 20A
Simple Harmonic Motion—Mass on a Spring Using a Motion Sensor
Observe position, velocity, and acceleration of mass on a spring and determine the dependence of the period of motion on mass and amplitude

● LABORATORY 21

Standing Waves on a String 217

Demonstration of the relationship between the string tension T, the wavelength λ, frequency f, and mass per unit length of the string ρ

LABORATORY 22

Speed of Sound—Resonance Tube 225

Speed of sound using a tuning fork for resonances in a tube closed at one end

LABORATORY 23

Specific Heat of Metals 235

Determination of the specific heat of several metals by calorimetry

LABORATORY 24

Linear Thermal Expansion 243

Determination of the linear coefficient of thermal expansion for several metals by direct measurement of their expansion when heated

● LABORATORY 25

The Ideal Gas Law 251

Demonstration of Boyle's law and Charles' law using a homemade apparatus constructed from a plastic syringe

● LABORATORY 26

Equipotentials and Electric Fields 259

Mapping of equipotentials around charged conducting electrodes painted on resistive paper, construction of electric field lines from the equipotentials, dependence of the electric field on distance from a line of charge

◆ ● LABORATORY 27

Capacitance Measurement with a Ballistic Galvanometer 269

Ballistic galvanometer calibrated by known capacitors charged to known voltage, unknown capacitors measured, series and parallel combinations of capacitance

◆ ● LABORATORY 28

Measurement of Electrical Resistance and Ohm's Law 279

Relationship between voltage V, current I, and resistance R, dependence of resistance on length and area, series and parallel combinations of resistance

◆ LABORATORY 29

Wheatstone Bridge 289

Demonstration of bridge principles, determination of unknown resistors, introduction to the resistor color code

LABORATORY 30

Bridge Measurement of Capacitance 299

Alternating current bridge used to determine unknown capacitance in terms of a known capacitor, series and parallel combinations of capacitors

LABORATORY 31

Voltmeters and Ammeters 307

Galvanometer characteristics, voltmeter and ammeter from galvanometer, and comparison with standard voltmeter and ammeter

LABORATORY 32

Potentiometer and Voltmeter Measurements of the emf of a Dry Cell 319

Principles of the potentiometer, comparison with voltmeter measurements, internal resistance of a dry cell

LABORATORY 33

The RC Time Constant 329

RC time constant using a voltmeter as the circuit resistance *R*, determination of an unknown capacitance, determination of unknown resistance

www LABORATORY 33A

RC Time Constant with Positive Square Wave and Voltage Sensors

Determine the time constant, and time dependence of the voltages across the capacitor and resistor in an *RC* circuit using voltage sensors

LABORATORY 34

Kirchhoff's Rules 339

Illustration of Kirchhoff's rules applied to a circuit with three unknown currents and to a circuit with four unknown currents

LABORATORY 35

Magnetic Induction of a Current Carrying Long Straight Wire 349

Induced emf in a coil as a measure of the *B* field from an alternating current in a long straight wire, investigation of *B* field dependence on distance *r* from wire

www LABORATORY 35A

Magnetic Induction of a Solenoid

Determination of the magnitude of the axial *B* field as a function of position along the axis using a magnetic field sensor

LABORATORY 36

Alternating Current LR *Circuits* 359

Determination of the phase angle ϕ, inductance *L*, and resistance *r* of an inductor

www LABORATORY 36A

Direct Current LR *Circuits*

Determination of the phase relationship between the circuit elements and the time constant for an *LR* circuit

LABORATORY 37

Alternating Current RC *and* LCR *Circuits* 369

Phase angle in an *RC* circuit, determination of unknown capacitor, phase angle relationships in an *LCR* circuit

LABORATORY 38

Oscilloscope Measurements 379

Introduction to the operation and theory of an oscilloscope

LABORATORY 39

Joule Heating of a Resistor 391

Heat (calories) produced from electrical energy dissipated in a resistor (joules), comparison with the expected ration of 4.186 joules/calorie

LABORATORY 40

Reflection and Refraction with the Ray Box 401

Law of reflection, Snell's law of refraction, focal properties of each

LABORATORY 41

Focal Length of Lenses 413

Direct measurement of focal length of converging lenses, focal length of a converging lens with converging lens in close contact

LABORATORY 42

Diffraction Grating Measurement of the Wavelength of Light 421

Grating spacing from known wavelength, wavelengths from unknown heated gas, wavelength of colors from continuous spectrum

www LABORATORY 42A

Single-Slit Diffraction and Double-Slit Interference of Light

Light sensor and motion sensor measurement of the intensity distribution of laser light for both a single slit and a double slit

LABORATORY 43

Bohr Theory of Hydrogen—The Rydberg Constant 431

Comparison of the measured wavelengths of the hydrogen spectrum with Bohr theory to determine the Rydberg constant

www LABORATORY 43A

Light Intensity versus Distance with a Light Sensor

Investigate the dependence of light intensity versus distance from a light source using a light sensor

LABORATORY 44

Simulated Radioactive Decay Using Dice "Nuclei" 441

Measurement of decay constant and half-life for simulated radioactive decay using 20-sided dice as "nuclei"

LABORATORY 45

Geiger Counter Measurement of the Half-Life of ^{137}Ba 451

Geiger counter plateau, half-life from activity versus time measurements

LABORATORY 46

Nuclear Counting Statistics 463

Distribution of series of counts around the mean, demonstration that $\sqrt{N}$ is a measure of the uncertainty in the count N

LABORATORY 47

Absorption of Beta and Gamma Rays 473

Comparison of absorption of beta and gamma radiation by different materials, determination of the absorption coefficient for gamma rays

Appendix I 483

Appendix II 485

Appendix III 487

Preface

This laboratory manual is intended for use with a two-semester introductory physics course, either calculus-based or noncalculus-based. For the most part, the manual includes the standard laboratories that have been used by many physics departments for years. However, in this edition there are available some laboratories that use the newer computer-assisted data-taking equipment that has recently become popular. The major change in the current addition is an attempt to be more concise in the Theory section of each laboratory to include only what is required to prepare a student to take the needed measurements. As before, the Instructor's Manual gives examples of the best possible experimental results that are possible for the data for each laboratory. Complete solutions to all portions of each laboratory are included. All of the laboratories are written in the same format that is described below in the order in which the sections occur.

OBJECTIVES

Each laboratory has a brief description of what subject is to be investigated. The current list of objectives has been condensed compared to the previous edition.

EQUIPMENT

Each laboratory contains a brief list of the equipment needed to perform the laboratory.

THEORY

This section is intended to be a description of the theory underlying the laboratory to be performed, particularly describing the variables to be measured and the quantities to be determined from the measurements. In many cases, the theory has been shortened significantly compared to previous editions.

EXPERIMENTAL PROCEDURE

The procedure given is usually very detailed. It attempts to give very explicit instructions on how to perform the measurements. The data tables provided include the units in which the measurements are to be recorded. With few exceptions, SI units are used.

CALCULATIONS

Very detailed descriptions of the calculations to be performed are given. When practical, actual data are recorded in a data table, and calculated quantities are recorded in a calculations table. This is the preferred option because it emphasizes the distinction between measured quantities and quantities calculated from the measured quantities. In some cases it is more practical to combine the two into a data and calculations table. That has been done for some of the laboratories.

Whenever it is feasible, repeated measurements are performed, and the student is asked to determine the mean and standard error of the measured quantities. For data that are expected to show a linear relationship between two variables, a linear least squares fit to the data is required. Students are encouraged to do these statistical calculations with a spreadsheet program such as Excel. It is also acceptable to do them on a handheld calculator capable of performing them automatically. Use of the statistical calculations is included in 35 of the 47 laboratories.

GRAPHS

Any graphs required are specifically described. All linear data are graphed and the least squares fit to the data is shown on the graph along with the data.

PRE-LABORATORY

Each laboratory includes a pre-laboratory assignment that is based upon the laboratory description. We intend to prepare students to perform the laboratory by having them answer a series of questions about the theory and working numerical problems related to the calculations in the laboratory. The questions in the pre-laboratory have been changed somewhat to include more conceptual questions about the theory behind the laboratory. However, there remains an emphasis on preparing students for the quantitative processes needed to perform the laboratory.

LABORATORY REPORT

The laboratory includes the data and calculations tables, a sample calculations section, and a list of questions. Usually the questions are related to the actual data taken by the student. They attempt to require the student to think critically about the significance of the data with respect to how well the data can be said to verify the theoretical concepts that underlie the laboratory.

COMPUTER-ASSISTED LABORATORIES

The Table of Contents lists 10 laboratories, prefaced by a symbol www that use computer-assisted data collection and analysis. *DataStudio* software and compatible sensors are to be used for these laboratories. The laboratories are available to purchasers of this manual at academic.cengage.com/physics/loyd. Options for including these computer-assisted laboratories in a customized version of the lab manual are available through Cengage Learning's digital library, Textchoice. Visit www.textchoice.com or contact your local Cengage Learning representative.

CONTACT INFORMATION FOR AUTHOR

Please contact me at david.loyd@angelo.edu if you find any errors or have any suggestions for improvements in the laboratory manual. I will keep an updated list of errors and suggestions at the Cengage Learning website.

Acknowledgements

I wish to acknowledge the mutual exchange of ideas about laboratory instruction that occurred among H. Ray Dawson, C. Varren Parker and myself for over 30 years at Angelo State University. I also thank the following users of previous editions of the manual for helpful comments: (1) Charles Allen, Angelo State University (2) William L. Basham, University of Texas at Permian Basin (3) Gerry Clarkson, Howard Payne University (4) Carlos Delgado, College of Southern Nevada (5) Poovan Murgeson, San Diego City.

I am grateful to all the highly professional and talented people of Brooks/Cole for their excellent work to improve this third edition of the laboratory manual. I especially want to acknowledge the help and encouragement of Rebecca Heider and Chris Hall in this rather lengthy process. Their comments and suggestions about the changes and additions that were needed were very beneficial.

I wish to thank the Literary Executor of the late Sir Ronald A. Fisher, F.R.S., to Dr. Frank Yates, F.R.S., and to Longman Group Ltd., London, for permission to reprint the table in Appendix I from their book *Statistical Tables for Biological, Agricultural and Medical Research*. (6th edition, 1974)

I thank Melissa Vigil, Marquette University and Marllin Simon, Auburn University for conversations we have had about laboratory instruction. I am particularly indebted to Marllin Simon for his permission to use the procedures and other aspects from several of his laboratories that use computer assisted data acquisition techniques.

My final and most important acknowledgement is to my wife of 47 years, Judy. Her encouragement and help with proof-reading have been especially important during this project. Her good humor and practical advice are always appreciated.

David H. Loyd

General Laboratory Information

PURPOSE OF LABORATORY

The laboratory provides a unique opportunity to validate physical theories in a quantitative manner. Laboratory experience demonstrates the limitations in the application of physical theories to real physical situations. It teaches the role that experimental uncertainty plays in physical measurements and introduces ways to minimize experimental uncertainty. In general, the purpose of these laboratory exercises is both to demonstrate some physical principle and to teach techniques of careful measurement.

DATA-TAKING PROCEDURES

Original data should always be recorded directly in the data tables provided. Avoid the habit of recording the original data on scratch sheets and transferring them to the data tables later.

When working in a group, all partners should contribute to the actual process of taking the measurements. If time and other considerations permit, each partner should perform a separate set of measurements as a check on the procedure. Each partner should record data separately even if only one set of data is taken by the group.

SIGNIFICANT FIGURES

The number of significant figures means the number of digits known in some number. The number of significant figures does not necessarily equal the total digits in the number because zeros are used as place keepers when digits are not known. For example, in the number 123 there are three significant figures. In the number 1230, although there are four digits in the number, there are only three significant figures because the zero is assumed to be merely keeping a place. Similarly, the numbers 0.123 and 0.0123 both have only three significant figures. The rules for determining the number of significant figures in a number are:

- The most significant digit is the leftmost nonzero digit. In other words, zeros at the left are never significant.
- In numbers that contain no decimal point, the rightmost nonzero digit is the least significant digit.
- In numbers that contain a decimal point, the rightmost digit is the least significant digit, regardless of whether it is zero or nonzero.
- The number of significant digits is found by counting the places from the most significant to the least significant digit.

As an example, the numbers in the following list of numbers all have four significant figures. An explanation for each is given.

- 3456: All four nonzero digits are significant.
- 135700: The two rightmost zeros are not significant because there is no decimal point.
- 0.003043: Zeros at the left are never significant.
- 0.01000: The zero at the left is not significant, but the three zeros at the right are significant because there is a decimal point.
- 1030.: There is a decimal point, so all four numbers are significant.
- 1.057: Again, there is a decimal point, so all four are significant.
- 0.0002307: Zeros at the left are never significant.

READING MEASUREMENT SCALES

For the measurement of any physical quantity such as mass, length, time, temperature, voltage, or current, some appropriate measuring device must be chosen. Despite the diverse nature of the devices used to measure the various quantities, they all have in common a measurement scale, and that scale has a smallest marked scale division. All measurements should be done in the following very specific manner. All meters and measuring devices should be read by interpolating between the smallest marked scale division. Generally the most sensible interpolation is to attempt to estimate 10 divisions between the smallest marked scale division. Consider the section of a meter stick pictured in Figure 1 that shows the region between 2 cm and 5 cm. The smallest marked scale divisions are 1 mm apart. The location of the arrow in the figure is to be determined. It is clearly between 3.4 cm and 3.5 cm, and the correct procedure is to estimate the final place. In this case a reading of 3.45 cm is estimated. For this measurement the first two digits are certain, but the last digit is estimated. This measurement is said to contain three significant figures. Much of the data taken in this laboratory will have three significant figures, but occasionally data may contain four or even five significant figures.

MISTAKES OR PERSONAL ERRORS

All measurements are subject to errors. There are three types of errors, which are classified as personal, systematic, or random. Random errors are sometimes called statistical errors. This section deals with personal errors. Systematic and random errors will be discussed later. In fact, personal errors are not really errors in the same sense as the other two types of errors. Instead, they are merely mistakes made by the experimenter. Mistakes are fundamentally different from the other two types of errors because mistakes can be completely eliminated if the experimenter is careful. Mistakes can be made either while taking the data or later in calculations done with the original data. Either type of mistake is bad, but a mistake made in the data-taking process is probably worse because often it is not discovered until it is too late to correct it.

The correct attitude toward all data-taking processes is one of skepticism about all the procedures that are carried out in the laboratory. Essentially, this amounts to assuming that things will go wrong unless

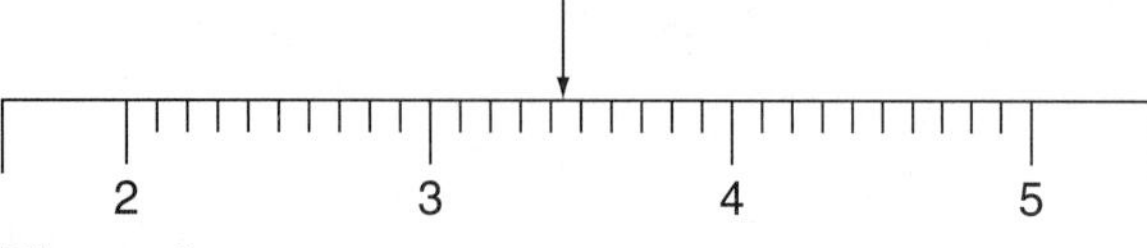

Figure 1

constant attention is given to making sure that no mistakes are made. For every measurement taken, all aspects of the process must be checked and rechecked. Everyone in the group must be convinced that they know exactly what is supposed to be measured, what the correct procedure is to measure it, and that the group is making no mistakes in carrying out that procedure.

ACCURACY AND PRECISION

The central point to experimental physical science is the measurement of physical quantities. It is assumed that there exists a true value for any physical quantity, and the measurement process is an attempt to discover that true value. It is expected that there will be some difference between the true value and the measured value. The terms accuracy and precision are used to describe different aspects of the difference between the measured value and the true value of some quantity.

The accuracy of a measurement is determined by how close the result of the measurement is to the true value. For example, in several of the experiments, we will determine a value for the acceleration due to gravity. For this case, the accuracy of the result is decided by how close it is to the true value of $9.80\ m/s^2$. For many laboratory experiments, the true value of the measured quantity is not known, and we cannot determine the accuracy of the experiment from the available data.

The precision of a measurement refers essentially to how many digits in the result are significant. It indicates also how reproducible the results are when measurements of some quantity are repeated. The smaller the variations of the individual repeated measurements of a quantity, the more precise the quoted value of the measurement is considered to be. We will elaborate upon and quantify this idea about the relationship between the size of the variations in the measurements and the precision of the measurement in a later section on statistical methods.

SYSTEMATIC ERRORS

Systematic errors are errors that tend to be in the same direction for repeated measurements, giving results that are either consistently above the true value or consistently below the true value. In many cases such errors are caused by some flaw in the experimental apparatus. For example, a voltmeter could be incorrectly calibrated in such a way that it consistently gives a reading that is 80% of the true voltage across its input terminals. It is also possible to have a voltmeter with a zero offset on its scale, which is assumed for this discussion to be 0.50 volts. In the first case, the error is a constant fraction of the true value (in this case, 20%), and in the second case, the error is a constant absolute voltage. Either of these is a systematic error, and the answer to the question of which one is worse depends upon the magnitude of the voltage to be measured. If the voltage to be measured is 1.00 volts, then the meter with absolute error of 0.50 volts causes an error of 50%, whereas the meter with relative error causes an error of 20%. On the other hand, if the voltage to be measured is 100 volts, the meter with absolute error of 0.50 volts causes only a 0.5% error, and the other meter still causes a 20% error, or in this case, 20 volts. If this measured voltage is used to calculate some other quantity, it too will show a systematic error in the results.

A second common type of systematic error is failure to consider all of the variables that are important in the experiment. In some cases one may be aware that some other factors need to be considered, but might not have the ability to do so quantitatively. For example, when using an air table to validate Newton's Laws, it is common to ignore friction. This is done because friction is assumed to be small, but also because often there is no easy way to determine its contribution. It is expected, therefore, that neglecting friction might introduce a systematic error.

For purposes of this laboratory, the concern with systematic errors will usually be twofold—to attempt to eliminate any obvious systematic errors to the extent possible, and to attempt to identify any data that show systematic error, and suggest possible reasonable causes for such error.

RANDOM ERRORS

The final class of errors is those that are produced by unpredictable and unknown variations in the total experimental process even when one does the experiment as carefully as is humanly possible. The variations caused by an observer's inability to estimate the last digit the same way every time will definitely be one contribution. Other variations can be caused by fluctuations in line voltage, temperature changes, mechanical vibrations, or any of the many physical variations that may be inherent in the equipment or any other aspect of the measurement process. It is important to realize the following difference between random errors and personal and systematic errors. In principle all personal and systematic errors can be eliminated, but there will always remain some random errors in any measurement. Even in principle the random errors can never be completely eliminated.

Random errors, on the other hand, can be determined in a prescribed way. It has been found empirically that random errors often are distributed according to a particular statistical distribution function called the Gauss distribution function, which is also referred to as the normal error function. Random measurement errors are said to be normally distributed when a histogram of the frequency distribution of the results of a large number of repeated measurements produces a bell-shaped curve with a peak at the mean of the measurements. The histogram of the frequency distribution is simply a graph of the number of times the measurements fall within a certain range versus the measured values.

MEAN AND STANDARD DEVIATION

Assume a series of repeated measurements is made in which there are no systematic or personal errors, and thus only random errors are present. Assume that there are n measurements made of some quantity x, and the ith value obtained is x_i where i varies from 1 to n. If it is true that the errors are normally distributed, statistical theory says that the mean is the best approximation to the true value. In formal mathematical terms, the mean (which has a symbol of $\bar{x}$) is given by the equation

$$\bar{x} = \left[\frac{1}{n}\right]\sum_{1}^{n} x_i \qquad \text{(Eq. 1)}$$

For example, assume that four measurements are made of some quantity x, and that the four results are 18.6, 19.3, 17.7, and 20.4. Equation 1 is simply shorthand notation for the averaging process given by

$$\bar{x} = (1/4)\ (18.6 + 19.3 + 17.7 + 20.4) = 19.0 \qquad \text{(Eq. 2)}$$

It is not surprising that the mean is the best approximation to the true value. It seems intuitively reasonable. We can prove mathematically that the mean is indeed the proper choice by something called the principle of least squares, which we state in the following way. The most probable value for some quantity determined from a series of measurements is that value that minimizes the sum of the squares of the deviations between the chosen value and the measured values. We can demonstrate that the proper choice to produce this minimum sum of deviations is simply the mean of the measurements. This idea can be usefully generalized later for the case of two variables.

Statistical theory, furthermore, states that the precision of the measurement can be determined by calculating a quantity called the standard deviation from the mean of the measurements. The symbol for standard deviation from the mean is σ_{n-1}, and it is defined by the equation

$$\sigma_{n-1} = \sqrt{\frac{1}{n-1}\sum_{1}^{n}[x_i - \bar{x}]^2} \qquad \text{(Eq. 3)}$$

For the data given, the standard deviation is calculated from Equation 3 to be the following:

$$\sigma_{n-1} = \sqrt{\frac{1}{4-1}((18.6-19.0)^2 + (19.3-19.0)^2 + (17.7-19.0)^2 + (20.4-19.0)^2)} = 1.1$$

The quantity σ_{n-1}, which is actually called the sample standard deviation, is a measure of the precision of the measurement in the following statistical sense. It gives the probability that the measurements fall within a certain range of the measured mean. From the sample standard deviation and tables of the standard error function, it is possible to determine the probability that the measurements fall within any desired range about the mean. The common range to be quoted is the range of one standard deviation as calculated by Equation 3.

Probability theory states that approximately 68.3% of all repeated measurements should fall within a range of plus or minus σ_{n-1} from the mean. Furthermore, 95.5% of all repeated measurements should fall within a range of $2\sigma_{n-1}$ from the mean. For the example given above, 68.3% should fall in the range 19.0 ± 1.1 (from 17.9 to 20.1), and 95.5% should fall in the range 19.0 ± 2.2 (from 16.8 to 21.2).

As a final note on the expected distribution for measurements that follow a normal error curve, 99.73% of all measurements should fall within $3\sigma_{n-1}$ of the mean. This implies that if one of the measurements is $3\sigma_{n-1}$ or farther from the mean, it is very unlikely that it is a random error. It is much more likely to be the result of a personal error.

A second issue that can be addressed by these repeated measurements is the precision of the mean. After all, this is what is really of concern, because the mean is the best estimate of the true value. The precision of the mean is indicated by a quantity called the standard error. The standard error, which has a symbol of α, is defined by

$$\alpha = \frac{\sigma_{n-1}}{\sqrt{n}} \qquad \text{(Eq. 4)}$$

For the example given above with $\sigma_{n-1} = 1.1$ and $n = 4$, the value is $\alpha = 0.55$. The significance of α is that if several groups of n measurements are made, each producing a value for the mean, 68.3% of the means should fall in the range 19.0 ± 0.6. In other words, there is a 68.3% probability that the true value lies in this range. Of course, all these statements are valid only if there are no other errors present other than random errors.

In this laboratory, students will often be asked to make repeated measurements of some quantity and to determine the mean. Assuming that α represents the uncertainty in the value of the mean, a crucial question is the appropriate number of significant figures to retain in α. In this laboratory, the convention to be followed is to retain *one significant figure* in α and to make the least significant figure in the mean be in the same *decimal place* as α. In this context the appropriate procedure is to originally calculate the mean and σ_{n-1} to more significant figures than it is assumed are needed, and then allow the value of α to determine the significant figures to be retained in the mean. In the example given above, the result should be stated as 19.0 ± 0.6. Notice that as described above, only one significant figure has been retained in α, and the mean has its least significant digit in the same decimal place as α.

To illustrate how the concepts of the mean and standard error apply to accuracy and precision, consider the following sets of three measurements of the acceleration due to gravity made by four students named Alf, Beth, Carl, and Dee. The results for each measurement, the means, the sample standard deviations σ_{n-1}, and the standard errors α are given for each student.

The accuracy of each student's data is determined by comparing the mean with the true value of 9.80. Dee's value of 9.76 is the most accurate, Alf's value of 9.43 is second, Beth's value of 9.26 is third, and Carl's value of 8.74 is the least accurate. Using the values of the standard errors of the mean as a criterion for precision, Carl's value is the most precise, Dee's is second, Beth's is third, and Alf's value is the least precise.

In fact, the situation is not quite so simple as has been presented. There is an interplay between the concepts of accuracy and precision that we must consider. If a measurement appears to be very accurate,

Table 1

	Alf	*Beth*	*Carl*	*Dee*
Measurement 1	7.83	9.53	8.70	9.72
Measurement 2	11.61	9.38	8.75	9.86
Measurement 3	8.85	8.87	8.77	9.70
Mean	9.43	9.26	8.74	9.76
Standard Deviation	1.96	0.35	0.036	0.087
Standard Error	1	0.2	0.02	0.05

but the precision is poor, we do not know if the results are meaningful. Consider Alf's mean of 9.43, which differs from the true value of 9.80 by only 0.37, and thus appears to be quite accurate. However, all of his measurements have large deviations from the true value, and his standard error is very large. It seems much more likely, then, that Alf's mean of 9.43 is due to luck rather than to a careful measurement. In contrast, it seems likely that Dee's mean of 9.76 is meaningful because the value of her standard error is small.

Carl's results are an example of a situation that is common in the interplay between accuracy and precision. Carl's precision is extremely high, yet his accuracy is not very good. When a measurement has high precision but poor accuracy, it is often the sign of a systematic error, and in this case it seems very likely that Carl has some systematic error in his measurements.

PROPAGATION OF ERRORS

Consider the following set of data that was taken by measuring the coordinate position d of some object as a function of time t.

Table 2

d (m)	t (s)
7.57	1.00
11.97	2.00
16.58	3.00
21.00	4.00
25.49	5.00

From these data the average speed over each time interval can be calculated. The average speed $\overline{v}$ over some time interval Δt during which a distance interval Δd was traveled is given by

$$\overline{v} = \frac{\Delta d}{\Delta t} \qquad \text{(Eq. 5)}$$

For the data of Table 2 there are four intervals for the five data points, and for the first two intervals the results are:

$$\overline{v}_1 = \frac{11.97 - 7.57}{2.00 - 1.00} = 4.40 \text{ m/s} \quad \overline{v}_2 = \frac{16.58 - 11.97}{3.00 - 2.00} = 4.61 \text{ m/s}$$

The other two intervals give average speeds of 4.42 m/s and 4.49 m/s. A basic question is on what basis was the decision made to express $\overline{v}_1$, for example, as 4.40 rather than 4.4 or 4.400? We derive the answer by further extending the rules for significant figures to include calculations. Use the following rules to determine the number of significant figures to retain at the end of a calculation:

- When adding or subtracting, figures to the right of the last column in which all figures are significant should be dropped.
- When multiplying or dividing, retain only as many significant figures in the result as are contained in the least precise quantity in the calculation.
- The last significant figure is increased by 1 if the figure beyond it (which is dropped) is 5 or greater.

These rules apply only to the determination of the number of significant figures in the final result. In the intermediate steps of a calculation, one more significant figure should be kept than is kept in the final result.

Consider these examples of addition, multiplication, and division of numbers:

$$\begin{array}{r} 753.1 \\ 37.08 \\ 0.697 \\ \underline{56.3} \\ 847.177 \end{array} \qquad \begin{array}{r} 753.1 \\ 37.1 \\ 0.7 \\ \underline{56.3} \\ 847.2 \end{array} \qquad \begin{array}{r} 327.23 \\ \underline{\times \quad 36.73} \\ 12019.158 \end{array} \qquad \begin{array}{r} 8.90906 \\ 36.73 \overline{)327.23} \end{array}$$

Following the above rules for addition strictly implies rewriting each number as shown in the second addition where the first digit beyond the decimal is the least significant digit. This is true because that column is the rightmost column in which all digits are significant. Note that one gets the same result if the numbers are added on the calculator (as done at the left), and then it is noted that the first digit beyond the decimal is the last one that can be kept. Therefore 847.177 is rounded off to 847.2. A similar process is used for multiplication and division, as shown in the third and fourth part above. In each case, the result is rounded to four significant figures because the least significant number in each calculation (36.73) has only four significant figures. For the multiplication the result is 12020, and for the division it is 8.909.

LINEAR LEAST SQUARES FITS

Often measurements are taken by changing one variable (call it x) and measuring how a second variable (call it y) changes as a function of the first variable. In many cases of interest it is assumed that there exists a linear relationship between the two variables. In mathematical terms one can say that the variables obey an equation of the form

$$y = mx + b \qquad \text{(Eq. 6)}$$

where m and b are constants. This also implies that if a graph is made with x as the horizontal axis and y as the vertical axis, it will be a straight line with m equal to the slope (defined as $\Delta y/\Delta x$) and b equal to the y intercept (the value of y at $x = 0$).

The question is how to best verify that the data do indeed obey Equation 6. One way is to make a graph of the data, and then try to draw the best straight line possible through the data points. This will give a qualitative answer to the question, but it is possible to give a quantitative answer to the question by the process described below.

The measurements are repeated measurements in the sense that they are to be considered together in the attempt to determine to what extent the data obey Equation 6. It is possible to generalize the idea of minimizing the sum of squares of the deviations described earlier for the mean and standard deviation to the present case. The result of the generalization to two-variable linear data is called a linear least squares fit to the data. It is also sometimes referred to as a linear regression.

The aim of the process is to determine the values of m and b that produce the best straight-line fit to the data. Any choice of values for m and b will produce a straight line, with values of y determined by the choice of x. For any such straight line (determined by a given m and b) there will be a deviation between each of the measured y's and the y's from the straight-line fit at the value of the measured x's. The least squares fit is that m and b for which the sum of the squares of these deviations is a minimum. Statistical theory states that the appropriate values of m and b that will produce this minimum sum of squares of the deviations are given by the following equations:

$$m = \frac{n\sum_{1}^{n} x_i y_i - \left(\sum_{1}^{n} x_i\right)\left(\sum_{1}^{n} y_i\right)}{n\sum_{1}^{n} x_i^2 - \left(\sum_{1}^{n} x_i\right)^2} \qquad \text{(Eq. 7)}$$

$$b = \frac{\left(\sum_{1}^{n} y_i\right)\left(\sum_{1}^{n} x_i^2\right) - \left(\sum_{1}^{n} x_i y_i\right)\left(\sum_{1}^{n} x_i\right)}{n\sum_{1}^{n} x_i^2 - \left(\sum_{1}^{n} x_i\right)^2} \qquad \text{(Eq. 8)}$$

Refer again to the data of Table 2 for coordinate position versus time. The question to be answered is whether or not the data are consistent with constant velocity. If the speed v is constant, the data can be fit by an equation of the form

$$d = vt + d_o \qquad \text{(Eq. 9)}$$

Equation 9 is of the form of Equation 6 with d corresponding to y, t corresponding to x, v corresponding to m, and d_o corresponding to b. Thus v will be the slope of a graph of d versus t, and d_o will be the intercept, which is the coordinate position at the arbitrarily chosen time $t = 0$.

Calculating some of the individual terms gives:

$$\sum t_i = 1.00 + 2.00 + 3.00 + 4.00 + 5.00 = 15.00$$
$$\sum (t_i)^2 = (1.00)^2 + (2.00)^2 + (3.00)^2 + (4.00)^2 + (5.00)^2 = 55.00$$
$$\sum d_i = 7.57 + 11.97 + 16.58 + 21.00 + 25.49 = 82.61$$
$$\sum t_i d_i = (1.00)(7.57) + (2.00)(11.97) + (3.00)(16.58) + (4.00)(21.00) + (5.00)(25.49) = 292.70$$
$$\sum (d_i)^2 = (7.57)^2 + (11.97)^2 + (16.58)^2 + (21.00)^2 + (25.49)^2 = 1566.22$$

Using these values in Equations 7 and 8 with the appropriate correspondence of variables gives $v = 4.49$ and $d_o = 3.06$. Thus the velocity is determined to be 4.49 m/s, and the coordinate at $t = 0$ is found to be 3.06 m.

At this point, the best possible straight-line fit to the data has been determined by the least squares fit process. A second goal remains, to determine how well the data actually fit the straight line that we have obtained. Again, we derive a qualitative answer to this question by making a graph of the data and the straight line and qualitatively judging the agreement between the line and the data.

There is, however, a quantitative measure of how well the data follow the straight line obtained by the least squares fit. It is given by the value of a quantity called the correlation coefficient, r. This quantity is a measure of the fit of the data to a straight line with $r = 1.000$ exactly signifying a perfect correlation, and $r = 0$ signifying no correlation at all. The equation to calculate r in terms of the general variables x and y is given by

$$r = \frac{n\sum_1^n x_i y_i - \left(\sum_1^n x_i\right)\left(\sum_1^n y_i\right)}{\sqrt{n\sum_1^n x_i^2 - \left(\sum_1^n x_i\right)^2}\sqrt{n\sum_1^n y_i^2 - \left(\sum_1^n y_i\right)^2}} \qquad \text{(Eq. 10)}$$

Making the substitutions for the variables of the problem of the fit to the displacement versus time by substituting t for x and d for y in the above equation and using the appropriate numerical values calculated earlier gives $r = 0.99998$. Thus the data show an almost perfect linear relationship because r is so close to 1.000. In calculations of r keep either three significant figures, or else enough until the last place is not a 9.

When performing a least squares fit to data, particularly when a small number of data points are involved, there is some tendency to obtain a surprisingly good value for r even for data that do not appear to be very linear. For those cases, we can determine the significance of a given value of r by comparing the obtained value of r with the probability that that value of r would be obtained for n values of two variables that are unrelated. A table for such comparisons is given in Appendix I in a table entitled Correlation Coefficients.

STATISTICAL CALCULATIONS

A very high percentage of the laboratories in this course will involve two variables that are linearly related. These cases usually will require a least squares fit to the data. Although the least squares fit calculations and mean and standard deviation calculations are not difficult in principle, they are tedious and time-consuming. The use of a spreadsheet computer program such as Excel is highly recommended. As an alternative, many handheld calculators have automatic routines built in that allow the calculation of these quantities simply by inputting the data points one after another. Note that most calculators will calculate two different standard deviations. The one needed is usually denoted σ_{n-1}, and it is the sample standard deviation. Also available on most calculators is a quantity that is usually denoted as σ_n. It applies to the case when the population is known, and it will never be appropriate for data taken in this laboratory. Always be sure to choose the quantity σ_{n-1}, which is the one defined by Equation 3.

PERCENTAGE ERROR AND PERCENTAGE DIFFERENCE

In several of the laboratory exercises, the true value of the quantity being measured will be considered to be known. In those cases, the accuracy of the experiment will be determined by comparing the experimental result with the known value. Normally this will be done by calculating the percentage error of your measurement compared to the given known value. If E stands for the experimental value, and K stands for the known value, then the percentage error is given by

$$\text{Percentage error} = \frac{|E - K|}{K} \times 100\% \qquad \text{(Eq. 11)}$$

In other cases we will measure a given quantity by two different methods. There will then be two different experimental values, E_1 and E_2, but the true value may not be known. For this case, we will calculate the percentage difference between the two experimental values. Note that this tells nothing about the accuracy of the experiment, but will be a measure of the precision. The percentage difference between the two measurements is defined as

$$\text{Percentage difference} = \frac{|E_2 - E_1|}{[E_1 + E_2]/2} \times 100\% \qquad \text{(Eq. 12)}$$

PREPARING GRAPHS

It is helpful to represent data in the form of a graph when interpreting the overall trend of the data. Most of the graphs for this laboratory will use rectangular Cartesian coordinates. Note that it is customary to denote the horizontal axis as x and the vertical axis as y when developing general equations, as was done in the development of the equations for a linear least squares fit. However, any two variables can be plotted against each other.

When preparing a graph, first choose a scale for each of the axes. It is not necessary to choose the same scale for both axes. In fact, rarely will it be convenient to have the same scale for both axes. Instead, choose the scale for each axis so that the graph will range over as much of the graph paper as possible, consistent with a convenient scale. Choose scales that have the smallest divisions of the graph paper equal to multiples of 2, 5, or 10 units. This makes it much easier to interpolate between the divisions to locate the data points when graphing.

The student is expected to bring to each laboratory a supply of good quality linear graph paper. A very good grade of centimeter by centimeter graph paper with one division per millimeter is the best choice. Do not, for example, ever use 1/4 inch by 1/4 inch sketch paper or other such coarse scaled paper as graph paper. In some cases special graph paper like semilog or log-log graph paper may be required.

Figure 2 is a graph of the data for displacement versus time from Table 2 for which the least squares fit was previously made. Note that scales for each axis have been chosen, to spread the graph over a reasonable portion of the page. Also note that because the data have been assumed linear, a straight line has been drawn through the data points. The straight line is the one obtained from the least squares fit to the data.

For most experiments, the variables will take on only positive values. For that case the scales should range from zero to greater than the largest value for any data point. For example, in Figure 2 the displacement is chosen to range from 0 to 30 meters because the largest displacement is 25.49, and the time scale has been chosen to range from 0 to 6 seconds because the largest time is 5.00 seconds. Also note that the scales should not be suppressed as a means to stretch out the graph. For example, if a set of data contains ordinates that range from 60 to 90, do not choose a scale that shows only that range. Instead a scale from 0 to 100 should be chosen, and there is nothing that can be done in that case to make the graph range over more than about 30% of the graph paper. Scales should always be chosen to increase to the right of the origin and to increase above the origin.

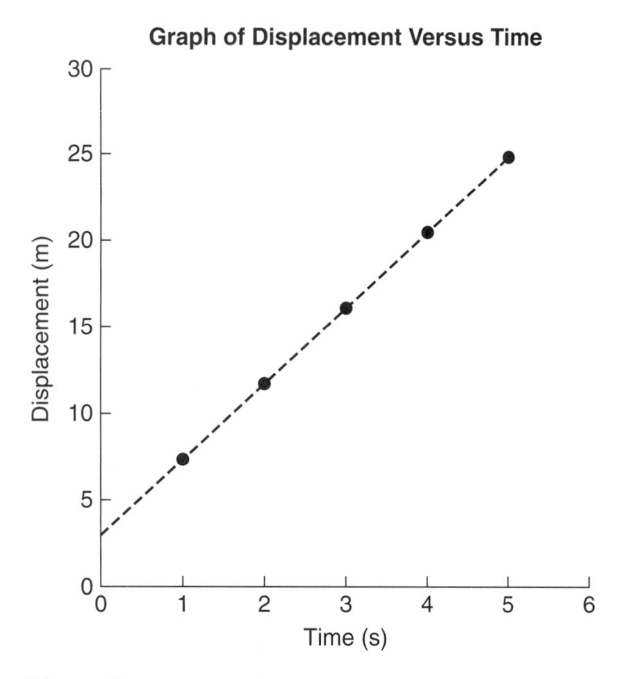

Figure 2

Graphs should always have the scales labeled with the name and units of each variable along each axis. Major scale divisions should be labeled with the appropriate numbers defining the scale. Always include a title for each graph, keeping in mind that it is customary to state the vertical axis versus the horizontal axis.

All graphs should be plotted as points with no attempt to connect the data with a smooth curve. Do not write the coordinates on the graph next to the data point, as is common practice in mathematics classes. The only time it is appropriate to draw any continuous line to represent the trend of the data is when it is assumed that the mathematical form of the data is known. In practice, the only time this will be true will be when linearity is assumed, and in that case, it is appropriate to draw the straight line that has been obtained by the least squares fitting procedure.

Measurement of Length

OBJECTIVES

- ❑ Demonstrate the specific knowledge gained by repeated measurements of the length and width of a table.
- ❑ Apply the statistical concepts of mean, standard deviation from the mean, and standard error to these measurements.
- ❑ Demonstrate propagation of errors by determining the uncertainty in the area calculated from the measured length and width.

EQUIPMENT LIST

- 2-meter stick
- Laboratory table

THEORY

In this laboratory it is assumed that the uncertainty in the measurement of the length and width of the table is due to random errors. If this assumption is valid, then the mean of a series of repeated measurements represents the most probable value for the length or width.

Consider the general case in which n measurements of the length and width of the table are made. We will make 10 measurements, so $n = 10$ for this case, but we will develop equations for the case in which n can be any chosen value. If L_i and W_i stand for the individual measurements of the length and width, and $\overline{L}$ and $\overline{W}$ stand for the **mean** of those measurements, the equations relating them are

$$\overline{L} = \left(\frac{1}{n}\right)\sum_{i}^{n} L_i \qquad \overline{W} = \left(\frac{1}{n}\right)\sum_{i}^{n} W_i \qquad \text{(Eq. 1)}$$

We get information about the precision of the measurement from the variations of the individual measurements using the statistical concept of the standard deviation. The values of the **standard deviation** from the mean for the length and width of the table, σ_{n-1}^{L} and σ_{n-1}^{W}, are given by the equations:

$$\sigma_{n-1}^{L} = \sqrt{\frac{1}{n-1}\sum_{1}^{n}(L_i - \overline{L})^2} \qquad \sigma_{n-1}^{W} = \sqrt{\frac{1}{n-1}\sum_{1}^{n}(W_i - \overline{W})^2} \qquad \text{(Eq. 2)}$$

If errors are only random, it should be true that approximately 68.3% of the measurements of length should fall in the range $\overline{L} \pm \sigma_{n-1}^{L}$, and that approximately 68.3% of the measurements of width should fall within the range $\overline{W} \pm \sigma_{n-1}^{W}$. Furthermore, 95.5% of the measurements of both length and width should fall within 2 σ_{n-1} of the mean, and 99.73% should fall within 3 σ_{n-1} of the mean.

The precision of the mean for $\overline{L}$ and $\overline{W}$ is given by quantities called the **standard error**, α_L and α_W. These quantities are defined by the following equations:

$$\alpha_L = \frac{\sigma_{n-1}^{L}}{\sqrt{n}} \qquad \alpha_W = \frac{\sigma_{n-1}^{W}}{\sqrt{n}} \qquad \text{(Eq. 3)}$$

The meaning of α_L and α_W is that, if the errors are only random, there is a 68.3% chance that the true value of the length lies within the range $\overline{L} \pm \alpha_L$ and the true value of the width lies within the range $\overline{W} \pm \alpha_W$.

An important problem in experimental physics is to determine the uncertainty in some quantity that is derived by calculations from other directly measured quantities. For this experiment, consider the area A of the table as calculated from the measured values of the length and width $\overline{L}$ and $\overline{W}$ by the following:

$$A = \overline{L} \times \overline{W} \qquad \text{(Eq. 4)}$$

For the case of an area that is the product of two measured quantities, the uncertainty in the area is related to the uncertainty of the length and width by:

$$\alpha_A = \sqrt{(\overline{L})^2(\alpha_W)^2 + (\overline{W})^2(\alpha_L)^2} \qquad \text{(Eq. 5)}$$

EXPERIMENTAL PROCEDURE

1. Place the 2-meter stick along the length of the table near the middle of the width and parallel to one edge of the length. Do not attempt to line up either edge of the table with one end of the meter stick or with any certain mark on the meter stick.
2. Let X stand for the coordinate position in the length direction. Read the scale on the 2-meter stick that is aligned with one end of the table and record that measurement in Data and Calculations Table 1 as X_1. Read the scale that is aligned at the other end of the table and record that measurement in Data and Calculations Table 1 as X_2. A 3×5 note card held next to the edge of the table may help to determine where the 2-meter stick is aligned with the table for each measurement. Note that the stick has 1 millimeter as the smallest marked scale division. *Therefore, each coordinate should be estimated to the nearest 0.1 millimeter (nearest 0.0001 m).*
3. Repeat Steps 1 and 2 nine more times for a total of 10 measurements of the length of the table. For each measurement place the 2-meter stick on the table with no attempt to align either end of the stick or any particular mark on the stick with either end of the table. Make the measurements at 10 different places along the width of the table so that any variation in the length of the table is included in the measurements.
4. Perform Steps 1 through 3 for 10 measurements of the width of the table. Let the coordinate for the width be given by Y and record the 10 values of Y_1 and Y_2 in Data and Calculations Table 2. Again place the stick along the different lines each time, but make no attempt to align any particular mark on the stick with either edge of the table.

CALCULATIONS

1. After all measurements are completed, perform the subtractions of the coordinate positions to determine the 10 values of the length L_i, and the 10 values of width W_i. Record the 10 values of L_i and W_i in the appropriate table.
2. Use Equations 1 to calculate the mean length $\overline{L}$ and the mean width $\overline{W}$ and record their values in the appropriate table. Keep five decimal places in these results. For example, typical values might be $\overline{L} = 1.37157$ m and $\overline{W} = 0.76384$ m.
3. For each measurement of length and width, calculate the values of $L_i - \overline{L}$ and $W_i - \overline{W}$ and record them in the appropriate table. Then for each value of the length and width, calculate and record the values of $(L_i - \overline{L})^2$ and $(W_i - \overline{W})^2$ in the appropriate table.
4. Perform the summations of the values of $(L_i - \overline{L})^2$ and the summations of the values of $(W_i - \overline{W})^2$ and record them in the appropriate box in the tables.
5. Use the values of the summations of $(L_i - \overline{L})^2$ and of $(W_i - \overline{W})^2$ in Equations 2 to calculate the values of σ_{n-1}^{L} and σ_{n-1}^{W} and record them in the appropriate table.
6. Calculate $\overline{L} - \sigma_{n-1}^{L}$, $\overline{L} + \sigma_{n-1}^{L}$, $\overline{W} - \sigma_{n-1}^{W}$, and $\overline{W} + \sigma_{n-1}^{W}$ and record the values in the appropriate table.
7. Use the values of σ_{n-1}^{L} and σ_{n-1}^{W} in Equations 3 to calculate the values of α_L and α_W and record them in the appropriate table.
8. Use the values of $\overline{L}$ and $\overline{W}$ in Equation 4 to calculate the value of A, the area of the table, and record it in the appropriate table. Use Equation 5 to calculate the value of α_A and record it in the appropriate table.

Name ______________________ Section __________ Date __________

1 LABORATORY 1 *Measurement of Length*

PRE-LABORATORY ASSIGNMENT

1. State the number of significant figures in each of the following numbers and explain your answer.

(a) 37.60________

(b) 0.0130________

(c) 13000________

(d) 1.3400________

2. Perform the indicated operations to the correct number of significant figures using the rules for significant figures.

(a) $\begin{array}{r} 37.60 \\ \times\ 1.23 \\ \hline \end{array}$ **(b)** $6.7\overline{)8.975}$ **(c)** $\begin{array}{r} 3.765 \\ +\ 1.2 \\ +37.21 \\ \hline \end{array}$

3–6. Three students named Abe, Barb, and Cal make measurements (in m) of the length of a table using a meter stick. Each student's measurements are tabulated in the table below along with the mean, the standard deviation from the mean, and the standard error of the measurements.

Student	L_1	L_2	L_3	L_4	$\bar{L}$	σ_{n-1}	α
Abe	1.4717	1.4711	1.4722	1.4715	1.4716	0.00046	0.0002
Barb	1.4753	1.4759	1.4756	1.4749	1.4754	0.00043	0.0002
Cal	1.4719	1.4723	1.4727	1.4705	1.4719	0.00096	0.0005

Note that in each case only one significant figure is kept in the standard error α, and this determines the number of significant figures in the mean. The actual length of the table is determined by very sophisticated laser measurement techniques to be 1.4715 m.

3. State how one determines the accuracy of a measurement. Apply your idea to the measurements of the three students above and state which of the students has the most accurate measurement. Why is that your conclusion?

4. Apply Equations 1, 2, and 3 to calculate the mean, standard deviation, and standard error for Abe's measurements of length. Confirm that your calculated values are the same as those in the table. Show your calculations explicitly.

5. State the characteristics of data that indicate a systematic error. Do any of the three students have data that suggest the possibility of a systematic error? If so, state which student it is, and state how the data indicate your conclusion.

6. Which student has the best measurement considering both accuracy and precision? State clearly what the characteristics are of the student's data on which your answer is based.

Name

Section

Date

Lab Partners

LABORATORY 1 Measurement of Length

LABORATORY REPORT

Data and Calculations Table 1 (nearest 0.0001 m, which is 0.1 mm)

Trial	X_1 (m)	X_2 (m)	$L_i = X_2 - X_1$ (m)	$L_i - \overline{L}$ (m)	$(L_i - \overline{L})^2$ (m^2)
				$\sum_{1}^{n}(L_i - \overline{L})^2 =$	

$\overline{L} =$ ________ $\sigma_{n-1}^{L} =$ ________ $\overline{L} - \sigma_{n-1}^{L} =$ ________ $\overline{L} + \sigma_{n-1}^{L} =$ ________ $\alpha_L =$ ________

Data and Calculations Table 2 (nearest 0.0001 m, which is 0.1 mm)

Trial	Y_1 (m)	Y_2 (m)	$W_i = Y_2 - Y_1$ (m)	$W_i - \overline{W}$ (m)	$(W_i - \overline{W})^2$ (m^2)
				$\sum_{1}^{n}(W_i - \overline{W})^2 =$	

$\overline{W} =$ ________ $\sigma_{n-1}^{W} =$ ________ $\overline{W} - \sigma_{n-1}^{W} =$ ________ $\overline{W} + \sigma_{n-1}^{W} =$ ________ $\alpha_W =$ ________

$A = \overline{L} \times \overline{W} =$ ____________________ $\sigma_A =$ ____________________

SAMPLE CALCULATIONS

1. $L_1 = X_2^1 - X_1^1 =$
2. $W_1 = Y_2^1 - Y_1^1 =$
3. $\overline{L} = \frac{1}{10}\sum_{1}^{10} L_i =$
4. $\overline{W} = \frac{1}{10}\sum_{1}^{10} W_i =$
5. $L_1 - \overline{L} =$
6. $(L_1 - \overline{L})^2 =$
7. $W_1 - \overline{W} =$
8. $(W_1 - \overline{W})^2 =$
9. $\sigma_{n-1}^{L} = \sqrt{\frac{1}{n-1}\sum_{1}^{n}(L_i - \overline{L})^2}$
10. $\overline{L} - \sigma_{n-1}^{L} =$
11. $\overline{L} + \sigma_{n-1}^{L} =$

12. $\sigma_{n-1}^{W} = \sqrt{\frac{1}{n-1}\sum_{1}^{n}(W_i - \overline{W})^2}$

13. $\overline{W} - \sigma_{n-1}^{W} =$

14. $\overline{W} + \sigma_{n-1}^{W} =$

15. $A = \overline{L} \times \overline{W} =$

16. $\sigma_A =$

QUESTIONS

1. According to statistical theory, 68% of your measurements of the length of the table should fall in the range from $\overline{L} - \sigma_{n-1}^{L}$ to $\overline{L} + \sigma_{n-1}^{L}$. About 7 of your 10 measurements should fall in this range. What is the range of these values for your data? From ________m to ________m. How many of your 10 measurements of the length of the table fall in this range? ________? State clearly the extent to which your data for the length agree with the theory. What is your evidence for your statement?

2. Answer the same question for the width. Range of $\overline{W} - \sigma_{n-1}^{W}$ to $\overline{W} + \sigma_{n-1}^{W}$ is from ________m to ________m. The number of measurements that fall in that range is ________. Do your data for the width of the table agree with the theory reasonably well? State your evidence for your opinion.

3. According to statistical theory, if any measurement of a given quantity has a deviation greater than $3\sigma_{n-1}$ from the mean of that quantity, it is very unlikely that it is statistical variation, but rather is more likely to be a mistake. Calculate the value of $3\sigma_{n-1}^{L}$. Do any of your measurements of length have a deviation from the mean greater than that value? If so, calculate how many times larger than σ_{n-1}^{L} it is. Do any of your measurements of the length appear to be a mistake, and, if so, which ones?

4. For the width measurements calculate $3\sigma_{n-1}^{W}$. Do any of your measurements of width have a deviation from the mean greater than that value? If so, calculate how many times larger than σ_{n-1}^{W} it is. Do any of your measurements of width appear to be a mistake, and, if so, which ones?

5. If possible, state the accuracy of your measurements of the length and width and give your reasoning. If this cannot be done, state why it is not possible. If possible, state the precision of your measurement of the length and width and give your reasoning. If this cannot be done, state why it is not possible.

Measurement of Density

OBJECTIVES

- ❑ Determine the mass, length, and diameter of three cylinders of different metals.
- ❑ Calculate the density of the cylinders and compare with the accepted values of the density of the metals.
- ❑ Determine the uncertainty in the value of the calculated density caused by the uncertainties in the measured mass, length, and diameter.

EQUIPMENT LIST

- Three solid cylinders of different metals (aluminum, brass, and iron)
- Vernier calipers
- Laboratory balance and calibrated masses

THEORY

The most general definition of **density** is mass per unit volume. Density can vary throughout the body if the mass is not distributed uniformly. If the mass in an object is distributed uniformly throughout the object, the density ρ is defined as the total mass M divided by the total volume V of the object. In equation form this is

$$\rho = \frac{M}{V} \qquad \text{(Eq. 1)}$$

For a cylinder the volume is given by

$$V = \frac{\pi d^2 L}{4} \qquad \text{(Eq. 2)}$$

where d is the cylinder diameter, and L is its length. Using Equation 2 in Equation 1 gives

$$\rho = \frac{4M}{\pi d^2 L} \qquad \text{(Eq. 3)}$$

We will determine the quantities M, d, and L by measuring each of them four times and calculating the mean and standard error for each quantity. Using the mean of each measured quantity in Equation 3 leads to the best value for the measured density ρ.

An important question in experimental physics is how the uncertainty in a quantity calculated from other measured quantities is related to the uncertainty in those measured quantities. For this laboratory, the uncertainty in the density (standard error) is related to the standard errors in the mass, length, and diameter by:

$$\alpha_\rho = \rho\sqrt{\left(\frac{\alpha_M}{M}\right)^2 + \left(\frac{\alpha_L}{L}\right)^2 + 4\left(\frac{\alpha_d}{d}\right)^2} \qquad \text{(Eq. 4)}$$

The form of this equation is stated here without proof, but it can be derived from the relationship between the measured quantities and the density described by Equation 3.

We determine the mass of the cylinders with a laboratory balance, which balances the weight of an unknown mass m against the weight of a known mass m_k. Although the balance is between two forces (the weight of the masses), the scales can be calibrated in terms of mass assuming that the force per unit mass is the same for both the known and unknown mass. The unknown mass on a pan at the left is balanced against the sum of all the known masses placed on the right pan plus the mass equivalent of the permanent sliding mass on the beam. Figure 2-1 shows a picture of a Harvard Trip balance, which has a calibrated beam along which a permanent sliding mass can be moved in units of 0.1 gram up to 10 grams.

The length and diameter of the metal cylinder will be measured with a vernier caliper. A **caliper** is actually any device used to determine thickness, the diameter of an object, or the distance between two surfaces. Often calipers are in the form of two legs fastened together with a rivet, so they can pivot about the fastened point. The vernier caliper used in this laboratory consists of a fixed rule that contains one jaw, and a second jaw with a vernier scale that slides along the fixed rule scale as shown in Figure 2-2. **Vernier** is the name given to any scale that aids in interpolating between marked divisions.

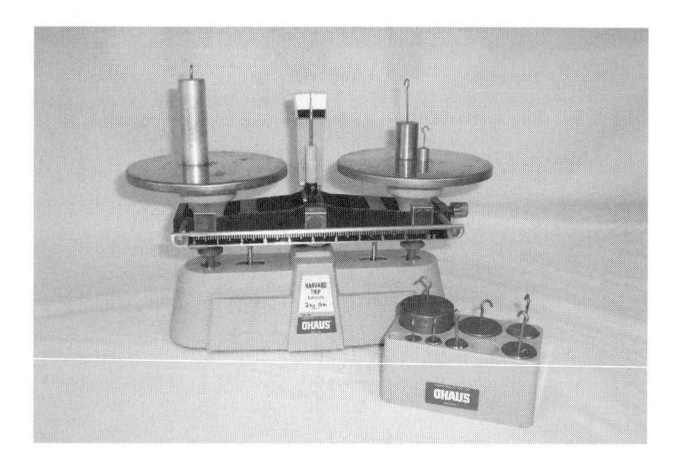

Figure 2-1 Harvard Trip balance.

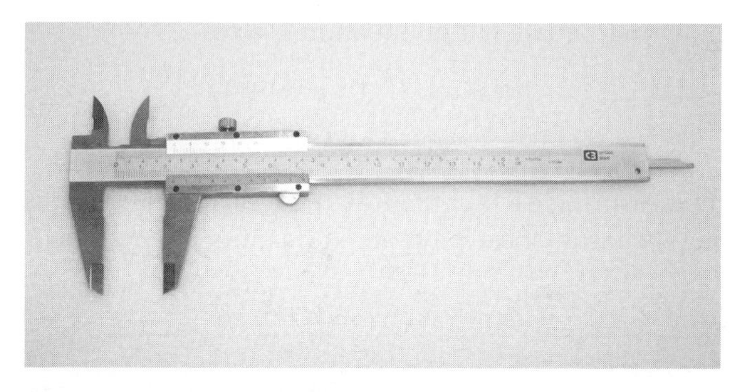

Figure 2-2 Vernier calipers.

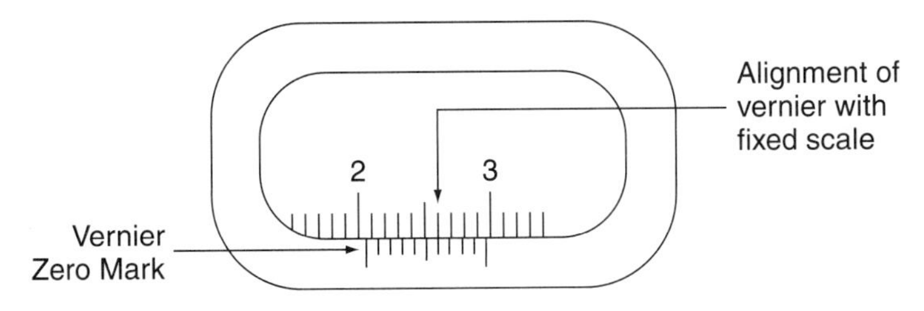

Figure 2-3 Illustration of vernier caliper reading of 2.06 cm.

The caliper has marked on the main scale major divisions of 1 cm for which there are both a mark and a number. On the main scale are also marked 10 divisions, each 1 mm apart between the 1 cm divisions. The 1 mm marks are not labeled with a number. This vernier is marked with a scale that, when aligned with different marks on the fixed rule scale, allows interpolation between the 1 mm marks on the fixed scale to 0.1 mm accurately. A vernier caliper can measure distances accurately to the nearest 0.01 cm.

A measurement is made by closing the jaws on some object and noting the position of the zero mark on the vernier and which one of the vernier marks is aligned with some mark on the fixed rule scale. This is illustrated in Figure 2-3. The position of the zero mark of the vernier scale gives the first two significant figures (2.0 cm in Figure 2-3). We derive the interpolation between 2.0 cm and 2.1 cm for this case from the fact that the sixth mark beyond the vernier zero is best aligned with a mark on the fixed rule scale. The reading in this example is 2.06 cm.

Before making any measurements, determine whether or not the vernier calipers read zero when the jaws are closed. If the calipers do not read zero when the jaws are closed, they are said to have a **zero error**. A correction is necessary for each measurement performed with the calipers. If the vernier zero is to the right of the fixed scale zero when the jaws are closed, the zero error is positive. Note the mark on the vernier scale that is aligned with the fixed scale, and subtract that number of units of 0.01 cm from each measurement. For example, if the third mark to the right of the vernier zero is aligned with the fixed scale when the jaws are closed, then each measurement should have 0.03 cm subtracted from it. If the vernier zero is to the left of the fixed scale zero, then the zero error is negative. In that case, find which vernier mark is aligned with the fixed scale. Then determine how far to the left of the 10 mark on the vernier scale the alignment occurs. For example, if the alignment occurs at the 7 mark on the vernier scale, you will add 0.03 cm to the reading.

EXPERIMENTAL PROCEDURE

1. Zero the laboratory balance according to directions given by your laboratory instructor.
2. Use the laboratory balance and calibrated masses to determine the mass of each of the three cylinders. Make four independent measurements for each of the cylinders and record the results in the Data Table.
3. Make four separate readings of the zero correction for the vernier calipers. Record the four values in the Data Table. Record the zero correction as positive if the vernier zero is to the right of the fixed scale zero and record it as negative if the vernier zero is to the left of the fixed scale zero.
4. Use the vernier calipers to measure the lengths of the three cylinders. Make four separate trials of the measurement of the length of each cylinder. Measure the length at different places on each cylinder for the four trials to sample any variation in length of the cylinders. Record the results in the Data Table.
5. Use the vernier calipers to measure the diameters of the three cylinders. Make four separate trials of the measurement of the diameter of each cylinder. Measure the diameter at four different positions along the length of the cylinders to sample any variation in diameter of the cylinders. Record the results in the Data Table.

CALCULATIONS

1. Calculate the mean $\overline{M}$, the standard deviation σ_{n-1}, and the standard error α_M for the four measurements of the mass of each cylinder and record the results in the Data Table. For this and all subsequent calculations keep *one significant figure* only for all standard errors, and then keep the number of *decimal places* in the mean that coincides with the *decimal place* of the standard error.
2. Determine the measured length and diameter for each trial by making the appropriate zero correction to each measurement and then calculating the means $\overline{L}$ and $\overline{d}$, the standard deviations, and the standard errors α_L and α_d for each cylinder. Record the results in the Data Table.
3. Use Equation 3 to calculate the density ρ of each of the cylinders. Use the mean values for the mass, diameter, and length. Use Equation 4 to calculate the standard error of the density. Record the results in the Data Table.
4. For purposes of this laboratory, assume that the density of aluminum is 2.70 gram/cm^3, the density of brass is 8.40 gram/cm^3, and the density of iron is 7.85 gram/cm^3. Calculate the percentage error in your results for the density of each of these metals.

Name _______________ Section _______ Date _______

2 LABORATORY 2 *Measurement of Density*

PRE-LABORATORY ASSIGNMENT

1. A cylinder has a length of 3.23 cm, a diameter of 1.75 cm, and a mass of 65.3 grams. What is the density of the cylinder? Based on its density, of what kind of material might it be made? Material is likely to be: ______________________ (Show your work.)

2. Figure 2-4 shows a vernier caliper scale set to a particular reading. What is the reading of the scale? Reading = ______________________ cm

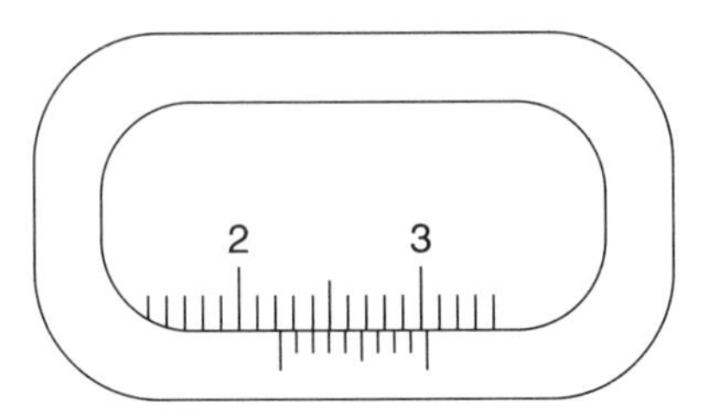

Figure 2-4 Example of a reading of a vernier caliper.

3. The caliper in Figure 2-5 has its jaws closed. If the caliper has a zero error, what is its value? Is it positive or negative? Error = _________ cm

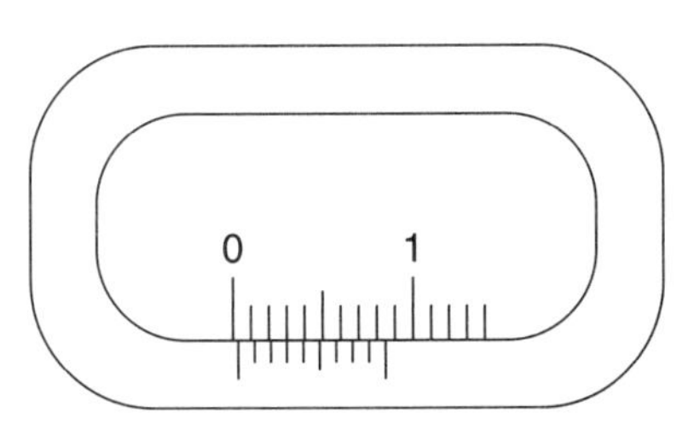

Figure 2-5 Vernier caliper with its jaws closed. Does it have a zero error?

4. A series of four measurements of the mass, length, and diameter are made of a cylinder. The results of these measurements are:

   Mass—20.6, 20.5, 20.6, and 20.4 grams

   Length—2.68, 2.67, 2.65, and 2.69 cm

   Diameter—1.07, 1.05, 1.06, and 1.05 cm

   Find the mean, standard deviation, and standard error for each of the measured quantities and tabulate them below. Keep only *one* significant figure in each standard error and then keep *decimal places* in the mean to coincide with the standard error.

   $\overline{M} =$ ________ $\sigma_{n-1} =$ ________ $\alpha_M =$ ________

   $\overline{L} =$ ________ $\sigma_{n-1} =$ ________ $\alpha_L =$ ________

   $\overline{d} =$ ________ $\sigma_{n-1} =$ ________ $\alpha_d =$ ________

   Calculate the density and the standard error of the density using Equations 3 and 4. Keep only *one significant figure* in the standard error and then keep *decimal places* in the density to coincide with the standard error.

   $\rho =$ ________ $\alpha_\rho =$ ________

5. Because α_ρ has only one digit, it determines the place of the least significant digit kept in the calculation of the density. From that information, how many significant figures are there in the density for the above calculation? State clearly the reasoning for your answer.

Name ______ Section ______ Date ______

Lab Partners ______

2 LABORATORY 2 *Measurement of Density*

LABORATORY REPORT

Mass Data and Calculations Table

	M_1 (kg)	M_2 (kg)	M_3 (kg)	M_4 (kg)	$\overline{M}$ (kg)	σ_{n-1} (kg)	α_M (kg)
Aluminum							
Brass							
Iron							

Zero Reading of the calipers ______

Length Data and Calculations Table

	L_1 (m)	L_2 (m)	L_3 (m)	L_4 (m)	$\overline{L}$ (m)	σ_{n-1} (m)	α_L (m)
Aluminum							
Brass							
Iron							

Diameter Data and Calculations Table

	d_1 (m)	d_2 (m)	d_3 (m)	d_4 (m)	$\overline{d}$ (m)	σ_{n-1} (m)	α_d (m)
Aluminum							
Brass							
Iron							

Density Data and Calculations Table

	ρ_{exp} (kg/m^3)	α_ρ (kg/m^3)	ρ_{known} (kg/m^3)	*Err* (kg/m^3)	% *Error*
Aluminum					
Brass					
Iron					

SAMPLE CALCULATIONS

1. $\overline{M} =$

2. $\sigma_{n-1} =$

3. $\alpha_M = \dfrac{\sigma_{n-1}}{\sqrt{n}} = \dfrac{\sigma_{n-1}}{\sqrt{4}} =$

4. $\rho = \dfrac{4M}{\pi d^2 L} =$

5. $\alpha_\rho = \rho\sqrt{\left(\dfrac{\alpha_M}{M}\right)^2 + \left(\dfrac{\alpha_L}{L}\right)^2 + 4\left(\dfrac{\alpha_d}{d}\right)^2} =$

6. $\text{Error} = \rho - \rho_{known} =$

7. $\%\ \text{Error} = \dfrac{\rho - \rho_{known}}{\rho_{known}} \times 100\% =$

QUESTIONS

1. Consider the uncertainty in the measured value of ρ to be given by α_ρ. Taking the one decimal place of α_ρ as the least significant digit in ρ, how many significant figures are indicated for each of the measurements of ρ?

2. What is the accuracy of your determination of the density for each metal? State clearly what quantity describes the accuracy of your measurements of the density.

3. Consider the value of the standard error α_ρ as an indication of the precision of your measurements. Express the standard error as a percentage of the measured value of the density, and relate it to the accuracy of each of your measurements.

4. Considering your answers to Question 3, is there evidence for a systematic error in any of your measurements of the density of any of the metals? State clearly your evidence either for or against the presence of a systematic error.

5. For the same percentage error in each of the three quantities, mass, diameter, and length, which would contribute the most to the error in the density? (Hint—Consider the form of Equation 4.)

Force Table and Vector Addition of Forces

OBJECTIVES

- ❑ Demonstrate the addition of several vectors to form a resultant vector using a force table.
- ❑ Demonstrate the relationship between the resultant of several vectors and the equilibrant of those vectors.
- ❑ Illustrate and practice graphical and analytical solutions for the addition of vectors.

EQUIPMENT LIST

- Force table with pulleys, ring, and string
- Mass holders and slotted masses
- Protractor and compass

THEORY

Physical quantities that can be completely specified by magnitude only are called **scalars**. Examples of scalars include temperature, volume, mass, and time intervals. Some physical quantities have both magnitude and direction. These are called **vectors**. Examples of vector quantities include spatial displacement, velocity, and force.

Consider the case of several forces with different magnitudes and directions that act at the same point. The single force, which is equivalent in its effect to the effect produced by the several applied forces, is called the **resultant force**. This resultant force can be found theoretically by a special addition process known as **vector addition**.

One process of vector addition is by graphical techniques. Figure 3-1(a) shows the case of two vectors, $\mathbf{F_1}$ of magnitude 20.0 N, and $\mathbf{F_2}$ of magnitude 30.0 N. A scale of 1.00 cm = 10.0 N is used, and these vectors are shown as 2.00 cm and 3.00 cm in length, respectively. The forces are assumed to act at the same point, but 60° different in direction as shown. Figure 3-1(b) shows the graphical addition process called the **parallelogram method**. Two lines are constructed, each one parallel to one of the vectors having the length of that vector as shown. The resultant $\mathbf{F_R}$ of the vector addition of $\mathbf{F_1}$ and $\mathbf{F_2}$ is found by constructing the straight line from the point at the tails of the two vectors to the opposite corner of the parallelogram formed by the original vectors and the constructed lines. A measurement of the length of $\mathbf{F_R}$ in Figure 3-1(b) shows it to be 4.35 cm in length, and a measurement of the angle between $\mathbf{F_R}$ and $\mathbf{F_1}$ shows it to be about 37°. Because the scale is 1.00 cm = 10.0 N, the value of the resultant $\mathbf{F_R}$ is 43.5 N, and it acts in a direction 37° with respect to the direction of $\mathbf{F_1}$.

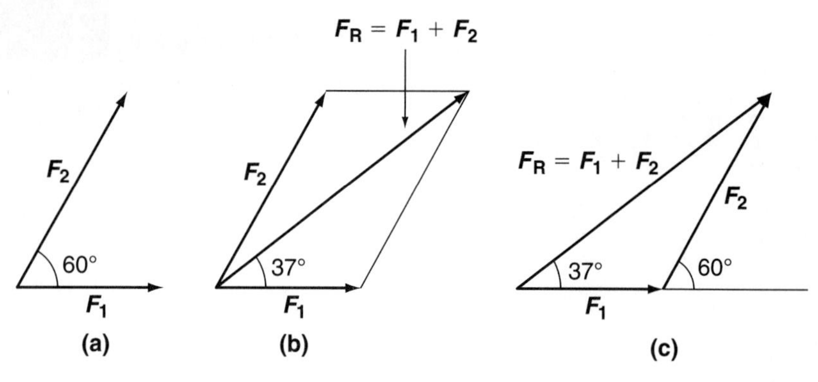

Figure 3-1 Illustration of the parallelogram and triangle addition of vectors.

In the graphical vector addition process known as the **polygon method** one of the vectors is first drawn to scale. Each successive vector to be added is drawn with its tail starting at the head of the preceding vector. The resultant vector is the vector drawn from the tail of the first arrow to the head of the last arrow. Figure 3-1(c) shows this process for the case of only two vectors (for which the polygon method is the triangle method). The second vector, $\mathbf{F_2}$, must be drawn at the proper angle relative to $\mathbf{F_1}$ by extending a line in the direction of $\mathbf{F_1}$ and constructing $\mathbf{F_2}$ relative to that line. In Figure 3-1(c) the length of $\mathbf{F_R}$ is 4.35 cm corresponding to 43.5 N, and it acts at 37° with respect to $\mathbf{F_1}$.

The polygon method is illustrated for the case of three vectors in Figure 3-2. Vector $\mathbf{F_1}$ is drawn, $\mathbf{F_2}$ is drawn at the proper angle α relative to $\mathbf{F_1}$, and $\mathbf{F_3}$ is drawn at the proper angle β relative to $\mathbf{F_2}$. The resultant $\mathbf{F_R}$ is the vector connecting the tail of $\mathbf{F_1}$ and the head of $\mathbf{F_3}$.

The analytical process of vector addition uses trigonometry to express each vector in terms of its components projected on the axes of a rectangular coordinate system. Figure 3-3 shows a vector, a coordinate system superimposed on the vector, and the components $|F|\cos\theta$ and $|F|\sin\theta$ into which the vector is resolved. When the analytical process for multiple vectors is used, each vector is resolved into components in that manner. The components along each axis are then added algebraically to produce the net components of the resultant vector along each axis. Those components are at right angles, and the magnitude of the resultant can be found from the Pythagorean theorem. The case of three vectors, $\mathbf{F_1}$, $\mathbf{F_2}$, and $\mathbf{F_3}$, is shown in Figure 3-4.

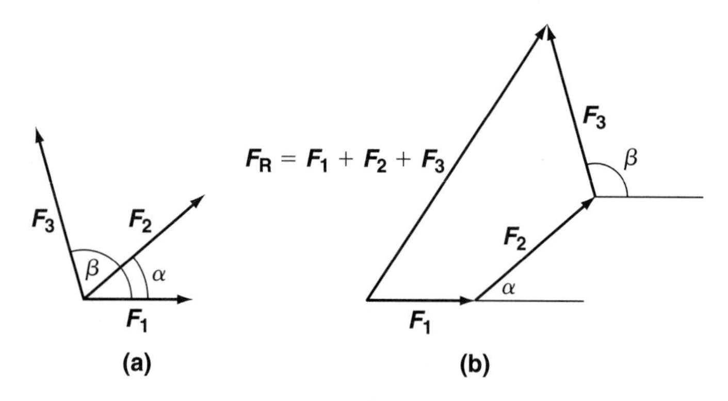

Figure 3-2 Illustration of the polygon method for vector addition.

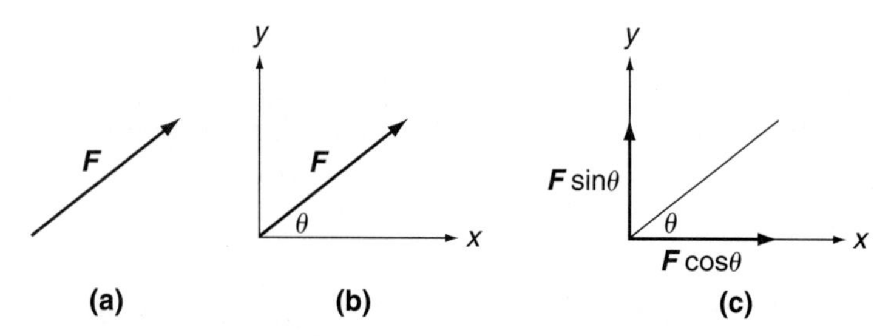

Figure 3-3 Illustration of analytical resolution of a vector.

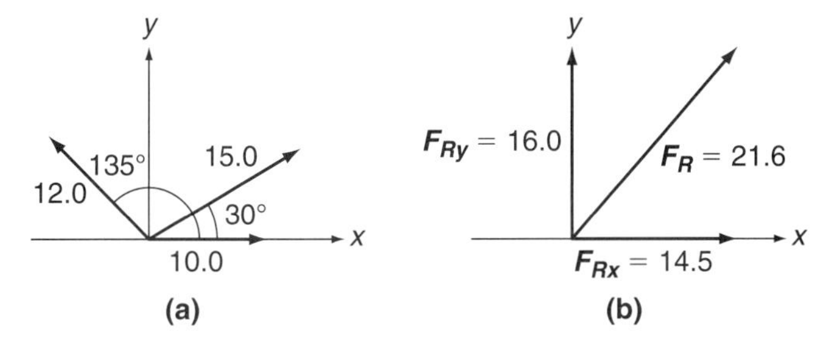

Figure 3-4 Illustration of the analytical addition of vectors.

Figure 3-5 Force Table. (Photo courtesy of Sargent-Welch Scientific Company)

Taking the algebraic sum of each of the components of the three vectors and combining the components to find the resultant and its direction leads to the following:

$$F_{Rx} = F_{1x} + F_{2x} + F_{3x} = 10.0\cos(0°) + 15.0\cos(30°) + 12.0\cos(135°) = 14.5$$

$$F_{Ry} = F_{1y} + F_{2y} + F_{3y} = 10.0\sin(0°) + 15.0\sin(30°) + 12.0\sin(135°) = 16.0$$

$$|\boldsymbol{F_R}| = F_R = \sqrt{(F_{Rx})^2 + (F_{Ry})^2} = \sqrt{(14.5)^2 + (16.0)^2} = 21.6$$

$$\theta = \arctan(F_{Ry}/F_{Rx}) = \arctan(16.0/14.5) = \text{arc tan }(1.10) = 47.7°$$

The force table (Figure 3-5) provides a force from the gravitational attraction on masses attached to a ring by a string passing over a pulley. Each force is applied over a separate pulley, and the pulley positions can be adjusted to any desired position around a circular plate. Experimentally the applied forces are balanced by the application of a single force that is equal to the magnitude of the resultant of the applied forces and acts opposite of the resultant. This balancing force (called the **equilibrant**) is what is determined by the measurements. The resultant is the same magnitude as the equilibrant and 180° different in direction.

EXPERIMENTAL PROCEDURE

Part 1. Two Applied Forces

1. Place a pulley at the 20.0° mark on the force table and place a total of 0.100 kg (including the mass holder) on the end of the string. Calculate the magnitude of the force (in N) produced by the mass.

Assume that $g = 9.80\ m/s^2$. Assume three significant figures for this and for all other calculations of force. Record the value of this force as F_1 in Data Table 1.

2. Place a second pulley at the 90.0° mark on the force table and place a total of 0.200 kg on the end of the string. Calculate the force produced and record as F_2 in Data Table 1.
3. Determine by trial and error the magnitude of mass needed and the angle at which it must be located for the ring to be centered on the force table. Jiggle the ring slightly to be sure that this equilibrium condition is met. Attach all strings to the ring so that they are directed along a line passing through the center of the ring. All the forces will then act through the point at the center of the table. Record this value of mass in Data Table 1 in the row labeled Equilibrant F_{E1}.
4. Calculate the force produced (mg) on the experimentally determined mass. Record the magnitude and direction of this equilibrant force F_{E1} in Data Table 1.
5. The resultant F_{R1} is equal in magnitude to F_{E1}, and its direction is 180° from F_{E1}. Record the magnitude of the force F_{R1}, the mass equivalent of this force, and the direction of the force in Data Table 1 in the row labeled Resultant F_{R1}.

Part 2. Three Applied Forces

1. Place a pulley at 30.0° with 0.150 kg on it, one at 100.0° with 0.200 kg on it, and one at 145.0° with 0.100 kg on it.
2. Calculate the force produced by those masses and record them as F_3, F_4, and F_5 in Data Table 2.
3. Determine the equilibrant force and the resultant force by following a procedure like that in Part 1, Steps 3 through 5 above. Record the magnitudes of the forces, the associated values of mass, and the directions in Data Table 2 in the rows labeled F_{E2} and F_{R2}.

CALCULATIONS

Part 1. Two Applied Forces

1. Find the resultant of these two applied forces by scaled graphical construction using the parallelogram method. Use a ruler and protractor to construct vectors with scaled length and direction that represent F_1 and F_2. A convenient scale might be 1.00 cm = 0.100 N. All directions are given relative to the force table. Account for this in the graphical construction to ensure the proper angle of one vector to another. Determine the magnitude and direction of the resultant from your graphical solution and record them in the appropriate section of Calculations Table 1.
2. Use trigonometry to calculate the components of F_1 and F_2 and record them in the analytical solution portion of Calculations Table 1. Add the components algebraically and determine the magnitude of the resultant by the Pythagorean theorem. Determine the angle of the resultant from the arc tan of the components. Record those results in Calculations Table 1.
3. Calculate the percentage error of the magnitude of the experimental value of F_R compared to the analytical solution for F_R. Also calculate the percentage error of the magnitude of the graphical solution for F_R compared to the analytical solution for F_R. For each of those comparisons, also calculate the magnitude of the difference in the angle. Record all values in Calculations Table 1.

Part 2. Three Applied Forces

1. Use the polygon scaled graphical construction method to find the resultant of the three applied forces. Determine the magnitude and direction of the resultant from your graphical solution and record them in Calculations Table 2.
2. Use trigonometry to calculate the components of all three forces, the components of the resultant, and the magnitude and direction of the resultant, and record them all in Calculations Table 2.
3. Make the same error calculations for this problem as described in Step 3 of Part 1 above. Record the values in Calculations Table 2.

Name ______________________ Section __________ Date __________

3 LABORATORY 3 *Force Table and Vector Addition of Forces*

PRE-LABORATORY ASSIGNMENT

1. Scalars are physical quantities that can be completely specified by their ______________________.

2. A vector quantity is one that has both ______________________ and ______________________.

3. Classify each of the following physical quantities as vectors or scalars:

   (a) Volume __________

   (b) Force __________

   (c) Density __________

   (d) Velocity __________

   (e) Acceleration __________

Answer Questions 4–7 with reference to Figure 3-6 below.

4. If $\mathbf{F_1}$ stands for a force vector of magnitude 30.0 N and $\mathbf{F_2}$ stands for a force vector of magnitude 40.0 N acting in the directions shown in Figure 3-6, what are the magnitude and direction of the resultant obtained by the vector addition of these two vectors using the analytical method? Show your work.

   Magnitude = __________N Direction(relative to x axis) = __________degrees

5. What is the equilibrant force that would be needed to compensate for the resultant force of the vectors $\mathbf{F_1}$ and $\mathbf{F_2}$ that you calculated in Question 4?

   Magnitude = __________N Direction(relative to x axis) = __________degrees

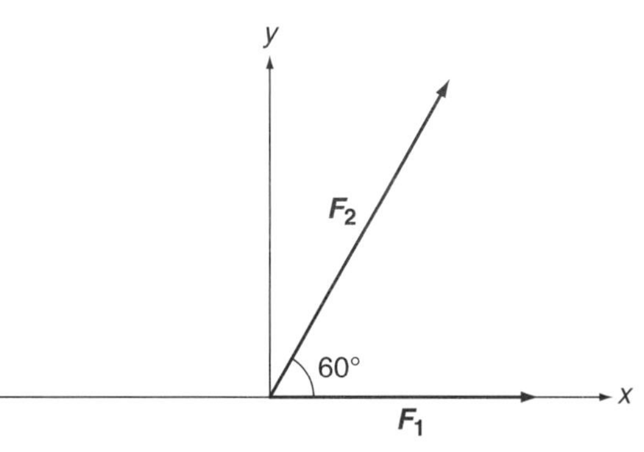

Figure 3-6 Addition of two force vectors.

6. Figure 3-6 has been constructed to scale with 1.00 cm = 10.0 N. Use the parallelogram graphical method to construct (on Figure 3-6) the resultant vector $\mathbf{F_R}$ for the addition of $\mathbf{F_1}$ and $\mathbf{F_2}$. Measure the length of the resultant vector and record it below. State the force represented by this length. Measure with a protractor the angle that the resultant makes with the x axis.

Resultant vector length = ________cm

Force represented by this length = ________N

Direction of resultant relative to x axis = ________degrees

7. Use the polygon method of vector addition to construct on the axes below a graphical solution to the problem in Figure 3-6. Use the scale 1.00 cm = 10.0 N.

Resultant vector length = ________cm

Force represented by this length = ________N

Direction of resultant relative to x axis = ________degrees

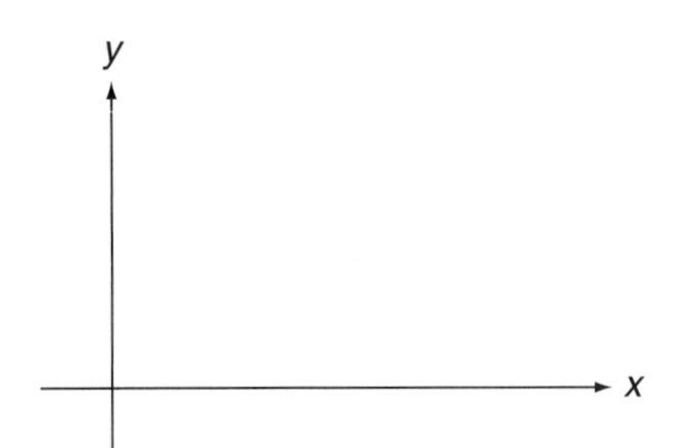

Name Section Date

Lab Partners

3 LABORATORY 3 *Force Table and Vector Addition of Forces*

LABORATORY REPORT

Data Table 1

Force	*Mass* (kg)	*Force* (N)	*Direction*
F_1	0.100		20.0°
F_2	0.200		90.0°
Equilibrant F_{E1}			
Resultant F_{R1}			

Data Table 2

Force	*Mass* (kg)	*Force* (N)	*Direction*
F_3	0.150		30.0°
F_4	0.200		100.0°
F_5	0.100		145.0°
Equilibrant F_{E2}			
Resultant F_{R2}			

Calculations Table 1

Graphical Solution			
Force	*Mass* (kg)	*Force* (N)	*Direction*
F_1	0.100		20.0°
F_2	0.200		90.0°
Resultant F_{R1}			

Analytical Solution					
Force	*Mass* (kg)	*Force* (N)	*Direction*	*x-component*	*y-component*
F_1	0.100		20.0°		
F_2	0.200		90.0°		
Resultant F_{R1}					

PART 1. ERROR CALCULATIONS

Percent Error magnitude Experimental compared to Analytical = ________%

Percent Error magnitude Graphical compared to Analytical = ________%

Absolute Error in angle Experimental compared to Analytical = ________degrees

Absolute Error in angle Graphical compared to Analytical = ________degrees

Calculations Table 2

Graphical Solution			
Force	*Mass* (kg)	*Force* (N)	*Direction*
F_3	0.150		30.0°
F_4	0.200		100.0°
F_5	0.100		145.0°
Resultant F_{R2}			

Analytical Solution					
Force	*Mass* (kg)	*Force* (N)	*Direction*	*x-component*	*y-component*
F_3	0.150		30.0°		
F_4	0.200		100.0°		
F_5	0.100		145.0°		
Resultant F_{R2}					

PART 2. ERROR CALCULATIONS

Percent Error magnitude Experimental compared to Analytical = ________%

Percent Error magnitude Graphical compared to Analytical = ________%

Absolute Error in angle Experimental compared to Analytical = ________degrees

Absolute Error in angle Graphical compared to Analytical = ________degrees

SAMPLE CALCULATIONS

1. $F = mg =$
2. $m = \frac{F}{g} =$
3. Direction F_E opposite F_R so direction F_R = direction $F_E - 180^\circ =$
4. $F_{1x} = F_1 \cos(20^\circ) =$
5. $F_{1y} = F_1 \sin(20^\circ) =$
6. $F_{R1} = \sqrt{(F_{Rx})^2 + (F_{Ry})^2} =$
7. $\theta = \tan^{-1}\left(\frac{F_y}{F_x}\right) =$
8. $\%\text{ Error Exp} = \frac{|\text{Experimental} - \text{Analytical}|}{\text{Analytical}} \times 100\% =$
9. Absolute Err $= \theta(\text{exp}) - \theta(\text{analytical}) =$

QUESTIONS

1. To determine the force acting on each mass it was assumed that $g = 9.80\ \text{m/sec}^2$. The value of g at the place where the experiment is performed may be slightly different from that value. State what effect (if any) it would have on the *percentage error* calculated for the comparisons. To test your answer to the question, leave g as a symbol in the calculation of the percentage error.

2. Two forces are applied to the ring of a force table, one at an angle of 20.0°, and the other at an angle of 80.0°. Regardless of the magnitudes of the forces, choose the correct response below.

   The equilibrant will be in the (a) first quadrant (b) second quadrant (c) third quadrant (d) fourth quadrant (e) cannot tell which quadrant from the available information.

The resultant will be in the (a) first quadrant (b) second quadrant (c) third quadrant (d) fourth quadrant (e) cannot tell which quadrant from the available information.

3. Two forces, one of magnitude 2 N and the other of magnitude 3 N, are applied to the ring of a force table. The directions of both forces are unknown. Which *best* describes the limitations on R, the resultant? Explain carefully the basis for your answer.

(a) $R \leq 5$ N **(b)** 2 N $\leq R \leq$ 3 N **(c)** $R \geq 3$ N **(d)** 1 N $\leq R \leq$ 5 N **(e)** $R \leq 2$ N.

4. Suppose the same masses are used for a force table experiment as were used in Part 1, but each pulley is moved 180° so that the 0.100 kg mass acts at 200°, and the 0.200 kg mass acts at 270°. What is the magnitude of the resultant in this case? How does it compare to the resultant in Part 1?

5. Pulleys introduce a possible source of error because of their possible friction. Given that they are a source of error, why are the pulleys used at all? What is the function of the pulleys?

Uniformly Accelerated Motion

OBJECTIVES

- ❑ Investigate how the displacement of a cart down the inclined plane of an air track is related to the elapsed time.
- ❑ Determine the acceleration of the cart from an analysis of the displacement versus time data.
- ❑ Determine an experimental value for g (the acceleration due to gravity) by interpretation of the cart's acceleration as a component of g. Compare with the accepted value of g.

EQUIPMENT LIST

- A 5-meter linear air track with a built-in 5-meter scale (If a shorter air track is used the suggested distances can be modified to fit the air track used.)
- Laboratory timer or stopwatch
- Block to raise the air track to form an inclined plane

THEORY

When an object undergoes one-dimensional uniformly accelerated motion its **velocity** increases linearly with time. If it is assumed that the initial velocity of the object is zero at time $t = 0$, then its velocity v at any later time t is given by

$$v = at \qquad \text{(Eq. 1)}$$

where a is the **acceleration,** which is constant in magnitude and direction.

Consider a time interval between $t = 0$ and any later time t. The average velocity $\overline{v}$ during the time interval is

$$\overline{v} = \frac{0 + v}{2} = \frac{v}{2} \qquad \text{(Eq. 2)}$$

The displacement x of the object during the time interval t is given by

$$x = \overline{v}t = \frac{vt}{2} \qquad \text{(Eq. 3)}$$

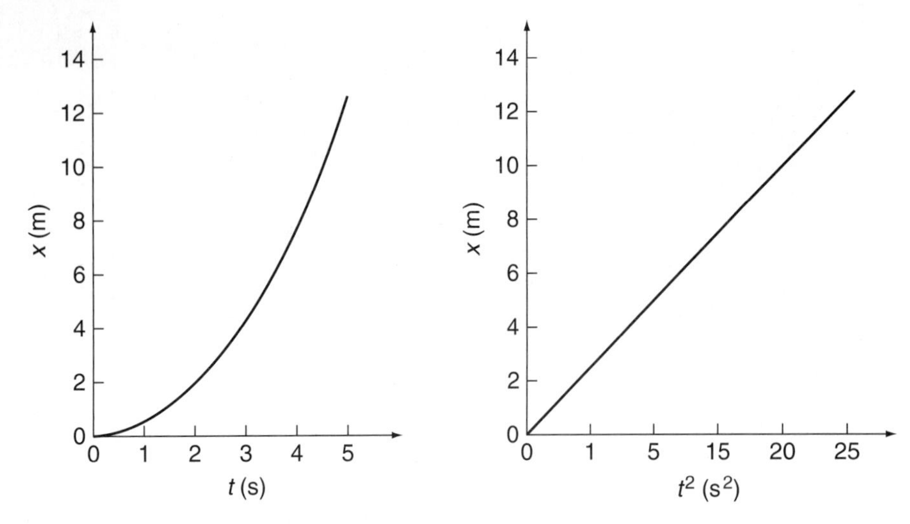

Figure 4-1 Graphs of displacement versus time and displacement versus the square of the time for uniformly accelerated motion. The displacement is linear with time squared.

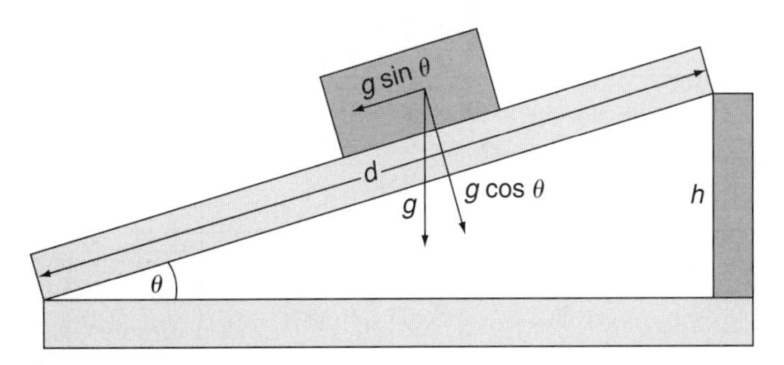

Figure 4-2 Components of g, the acceleration due to gravity on an air track.

Substituting Equation 1 for v in Equation 3 gives

$$x = \frac{at^2}{2} \qquad \text{(Eq. 4)}$$

Equation 4 states that if an object is released from rest, its displacement is directly proportional to the square of the elapsed time. Figure 4-1 shows graphs of both x versus t and x versus t^2 for uniformly accelerated motion.

A cart shown in Figure 4-2 is placed on an air track that is raised at one end to form an inclined plane with an angle of inclination of θ. The acceleration due to gravity points directly downward, but can be resolved into components that are perpendicular and parallel to the plane. The cart moves down the inclined plane with acceleration a, the component of the acceleration due to gravity g acting down the plane. In equation form this is:

$$a = g \sin\theta \qquad \text{(Eq. 5)}$$

EXPERIMENTAL PROCEDURE

1. Place a block approximately 10 cm in height under the support at one end of the air track. Measure the height h of the block to the nearest 0.1 mm. Have five members of the class perform this measurement independently and record those five trials in the Data Table.

2. Have five members of the class independently measure the distance d between the points of support of the air track as shown in Figure 4-2. Record the five trials for d in the Data Table.
3. Have one member of the class release the cart from rest at the top of the incline and simultaneously start a timer. Stop the timer when the cart has traveled 0.250 m down the track. Four other members of the class should repeat this measurement for a total of five trials at this distance. Record all times in the Data Table.
4. Repeat Step 3 for distances of 0.500, 0.750, 1.000, 1.500, 2.000, 3.000, and 4.000 m. At each distance have five different members of the class perform the measurement. Record all times in the Data Table.

CALCULATIONS

1. Calculate the mean $\bar{h}$, the standard deviation, and the standard error α_h for the five values of h, and record these results and all other calculated values in the Calculations Table. Keep only *one digit* in all standard errors and keep *decimal places* in all means that coincide with the *decimal place* of the standard error.
2. Calculate the mean $\bar{d}$, the standard deviation, and the standard error α_d for the five trials of d.
3. Calculate the mean time $\bar{t}$, the standard deviation, and the standard error α_t for the five trials of the time at each distance.
4. Calculate the square of each of the mean times $(\bar{t})^2$ for the eight distances.
5. Perform a linear least squares fit to the data with x as the vertical axis and $(\bar{t})^2$ as the horizontal axis. Determine the values of the slope and correlation coefficient r. In that calculation use (0,0) as one of the points. From the slope obtained in the least squares fit, calculate the acceleration a of the cart where $a = 2(\text{slope})$ according to Equation 4.
6. Calculate the value of $\sin\theta = h/d$ using the values of $\bar{h}$ and $\bar{d}$.
7. Use Equation 5 to calculate an experimental value for the acceleration due to gravity (g_{exp}) from the measured values of a and $\sin\theta$.
8. Calculate the percentage error in your value of g_{exp} compared to the accepted value of 9.80 m/sec^2.

GRAPHS

1. Construct a graph of the x versus $(\bar{t})^2$ data with x as the vertical axis and $(\bar{t})^2$ as the horizontal axis. Show on the graph the straight line from the linear least squares fit.

Name | Section | Date

LABORATORY 4 *Uniformly Accelerated Motion*

PRE-LABORATORY ASSIGNMENT

(a)

(b)

(c)

(d)

The carts pictured above are all moving in a straight line to the right. The pictures were taken 1.00 s apart. Choose which of the descriptions below matches which pictures.

1. These pictures show a cart that is moving at constant velocity.
   (a) (b) (c) (d)

2. These pictures show a cart that has a positive acceleration.
   (a) (b) (c) (d)

3. These pictures show a cart that travels at a constant velocity and then has a positive acceleration.
   (a) (b) (c) (d)

4. These pictures show a cart that has a negative acceleration.
   (a) (b) (c) (d)

5. A cart on a linear air track has a uniform acceleration of 0.172 m/s^2. Use Equation 1 to find the velocity of the cart 4.00 seconds after it is released from rest. Show your work.

6. How far does the cart in Question 5 travel in 4.00 seconds? Calculate the distance x two ways, first using Equation 3 and then using Equation 4. Show your work.

7. An air track like the one shown in Figure 4-2 has a block with a height $h = 12.0$ cm under one support. The other support is 3.50 m away. What is the angle of inclination θ? According to Equation 5, the component of acceleration parallel to the track is $a = g \sin\theta$ where $g = 9.80\,\text{m/s}^2$. For this value of θ what is a? Show your work.

Name | Section | Date

Lab Partners

4 LABORATORY 4 *Uniformly Accelerated Motion*

LABORATORY REPORT

Data Table

	Trial 1	*Trial 2*	*Trial 3*	*Trial 4*	*Trial 5*
h (m)					
d (m)					

x (m)	t_1 (s)	t_2 (s)	t_3 (s)	t_4 (s)	t_5 (s)

Calculations Tables

$\bar{h} =$ m	$\sigma_{n-1} =$ m	$\alpha_h =$ m
$\bar{d} =$ m	$\sigma_{n-1} =$ m	$\alpha_d =$ m
$\sin\theta = \bar{h}/\bar{d} =$		

x (m)								
$\bar{t}$ (s)								
σ_{n-1} (s)								
α_t (s)								
$\bar{t}^2$ (s^2)								

Slope =	a = m/s^2	g_{exp} = m/s^2	%Err =	r =

SAMPLE CALCULATIONS

1. $a = 2(\text{slope})$
2. $g_{exp} = a/(\sin\theta) =$
3. $\%\text{ error} = \dfrac{g_{exp} - g}{g} \times 100\% =$

QUESTIONS

1. The decimal place of the standard error coincides with the least significant digit and determines the number of significant figures in the values of $\bar{h}$ and $\bar{d}$. Because these are used to calculate the experimental value of g, they determine the number of significant figures in your value of g. How many significant figures are in your values of $\bar{h}$ and $\bar{d}$, and how many are in your experimental value of g?

2. Would friction tend to cause your experimental value for g to be greater or less than 9.80 m/sec^2? In which direction is your error for the value for g? Could friction be the cause of your observed error? State your reasoning.

3. What was the instantaneous velocity of the cart at $x = 4.00$ meters assuming your value of the acceleration a is correct? Show your work.

4. State how well the objectives of this laboratory were met. State your evidence for your opinion.

Uniformly Accelerated Motion on the Air Table

OBJECTIVES

- ❑ Determine the average velocity of a puck on an inclined air table from measured displacements and time intervals.
- ❑ Demonstrate the linear increase in the instantaneous velocity with time for the special case of constant acceleration, and determine the acceleration by a least squares fit to the velocity versus time data.
- ❑ Determine an experimental value for g (the acceleration due to gravity) by interpretation of the puck's acceleration as a component of g.

EQUIPMENT LIST

- Air table with sparktimer, air pump, foot switch, block, and level
- Recording paper and carbon paper
- Meter stick or measuring tape
- Balance and calibrated masses

THEORY

To investigate the one-dimensional motion of a puck on an air table, we choose an arbitrary point near the beginning of the motion as the origin of the coordinate system. We determine the location of the puck by the position x of sparks on the recording paper at a constant time interval between the marks on the paper. If the initial coordinate is x_i at some time t_i, and its later coordinate is x_f at some later time t_f, then the **displacement** Δx between those coordinates is given by

$$\Delta x = x_f - x_i \qquad \text{(Eq. 1)}$$

Although displacements are generally vector quantities, for the special case of one-dimensional motion in one direction, it is sufficient to consider only the magnitude of the displacements. All displacements for this laboratory will be positive because the puck moves in one direction.

The **average velocity** $\overline{v}$ is given by

$$\overline{v} = \frac{\Delta x}{\Delta t} = \frac{x_f - x_i}{t_f - t_i} \qquad \text{(Eq. 2)}$$

The average velocity is defined over some time interval Δt that is finite. The definition of instantaneous velocity, which is the velocity at some instant of time, is the limit of the average velocity as the time interval Δt approaches zero. The **average acceleration** $\overline{a}$ is the time rate of change of the instantaneous velocity and is given by

$$\overline{a} = \frac{\Delta v}{\Delta t} = \frac{v_f - v_i}{t_f - t_i} \qquad \text{(Eq. 3)}$$

In this laboratory the special case of constant acceleration will be the only case that is considered. When the initial velocity at $t = 0$ has a value of v_o, Equation 3 reduces to

$$v = v_o + a\,t \qquad \text{(Eq. 4)}$$

Equation 4 states that for constant acceleration, the instantaneous velocity increases linearly with time from its initial value of v_o at $t = 0$ to later values of v at any later time t. For constant acceleration there is a simple relationship between the average velocity and the instantaneous velocity. Because the velocity increases linearly with time, the average velocity during each time interval is equal to the instantaneous velocity at the middle of the time interval.

For constant acceleration, a determination of the average velocity as a function of time leads to a knowledge of the instantaneous velocity as a function of time. Assume that the coordinate position of some object is measured at fixed time intervals of 0.100 seconds. The average velocity during the interval from $t = 0$ to $t = 0.100$ seconds would equal the instantaneous velocity at $t = 0.050$ seconds (the middle of the time interval between $t = 0$ and $t = 0.100$ seconds). Similarly, the average velocity during any chosen interval is the same as the instantaneous velocity at the middle of that interval.

When the instantaneous velocity is known as a function of time, that data can be fit to Equation 4 by a linear least squares fit to obtain the initial velocity v_o at $t = 0$ and the acceleration a.

The acceleration of an object down an inclined plane is a component of the acceleration due to gravity g as shown in Figure 5-1. The component of g pointing parallel to the plane causing the acceleration a is

$$a = g \sin\theta \qquad \text{(Eq. 5)}$$

If θ is known and a is determined experimentally, Equation 5 provides an experimental value for g. If θ and g are assumed known, the expected value for the acceleration a can be determined from Equation 5.

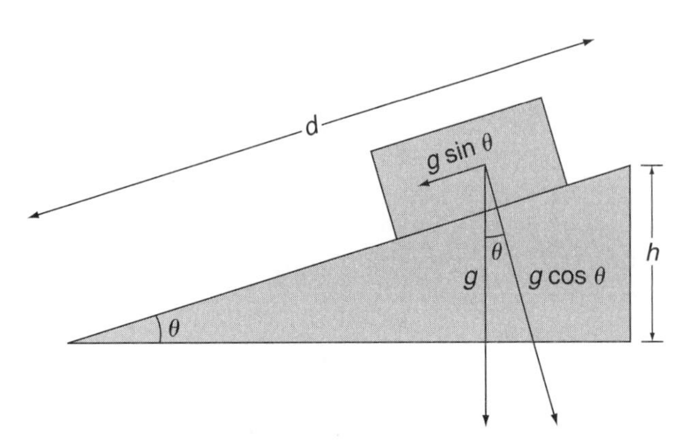

Figure 5-1 Components of g, the acceleration due to gravity.

Figure 5-2 Precision air table. (Photo courtesy of Central Scientific Company)

EXPERIMENTAL PROCEDURE

1. Read the instruction manual for the air table (Figure 5-2) or obtain instructions from your laboratory instructor. Become familiar with the operation of the air pump, the sparktimer, and the foot switch. Do not touch the conducting portions of the puck when the sparktimer is in operation.
2. Level the air table by means of the three adjustable legs until a puck placed near the center of the table is essentially motionless.
3. Place a small block under the air table single leg to make the table act as an inclined plane. The angle of the inclined plane is determined by the distance between the air table legs d and the height h of the block placed under one leg. Choose h to produce an angle of about 5°. Measure h and d to the nearest 0.0001 meters and record them in the Data Table.
4. Set the sparktimer to a spark rate of 10.0 Hertz. This will produce a spark every 0.100 seconds, and Δt between data points will be 0.100 seconds for all the data. Record this value of Δt in the Data Table. The values of time for x and for the center of the time interval have been included in the Data Table.
5. Place a piece of carbon paper and a sheet of recording paper on the air table with the recording paper on top. Place the two pucks on the table. To complete the circuit, both pucks must be attached to the sparktimer, but one puck can be left at rest at the bottom of the inclined plane. Simultaneously start the sparktimer and release the other puck from rest at the top of the inclined plane. Allow the puck to accelerate down the incline while the sparktimer is in operation, but stop the sparktimer as the puck hits the bottom rail. Release the puck with no initial velocity. Do not impart any sideways motion to the puck when releasing it.
6. Once a good spark record has been obtained, choose a point near the beginning of the track as the origin and label it as $x = 0$. Call the corresponding time $t = 0$. Measure to the nearest 0.0001 meters the displacement from the origin to each data point. Record in the Data Table the values of the coordinate x for 12 data points. For each data point, the coordinate x is to be measured from the origin to that point.
7. Determine the mass of the puck and record it in the Data Table in kilograms.

CALCULATIONS

1. Calculate the magnitude of the displacement Δx for each successive set of sparks by taking the differences between the coordinate values for successive data points. Record the values in the Calculations Table.
2. Use Equation 2 to calculate the average velocity during each time interval from the values of Δx and the value of $\Delta t = 0.100$ s. The average velocity for each interval is equal to the instantaneous velocity at

the middle of that interval. Record the instantaneous velocity at the time each velocity occurs in the Calculations Table.

3. Perform a linear least squares fit with v as the vertical axis and t as the horizontal axis. The slope obtained by that fit is equal to the acceleration a and the intercept obtained is equal to the initial velocity v_o. Record the acceleration, initial velocity, and correlation coefficient r in the Calculations Table.
4. Calculate the angle of the incline θ from the measured values of h and d using the equation $\sin\theta = h/d$. Record the value of θ in the Calculations Table.
5. Use the value of the acceleration a determined by the least squares fit and the measured value of θ to solve Equation 5 for g. Record this experimental value of g as g_{exp} in the Calculations Table. Calculate the percentage error for g_{exp} compared to the true value of $g = 9.80\,\text{m/s}^2$ and record it in the Calculations Table.

GRAPHS

1. Make a graph of the data for instantaneous velocity versus time with velocity as the vertical axis and time as the horizontal axis. Show the data as points. Draw on the graph the straight line that was obtained by the least squares fit procedure.

Name ______________________ Section ____________ Date ____________

5 **LABORATORY 5** *Uniformly Accelerated Motion on the Air Table*

PRE-LABORATORY ASSIGNMENT

1. The displacement is (a) always a vector, (b) a vector only if an object is at rest, (c) a vector only if an object is in motion, (d) always a scalar, (e) a scalar only if an object is in motion.

2. For one-dimensional motion, the instantaneous velocity is always defined as $\Delta x/\Delta t$.

   (a) true (b) false

3. The velocity of an object is proportional to elapsed time (a) always, (b) only for positive acceleration, (c) only for negative acceleration, (d) only for constant acceleration.

4. Suppose that an ideal frictionless inclined plane has an angle of inclination of 5.00° with respect to the horizontal. What is the acceleration of an object sliding down that plane? Assume that the acceleration due to gravity is $g = 9.80\,\text{m/s}^2$. Show your work.

5–7. The data below for the instantaneous velocity v versus the time t were obtained in a student experiment. Perform a linear least squares fit with v as the vertical axis and t as the horizontal axis. Determine the intercept (initial velocity v_o), slope (acceleration a), and correlation coefficient (r) of the straight line.

v (m/s)	0.352	0.496	0.655	0.808	0.939	1.073
t (s)	0.200	0.400	0.600	0.800	1.000	1.200

5. Intercept $= v_o =$ ____________ m/s

6. Acceleration $a =$ ____________ m/s^2

7. Correlation coefficient $r =$ ____________

8. There are six data points in the Question 7 least squares fit calculation. Statistical theory states that for six data points there is a 0.1% probability that a value of correlation coefficient $r \geq 0.974$ will be obtained for uncorrelated data. Compare the value of r obtained in Question 7 to 0.974. State your conclusion about the probability that these data show that velocity is proportional to time.

9. Assume that the data of Question 5–7 were taken for an inclined plane of angle 4.40°. Equation 5 states that the acceleration down the plane is $a = g \sin \theta$. Use this equation to determine an experimental value for g from the acceleration a determined in Question 6. Calculate the percentage error between this experimental value g_{exp} and the true value of 9.80 m/s^2.

g_{exp} = ______________ m/s^2 Percentage error = ______________ %

Name _______________ Section _______ Date _______

Lab Partners _______________

LABORATORY 5 Uniformly Accelerated Motion on the Air Table

LABORATORY REPORT

Data Table

Point	t (s)	x (m)
0	0.000	
1	0.100	
2	0.200	
3	0.300	
4	0.400	
5	0.500	
6	0.600	
7	0.700	
8	0.800	
9	0.900	
10	1.000	
11	1.100	
12	1.200	

Calculations Table

Δx (m)	v (m/s)	t (s)
		0.050
		0.150
		0.250
		0.350
		0.450
		0.550
		0.650
		0.750
		0.850
		0.950
		1.050
		1.150

$h =$	m	$a =$	m/s^2
$d =$	m	$v_o =$	m/s
$\Delta t =$	s	$\theta =$	degrees
Mass =	kg	$g_{exp} =$	m/s^2
$r =$		% Error =	

SAMPLE CALCULATIONS

1. $\Delta t = 1/f =$
2. $\Delta x = x_2 - x_1 =$
3. $v = \Delta x / \Delta t =$
4. $\sin\theta = h/d =$
5. $\theta = \sin^{-1}(h/d) =$
6. $t_{midpoint} = (t_i + t_{i+1})/2 =$
7. $g_{exp} = a/\sin\theta =$
8. % Error = (Exp-Known)/Known × 100% =

QUESTIONS

1. Suppose a different point had been chosen as the origin for the analysis of the data. Would this change significantly the value of the acceleration? State carefully your reasoning as to why it would or would not change.

2. Would choosing a different point as the origin for the analysis change significantly the value of v_o, the initial velocity? State carefully your reasoning as to why it would or would not change.

3. How long before the time you chose as zero for your analysis was the puck actually released? (Hint—Consider the time at which $v = 0$ in Equation 4.)

4. There are 12 data points in the linear least squares fit to Equation 4. Statistical theory states that there is only a 0.1% probability (1 chance in 1000) of achieving a value of $r \geq 0.823$ for 12 data points of uncorrelated data. State the most complete assessment you can make about how well your data confirm the theory.

Kinematics in Two Dimensions on the Air Table

OBJECTIVES

- ❑ Investigate the trajectory of a puck on a tilted air table.
- ❑ Determine the acceleration a_y in the y direction by taking the slope of a graph of the y component of velocity v_y versus time.
- ❑ Determine the very slight negative acceleration in the x direction due to friction from the very slight negative slope of v_x versus time.

EQUIPMENT LIST

- Air table with sparktimer, air pump, foot switch, level, and carbon paper
- Recording paper, ruler, square, and protractor

THEORY

In Figure 6-1 an object is launched at an angle θ relative to the horizontal (x axis) and moves in a two-dimensional path. If v_o is the magnitude of its velocity, then its initial velocity in the x direction is $v_o \cos\theta$, and its initial velocity in the y direction is $v_o \sin\theta$. Assuming no friction, there is no acceleration in the x direction, and the x component of velocity is constant in time. There is an acceleration in the negative y direction, opposite in direction to the initial y component of velocity. The two equations that describe the x and y components of velocity v_x and v_y as a function of time are

$$v_x = v_o \cos\theta \qquad \text{(Eq. 1)}$$

$$v_y = v_o \sin\theta - a_y t \qquad \text{(Eq. 2)}$$

In Equation 2 the direction of the acceleration has been shown by the negative sign, and a_y stands for the magnitude of the acceleration.

Assume that values of the x and y coordinates are known for some general two-dimensional motion. If x_i and x_f stand for the x coordinate positions at the initial time t_i and final time t_f, and y_i and y_f stand for the y coordinate positions at these same times, then the displacements Δx and Δy during the time interval Δt are given by

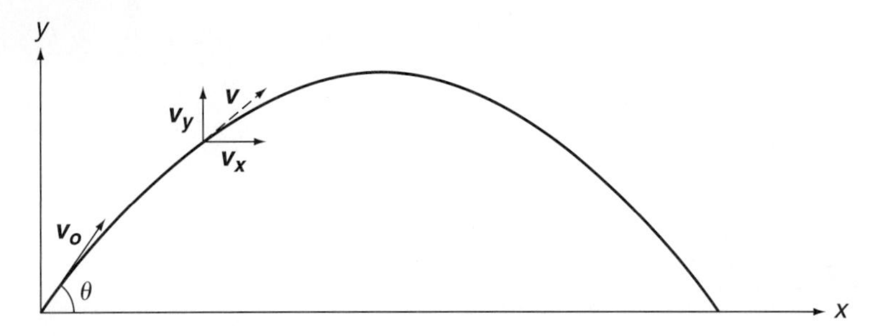

Figure 6-1 Two-dimensional motion with acceleration in the negative y direction.

$$\Delta x = x_f - x_i \quad \Delta y = y_f - y_i \quad \Delta t = t_f - t_i \tag{Eq. 3}$$

By definition the x and y components of the **average velocity** during a time interval Δt are given by the following equations:

$$\bar{v}_x = \frac{\Delta x}{\Delta t} \quad \text{and} \quad \bar{v}_y = \frac{\Delta y}{\Delta t} \tag{Eq. 4}$$

The velocities in Equations 4 are average velocities during a finite time interval Δt, and are not the same velocities as in Equations 1 and 2, which are instantaneous velocities at some particular time t. By definition the components of the **instantaneous velocities** are the limits of Equations 4 as Δt approaches zero.

For the special case of **constant acceleration,** the average velocity during the time interval Δt is equal to the instantaneous velocity at the time at the center of the interval Δt. For the other special case of **constant velocity,** the average velocity over the interval Δt is equal to the instantaneous velocity at any time during the time interval. Therefore, for the case being considered the instantaneous velocities at the center of the interval of time Δt are given by

$$v_x = \frac{\Delta x}{\Delta t} \quad \text{and} \quad v_y = \frac{\Delta y}{\Delta t} \tag{Eq. 5}$$

In this laboratory we use an air table to approximate a frictionless two-dimensional surface. We tilt the table to form an inclined plane to produce an acceleration in the negative y direction of a coordinate system as shown in Figure 6-1. A puck will be launched with an initial velocity having both x and y components. Measurements of Δx and Δy will allow the determination of v_x and v_y.

EXPERIMENTAL PROCEDURE

1. Read the instruction manual for the air table (Figure 6-2) to become familiar with its operation including the air pump, sparktimer, and foot switch. When the sparktimer is in operation, do not touch the pucks except by the insulated tubes.
2. Level the table by means of the three adjustable legs until a puck placed near the center of the table is essentially motionless.
3. Place a block with a height of about 5 cm under the single leg to tilt the table. Place a sheet of recording paper and a sheet of carbon paper on the table with the recording paper on top.
4. Choose a spark rate of 10.0 Hertz, which will produce a spark every 0.100 seconds. Record the value of $\Delta t = 0.100$ s in the Data Table.
5. Place the two pucks on the air table. Place one of them at the lower right-hand corner of the table and leave it there for all of the procedure. Launch the other puck by hand from a point near the lower

Figure 6-2 Precision air table. (Photo Courtesy of Central Scientific Company)

left-hand corner of the table. The puck should move upward and to the right and approach near the top center of the tilted table, and then move down to the vicinity of the lower right-hand corner. Several practice launches will probably be necessary to determine the correct launch speed and direction needed to achieve the desired trajectory.

6. When, after enough practice, you are sure that you can achieve the desired path, launch the puck and start the sparktimer simultaneously. Stop the sparktimer just before the puck arrives at the lower right-hand corner. Be sure that there are about 20 or so good data points on the record. *Do not remove the sheet of recording paper after making this measurement! Instead, record a second trace on this same paper to experimentally determine the vertical (y direction).* This vertical trace is obtained by releasing the puck from a point near the top of the table on the same side of the air table from which the puck is launched.
7. Remove the recording sheet for data analysis. Choose as the origin the lowest point on the trajectory that comes out clearly and call the corresponding time zero. Label that first point as 0 and then label the next 15 consecutive data points. Note that the times corresponding to these points are already given in the Data Table.
8. Draw a straight line through the vertical line formed by the sparktimer trace for the puck falling vertically. This defines the vertical direction.
9. Construct a straight line that passes through the origin chosen above in Step 7 and parallel to the vertical line drawn in Step 8. This is the y axis of the coordinate system. Construct a line through the origin and perpendicular to the y axis. This is the x axis of the coordinate system.
10. Measure the perpendicular distance of each data point from the y axis. Record these values in the Data Table as the values of the coordinate x.
11. Measure the perpendicular distance of each data point from the x axis. Record these values in the Data Table as the values of the coordinate y.

CALCULATIONS

1. Calculate the differences between the successive values of x and y and record them in the Calculations Table as the values of Δx and Δy. Note that Δy will change in magnitude, and it will become negative after the puck reaches its maximum height and starts back down.
2. Use Equations 5 to calculate the components of the instantaneous velocity v_x and v_y and record them in the Calculations Table. The time at which the instantaneous velocities apply are at the center of each interval, and those times have already been recorded in the Calculations Table for each interval.
3. Perform a linear least squares fit with v_x as the vertical axis and t as the horizontal axis. The slope of this fit should be very small and negative, and it is equal to the acceleration caused by friction.

The intercept is the initial velocity in the x direction $(v_x)_o$. Record the values of a_x, $(v_x)_o$, and r in the Calculations Table.

4. Perform a linear least squares fit with v_y as the vertical axis and time as the horizontal axis. The slope of this fit is equal to the acceleration in the y direction, and the intercept is the initial velocity in the y direction $(v_y)_o$. Record the values of a_y, $(v_y)_o$, and r in the Calculations Table.

GRAPHS

1. On one sheet of graph paper, graph both v_x and v_y as a function of time. Use different symbols for v_x and v_y. Because v_y can have both positive and negative values, choose the origin of the velocity scale in the center of the graph paper. Include the straight lines obtained by the least squares fit for each case.

Name | Section | Date

LABORATORY 6 *Kinematics in Two Dimensions on the Air Table*

PRE-LABORATORY ASSIGNMENT

Assume that the projectile motion shown in Figure 6-1 is ideal with no friction. For that assumption, choose the correct statements about the motion in 1 and 2 below.

1. (a) The value of v_x is always positive and decreases very slightly with time. (b) The value of v_x changes from positive to negative with time. (c) The value of v_x is always negative and increases very slightly with time. (d) The value of v_x is always positive and is constant with time. (e) The value of v_x changes from negative to positive with time.

2. (a) The value of v_y is always positive and decreases very slightly with time. (b) The value of v_y changes from positive to negative with time. (c) The value of v_y is always negative and increases very slightly with time. (d) The value of v_y is always positive and is constant with time. (e) The value of v_y changes from negative to positive with time.

Assume now for the projectile motion shown in Figure 6-1 that there is a very small frictional force acting on the puck. For that assumption, choose the correct statements about the motion in 3 and 4 below.

3. (a) The value of v_x is always positive and decreases very slightly with time. (b) The value of v_x changes from positive to negative with time. (c) The value of v_x is always negative and increases very slightly with time. (d) The value of v_x is always positive and is constant with time. (e) The value of v_x changes from negative to positive with time.

4. (a) The value of a_x is zero. (b) The value of a_x is very small and negative. (c) The value of a_x is very small and positive. (d) The value of a_x changes from positive to negative. (e) The value of a_x changes from negative to positive.

A particle moves in such a way that its coordinate positions x and y as a function of time are given by the following table.

x(m)	0.000	0.550	1.100	1.650	2.200	2.750	3.300	3.850	4.400
y(m)	0.000	1.425	2.700	3.825	4.800	5.625	6.300	6.825	7.200
t(s)	0.000	0.100	0.200	0.300	0.400	0.500	0.600	0.700	0.800

Answer the following questions (5 through 7) with respect to these data.

5. What is the particle's average velocity in the y direction $\overline{v}_y$ between $t = 0.200$ s and $t = 0.300$ s? What is $\overline{v}_y$ between $t = 0.400$ s and $t = 0.500$ s? What is $\overline{v}_y$ between $t = 0.700$ s and $t = 0.800$ s? Show your work.

6. What is the particle's average velocity in the x direction $\overline{v}_x$ between $t = 0.200$ s and $t = 0.300$ s? What is $\overline{v}_x$ between $t = 0.400$ s and $t = 0.500$ s? What is $\overline{v}_x$ between $t = 0.700$ s and $\mathrm{t} = 0.800$ s? Show your work.

7. What is the particle's instantaneous velocity in the y direction v_y at $t = 0.250$ s? What is v_y at $t = 0.450$ s? What is v_y at $t = 0.750$ s? (Hint—Assume that the average velocity during a time interval is equal to the instantaneous velocity at the center of the time interval.) Show your work.

Name | Section | Date

Lab Partners

6 LABORATORY 6 Kinematics in Two Dimensions on the Air Table

LABORATORY REPORT

Data Table

t(s)	x(m)	y(m)
0.000		
0.100		
0.200		
0.300		
0.400		
0.500		
0.600		
0.700		
0.800		
0.900		
1.000		
1.100		
1.200		
1.300		
1.400		
1.500		

Calculations Table

Δx(m)	Δy(m)	v_x(m/s)	v_y(m/s)	t(s)
				0.050
				0.150
				0.250
				0.350
				0.450
				0.550
				0.650
				0.750
				0.850
				0.950
				1.050
				1.150
				1.250
				1.350
				1.450

	$a_x =$ m/s^2	$a_y =$ m/s^2
$\Delta t =$ s	$(v_x)_o =$ m/s	$(v_y)_o =$ m/s
	$r =$	$r =$

SAMPLE CALCULATIONS

1. $\Delta t = 1/f =$
2. $\Delta x = x_{n+1} - x_n =$
3. $\Delta y = y_{n+1} - y_n =$
4. $v_x = \Delta x / \Delta t$
5. $v_y = \Delta y / \Delta t$

QUESTIONS

1. Suppose a different point had been chosen as the origin for the analysis of the data. In principle would this significantly change the value of the acceleration a_y? State clearly why it would or would not change the value.

2. Would choosing a different point as the origin for the analysis change significantly the value of the initial velocity $(v_y)_o$? State clearly why it would or would not change.

3. What is the equation for the magnitude of the initial velocity of the puck? Calculate the value of the initial velocity of the puck.

4. Calculate the ratio of the magnitude of a_x to the magnitude of a_y.

5. The small negative value of a_x is due to friction. If the ratio calculated in Question 4 is 0.02, it would indicate that friction is about a 2% effect. If the ratio were 0.01 it would indicate that friction is approximately a 1% effect. Based on your value of that ratio, estimate how large an effect friction was in your data.

Coefficient of Friction

OBJECTIVES

- ❑ Determine μ_s, the static coefficient of friction, and μ_k, the kinetic coefficient of friction, by sliding the block down a board that acts as an inclined plane.
- ❑ Determine μ_s and μ_k using the board with a pulley mounted on it in a horizontal position and applying known forces to the block.
- ❑ Compare the different values for μ_s and μ_k obtained by the two different techniques.
- ❑ Demonstrate that the coefficients are independent of the normal force, and that $\mu_s > \mu_k$.

EQUIPMENT LIST

- Smooth wooden board with pulley attached to one end
- Smooth wooden block (6-inch long 2×4 for example) with hook attached
- Second wooden block of known length to form inclined plane
- String, balance, calibrated masses, mass holder, and slotted masses

THEORY

Friction is a resisting force that acts along the tangent to two surfaces in contact when one body slides or attempts to slide across another. **Normal force** is the force that each body exerts on the other body, and it acts perpendicular to each surface. The frictional force is directly proportional to the normal force.

There are two different kinds of friction. **Static friction** occurs when two surfaces are still at rest with respect to each other, but an attempt is being made to cause one of them to slide over the other one. Static friction arises to oppose any force trying to cause motion tangent to the surfaces. The static frictional force f_s is given by

$$f_s \leq \mu_s N \qquad \text{(Eq. 1)}$$

where N stands for the normal force between the two surfaces, and μ_s is a constant called the **coefficient of static friction**. The meaning of Equation 1 is that the static frictional force varies in response to applied forces from zero up to a maximum value given by the equality in that equation. If the applied force is less than the maximum, then the frictional force that arises is equal to the applied force, and there is no motion. If the applied force is greater than the maximum, the object will begin to move, and static friction conditions are no longer valid.

The other kind of friction occurs when two surfaces are moving with respect to each other. It is called **kinetic friction,** and it is characterized by a constant μ_k called the **coefficient of kinetic friction**. The kinetic frictional force f_k is given by

$$f_k = \mu_k N \qquad \text{(Eq. 2)}$$

where N is again the normal force. Equation 2 states that the kinetic frictional force is a constant value any time the object is in motion. In fact, the coefficient of kinetic friction does vary somewhat with speed. It is assumed for this laboratory that at the slow speeds used, μ_k does not depend upon speed. To a good approximation both coefficients are independent of the apparent area of contact between the two surfaces.

An inclined plane with an angle θ that can be adjusted is shown in Figure 7-1. If a block is placed on the plane, and the angle is slowly increased, the block will begin to slip at some angle. The normal force N acts perpendicularly to the plane, and a component of the weight of the block $mg \cos \theta_s$ acts in the opposite direction. The block is in equilibrium for motion perpendicular to the plane, and these forces are equal and

$$N = mg \cos \theta_s \qquad \text{(Eq. 3)}$$

where θ_s is the angle at which the block just begins to slip on the inclined plane. Parallel to the plane there are also two forces. A component of the weight of the block $mg \sin \theta_s$ acts down the plane, and the frictional force f_s acts up the plane. At the point where the block just slips, the maximum frictional force is exerted, and these two forces are equal. In equation form

$$f_s = \mu_s N = mg \sin \theta_s \qquad \text{(Eq. 4)}$$

Combining Equations 3 and 4 leads to

$$\mu_s = \frac{mg \sin \theta_s}{N} = \frac{mg \sin \theta_s}{mg \cos \theta_2} = \tan \theta_s \qquad \text{(Eq. 5)}$$

Equation 5 can be used to determine μ_s by measuring $\tan \theta_s$ when the block just begins to slip on the inclined plane.

In a similar way the same inclined plane can be used to determine μ_k. By giving the block a slight push to get it started, the angle can be determined at which the block slides down the plane at constant velocity. A push is needed to get it started because generally $\mu_s > \mu_k$, and the static frictional force is greater than the

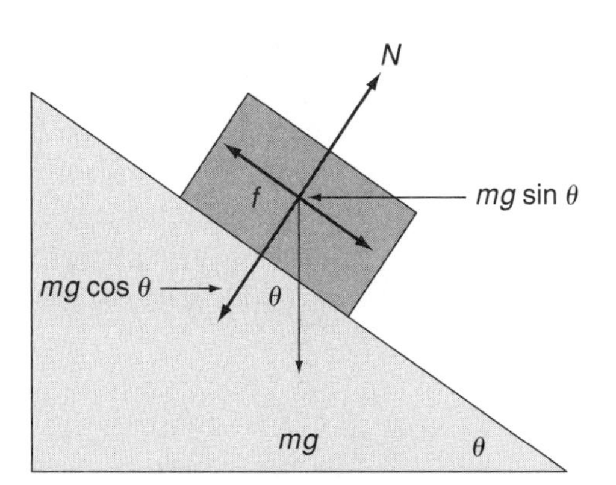

Figure 7-1 Forces acting on a block on an inclined plane.

kinetic frictional force. When the block is moving down the plane at constant velocity, the block is in equilibrium with the vector sum of forces on the block equal to zero. In equation form

$$N = mg\cos\theta_k \quad \text{and} \quad f_k = mg\sin\theta_k = \mu_k N \tag{Eq. 6}$$

Using algebra to combine these equations leads to

$$\mu_k = \frac{mg\sin\theta_k}{N} = \frac{mg\sin\theta_k}{mg\cos\theta_k} = \tan\theta_k \tag{Eq. 7}$$

Equation 7 can be used to determine μ_k from the angle at which the block slides down the inclined plane at constant velocity after it has been given a slight push to get it started.

If the inclined board is lowered to the horizontal position, a force can be applied to the block by means of a string running over a pulley and down to a mass as shown in Figure 7-2. For a given block mass M_1 it is possible to slowly add mass to M_2 until M_1 moves. The point at which the block just moves is the last point that the system is in equilibrium, and it occurs when the maximum static friction is acting. At that point the following conditions are met:

$$T = f_s \qquad T = M_2 g \qquad N = M_1 g \qquad f_s = \mu_s N \tag{Eq. 8}$$

In these equations T is the tension in the string, and the other symbols are the same as previously defined. Combining these four equations leads to

$$f_s = M_2 g = \mu_s N = \mu_s M_1 g \tag{Eq. 9}$$

Using the second and fourth terms above and canceling the common factor of g gives

$$M_2 = \mu_s M_1 \tag{Eq. 10}$$

Equation 10 can be used to determine μ_s by finding the minimum mass M_2 needed to cause the block of mass M_1 to just move.

The same procedure can also be used to determine the coefficient of kinetic friction. Refer to Figure 7-2 and imagine that f_s is replaced by f_k for the case when the block is in motion. When the system is moving at constant velocity it is also in equilibrium, and the following conditions are met:

$$T = f_k \qquad T = M_2 g \qquad N = M_1 g \qquad f_k = \mu_k N \tag{Eq. 11}$$

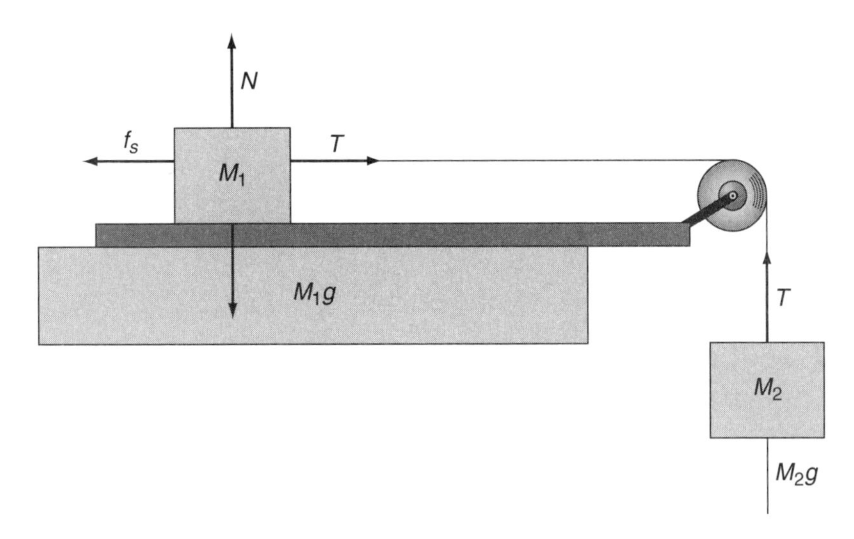

Figure 7-2 Force applied to a block on a horizontal plane.

Combining these equations leads to

$$M_2 = \mu_k M_1 \qquad \text{(Eq. 12)}$$

Equation 12 can be used to measure μ_k by finding the value of M_2 needed to cause M_1 to move at constant velocity.

EXPERIMENTAL PROCEDURE

The Inclined Plane

1. *Note which side of the block and which side of the board you are using. Continue to use the same surfaces for all the measurements made in this laboratory.*
2. Place the block with the hook attached with its large surface down on the board and incline the board until the block just begins to slide on its own. Incline the board by placing the block of known height (approximately 15.0 cm) under one end of the board. Record the value of the height of the block as Y (to the nearest 1 mm, which is 0.001 m) in both Data Table 1 and Data Table 2. Move the block of known height Y toward the line along which the board is resting on the table. This will increase the angle θ as shown in Figure 7-3.
3. When the block on top slides down the board because static friction can no longer hold it in place, record in Data Table 1 (to the nearest 0.001 m) the value of X, the distance from the pivot line of the board to the block as shown in Figure 7-3.
4. Repeat Step 3 three more times for a total of four trials with only the block itself on top of the board. Record the values of X associated with each of these trials in Data Table 1 in the column labeled 0 mass added.
5. Using light tape, attach a 0.200 kg mass to the top of the block and repeat the steps above, recording in Data Table 1 the values of X for the four trials in the column labeled 0.200 kg added.
6. Continue the process adding 0.400 kg and then 0.600 kg, and record the values of X for four trials in each case in the appropriate column in Data Table 1. The data taken in these first six steps will determine μ_s, from the fact that $\mu_s = \tan\,\theta_s = Y/X$.
7. Repeat all of the procedures above again, but instead of finding the point at which the block begins to move on its own, find the point at which the block moves at approximately constant speed after it is given a slight push to begin its motion. There will probably be more variation in the values of X obtained in this case because it is difficult to determine when the speed is constant. Again start with the block alone, and then add the same values of mass as above. In each case do four trials and record the values of X in the appropriate columns in Data Table 2. The data taken in these steps will be used to determine the coefficient of kinetic friction, μ_k.

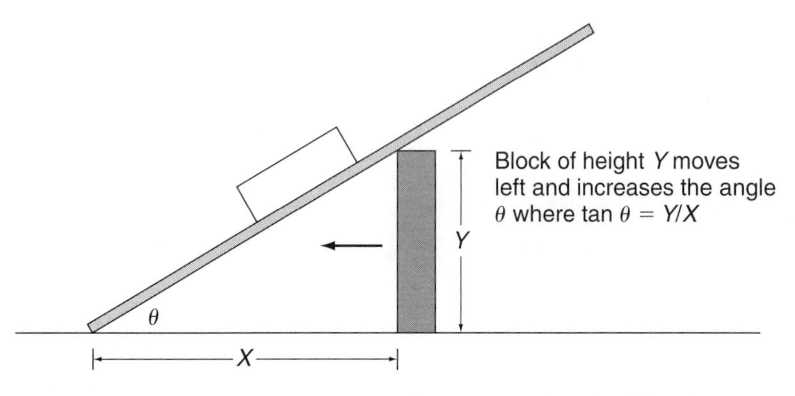

Figure 7-3 Geometry to increase θ by moving block of height Y.

Horizontal Plane with Pulley

1. Determine the mass of the block with hook attached using the balance. Record it in Data Table 3 as M_1 in the space labeled 0 kg added.
2. Place the board in a horizontal position on the laboratory table with the pulley beyond the edge of the table as shown in Figure 7-2.
3. Attach a piece of string to the hook in the block. Place it over the pulley and attach the mass holder to the other end of the string. *Be sure that the same surface of the block and board are in contact as in the first procedure.* Add mass to the mass holder to find the minimum mass needed to just cause the block to move. Record the value as M_2 in Data Table 3. Include the 0.050 kilogram mass of the holder in the total for M_2. Repeat the procedure two more times for a total of three trials.
4. Repeat Step 3 but add 0.200 kg to the top of the block. Record the value of the mass of the block plus 0.200 kg as M_1. Again determine the minimum mass needed to just cause the mass M_1 to move. Do three trials and record each as M_2 in Data Table 3.
5. Continue this process adding 0.400, 0.600, and finally 0.800 kg to the top of the block. In each case take three trials.
6. Perform a similar set of measurements as just described in Steps 1 through 5, but this time determine the mass M_2 needed to keep the block moving at constant velocity after it has been started with a small push. Again take three trials for each case and use values of M_1 beginning with the mass of the block and increasing in steps of 0.200 kg up to a total added mass of 0.800 kg. Record the values of M_2 and M_1 for all cases in Data Table 4.

CALCULATIONS

Board as an Inclined Plane

1. For the static friction data calculate the value of the mean $\overline{X}$ for the four trials of X at each value of M_1 and record them in Calculations Table 1 under the Static section. For each of the values of $\overline{X}$ in Calculations Table 1 calculate the value of μ_s as $\tan\theta_s = Y/\overline{X}$ and record them in Calculations Table 1 under the Static section. Calculate $\overline{\mu}_s$ and the standard error $\alpha_{\mu s}$ for the four measurements. Record those results.
2. For the kinetic friction data calculate the value of the mean $\overline{X}$ for the four trials of X at each value of M_1 and record them in Calculations Table 1 under the Kinetic section. For each of the values of $\overline{X}$ calculate the value of μ_k as $\tan\theta_k = Y/X$ and record them in Calculations Table 1 under the Kinetic section. Calculate $\overline{\mu_k}$ and the standard error $\alpha_{\mu k}$ for the four measurements. Record those results.

Horizontal Plane with Pulley

1. Calculate the mean $\overline{M_2}$ for the three trials of M_2 for each of the values of M_1 for both the static and kinetic friction cases. Record these values in Calculations Table 2 along with the value of M_1 for each value of $\overline{M_2}$.
2. According to Equation 10 there is a linear relationship between M_2 and M_1. A linear least squares fit to the static friction values of $\overline{M_2}$ versus M_1 should produce a straight line with a slope of μ_s. Perform such a fit with $\overline{M_2}$ as the vertical axis and M_1 as the horizontal axis. Record the value of the slope as μ_s in Calculations Table 2 under the Static section. Record the value of r.
3. Equation 12 states that there should also be a linear relationship between M_2 and M_1 for the kinetic friction data. Perform a linear least squares fit with $\overline{M_2}$ as the vertical axis and M_1 as the horizontal axis. Record the value of the slope as μ_k in Calculations Table 2 under the Kinetic section. Record the value of r.

GRAPHS

1. Graph the static friction data for the horizontal plane case with $\overline{M_2}$ as the vertical axis and M_1 as the horizontal axis. Show the straight line obtained from the fit.
2. Graph the kinetic friction data for the horizontal plane case with $\overline{M_2}$ as the vertical axis and M_1 as the horizontal axis. Show the straight line obtained from the fit.

Name ____________ Section ____________ Date ____________

7 LABORATORY 7 *Coefficient of Friction*

PRE-LABORATORY ASSIGNMENT

1. For kinetic friction the direction of the frictional force on a given object is always opposite the direction of that object's motion. (a) true (b) false

2. The two coefficients of friction discussed in this laboratory are static (μ_s) and kinetic (μ_k). Describe the conditions under which each kind is appropriate. Generally, which of the two is larger?

3. Suppose a block of mass 25.0 kg rests on a horizontal plane, and the coefficient of static friction between the surfaces is 0.220. (a) What is the maximum possible static frictional force that could act on the block? ____________N (b) What is the actual static frictional force that acts on the block if an external force of 25.0 N acts horizontally on the block? ____________N Assume $g = 9.80\ \text{m/s}^2$. Show your work and explain both answers.

4. To measure the coefficient of kinetic friction by sliding a block down an inclined plane the block must be in equilibrium. What experimental condition must you try to accomplish that will assure you that the block is in equilibrium?

5. A 5.00 kg block rests on a horizontal plane. A force of 10.0 N applied horizontally causes the block to move horizontally at constant velocity. What is the coefficient of kinetic friction between the block and the plane? Assume $g = 9.80\ \text{m/sec}^2$. Show your work.

6. Both types of coefficient of friction are dimensionless. Why is this true?

7. For either type of coefficient of friction, what is generally assumed about the dependence of the value of the coefficient on the area of contact between the two surfaces?

Name | Section | Date

Lab Partners

7 LABORATORY 7 *Coefficient of Friction*

LABORATORY REPORT

Data Table 1 Inclined Plane—Static Friction

Y = ______________ m

Added Mass	0.000 kg	0.200 kg	0.400 kg	0.600 kg
M_1 (kg)				
X Trial 1 (m)				
X Trial 2 (m)				
X Trial 3 (m)				
X Trial 4 (m)				

Data Table 2 Inclined Plane—Kinetic Friction

Y = ______________ m

Added Mass	0.000 kg	0.200 kg	0.400 kg	0.600 kg
M_1 (kg)				
X Trial 1 (m)				
X Trial 2 (m)				
X Trial 3 (m)				
X Trial 4 (m)				

Data Table 3 Horizontal Plane—Static Friction

Added Mass	0.000 kg	0.200 kg	0.400 kg	0.600 kg	0.800 kg
M_1 (kg)					
M_2 Trial 1					
M_2 Trial 2					
M_2 Trial 3					

Data Table 4 Horizontal Plane—Kinetic Friction

Added Mass	0.000 kg	0.200 kg	0.400 kg	0.600 kg	0.800 kg
M_1 (kg)					
M_2 Trial 1					
M_2 Trial 2					
M_2 Trial 3					

Calculations Table 1—Inclined Plane

Static			
$\overline{X}$ (m)	$\mu_s = Y/\overline{X}$	$\overline{\mu}_s$	$\alpha_{\mu s}$

Kinetic			
$\overline{X}$ (m)	$\mu_k = Y/\overline{X}$	$\overline{\mu}_k$	$\alpha_{\mu k}$

Calculations Table 2—Horizontal Plane

Static			
M_1 (kg)	$\overline{M_2}$ (kg)	μ_s	r

Kinetic			
M_1 (kg)	$\overline{M_2}$ (kg)	μ_k	r

SAMPLE CALCULATIONS

1. $\overline{X} = (X_1 + X_2 + X_3 + X_4)/4 =$
2. $\mu_s = Y/\overline{X} =$
3. Percent Difference for $\mu =$

QUESTIONS

1. Discuss the agreement between the two different measured values of μ_s. Calculate the percentage difference between them. Percentage difference = ____________%. Assume that the error in the horizontal plane value is approximately equal to the standard error for the inclined plane data. Under the assumption that both values of μ_s have that same error, do the two values of μ_s overlap within that assumed error? State the range of each measurement and quantitatively state the extent to which they overlap.

2. Answer the same questions as asked in Question 1, but now consider the kinetic friction data. Percentage difference = ____________%.

3. To what extent do your data confirm the expectation that the coefficients of friction, both static and kinetic, are independent of the normal force? What is the evidence for the inclined plane data? What is the evidence for the horizontal plane data? Give as quantitative an answer as possible in both cases.

4. Do your data confirm the expectation that $\mu_s \geq \mu_k$? Comment for both values of each coefficient and state your evidence.

5. State clearly in your own words what was to be accomplished in this laboratory. To what extent did your performance of the laboratory accomplish those goals?

Newton's Second Law on the Air Table

OBJECTIVES

- ❑ Investigate the variation in the acceleration produced when different forces are applied to a fixed mass.
- ❑ Demonstrate that the acceleration is proportional to the applied force.
- ❑ Determine the frictional force that acts on the system.

EQUIPMENT LIST

- Air table with pucks, sparktimer, air pump, foot switch, and level
- Carbon paper, recording paper, meter stick or tape
- Pulley, string, laboratory balance, and calibrated masses

THEORY

The relationship between the *net* force F exerted on a body of mass m and the acceleration a of the body is **Newton's Second Law.**

$$F = ma \qquad \text{(Eq. 1)}$$

A mass m_1 is shown in Figure 8-1 on a horizontal surface. It is connected to a second mass m_2 by a string running over a pulley. There is tension T in the string connecting the masses. Each mass is subject to several forces shown in a free body diagram in Figure 8-2. Four forces act on mass m_1, but only two of them lie along the direction of motion of the mass. The weight of m_1 and the normal force from the table are equal in magnitude and opposite in direction and sum to zero. The net force acting on m_1 is equal to $T - f$ and Equation 1 gives

$$T - f = m_1 a \qquad \text{(Eq. 2)}$$

The two forces acting on mass m_2 are its weight $m_2 g$ and the tension T. Because m_1 and m_2 are connected, the acceleration of m_2 has the same magnitude as the acceleration of m_1. In equation form

$$m_2 g - T = m_2 a \qquad \text{(Eq. 3)}$$

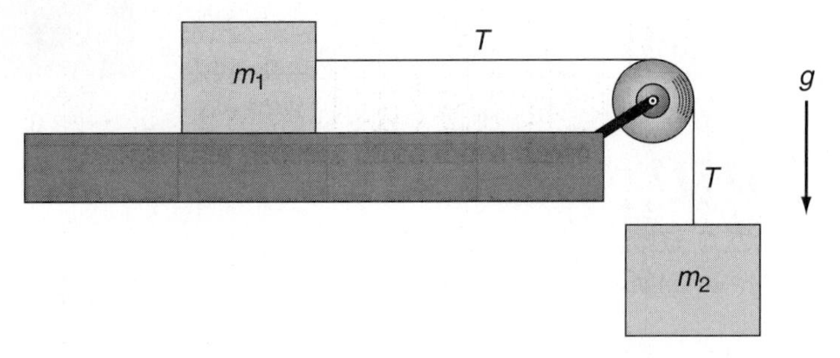

Figure 8-1 Force applied to mass m_1 by weight of mass m_2.

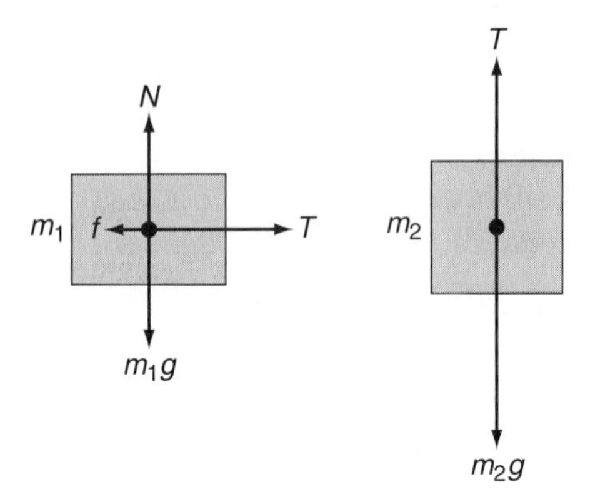

Figure 8-2 Free body diagram of the forces on masses m_1 and m_2.

If Equations 2 and 3 are combined and f added to both sides, the result is

$$m_2 g = (m_1 + m_2)a + f \qquad \text{(Eq. 4)}$$

Equation 4 assumes that f represents all the friction in the system including the pulley.

We will make a series of measurements varying the applied force $m_2 g$ while the mass $(m_1 + m_2)$ is kept constant. We will measure the resulting acceleration a of the masses as a function of the force $m_2 g$. The coordinate position x of the puck as a function of time t will be measured using a sparktimer to mark x at time intervals of 0.100 seconds. The average velocity $\overline{v}$ during any time interval Δt is

$$\overline{v} = \frac{\Delta x}{\Delta t} \qquad \text{(Eq. 5)}$$

where Δx is the displacement during the time interval Δt. Because the acceleration is constant, the average velocity $\overline{v}$ during time interval Δt is the instantaneous velocity v at the time t at the center of Δt. The equation

$$v = \frac{\Delta x}{\Delta t} \qquad \text{(Eq. 6)}$$

gives the instantaneous velocity v at the center of time interval Δt. From the resulting values of v as a function of time, we will determine the acceleration a.

EXPERIMENTAL PROCEDURE

1. Read the instruction manual for the air table (Figure 8-3) to become familiar with its operation, including the air pump, sparktimer, and foot switch. Do not touch the pucks when the sparktimer is in operation except by the insulated tubes.

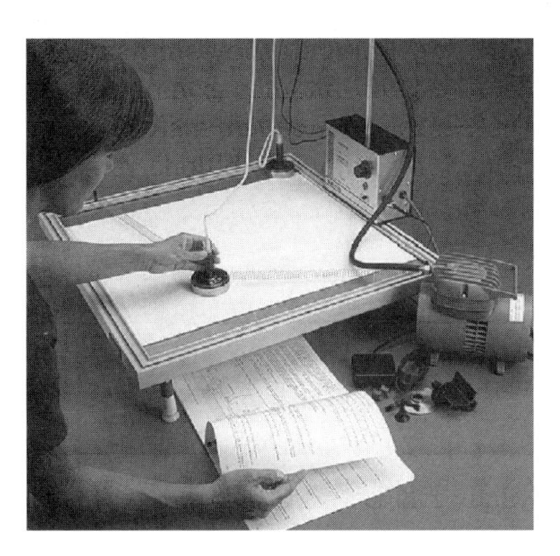

Figure 8-3 Precision air table. (Photo courtesy of Central Scientific Company)

2. Level the table by means of the three adjustable legs until a puck placed near the center of the table is essentially motionless. Place a sheet of carbon paper and a sheet of recording paper on the table with the recording paper on top.
3. Place one of the pucks in the corner of the table and leave it there for the rest of the procedure. Place the other puck on a laboratory balance and determine its mass m_p. Record the value of m_p in kg in Data and Calculations Table 1.
4. Place the puck for which mass has been determined on the air table. Attach a string to the puck. Run the string over the pulley as shown in Figure 8-1 and attach a mass of 0.010 kg on the end of the string to serve as mass m_2.
5. Place mass totaling 0.040 kg on top of the puck. The total mass $m_1 + m_2$ is equal to the puck mass m_p plus the 0.040 kg plus the 0.0100 kg on the end of the string. Record this value ($m_p + 0.050$ kg) in Data and Calculations Table 1 as $m_1 + m_2$.
6. Choose a spark rate of 20.0 Hertz. Release the puck near the edge of the table with zero velocity and simultaneously start the sparktimer. Stop the sparktimer just before the puck reaches the other side of the table. Release the puck with no initial velocity.
7. Remove 0.010 kg from the puck and add it to the 0.010 kg on the end of the string so that m_2 is now 0.020 kg, but $m_1 + m_2$ remains the same. Slide the recording paper over slightly, again release the puck with no initial velocity, and record this trace with the sparktimer.
8. Continue this process, each time shifting 0.010 kg from the puck to the end of the string to produce traces with values of m_2 equal to 0.030, 0.040, and 0.050 kg while the value of $m_1 + m_2$ remains fixed. Record all five traces on the same sheet of paper.
9. Choose an origin for each of the five traces and label every other point. This will produce data points with time intervals of $\Delta t = 0.100$ seconds. Measure the values of the coordinate position x as a function of time from $t = 0$ to $t = 0.600$ seconds for each of the traces. Record the values of x in Data and Calculations Table 1.

CALCULATIONS

1. Calculate the values of the displacement Δx between successive data points for each of the five traces. Record the values of Δx in Data and Calculations Table 1.
2. Use Equation 6 to calculate the instantaneous velocity at the times corresponding to the center of the measured time intervals. Record the values of v at the appropriate instantaneous times in Data and Calculations Table 1.

3. From the values of m_2 using $g = 9.80\,\text{m/s}^2$, calculate the values m_2g for each of the applied forces. Record the results in Data and Calculations Table 2.
4. Perform a linear least squares fit to each of the five sets of v versus t data with v as the vertical axis and t as the horizontal axis. Record the values of the slope for each fit as the acceleration a in Data and Calculations Table 2. Record the values of the intercept of each fit as the initial velocity in the table. Record the values of the correlation coefficient r in the table.
5. Equation 4 states that a is proportional to m_2g with $m_1 + m_2$ as the constant of proportionality. Perform a linear least squares fit with m_2g as the vertical axis and a as the horizontal axis. Record in Data and Calculations Table 2 the slope of the fit as $(m_1 + m_2)_{exp}$ the experimental value for the total mass. Record in Data and Calculations Table 2 the intercept of the fit as f the friction.
6. Calculate the percentage error in the value of $(m_1 + m_2)_{exp}$ compared to the known value of $m_1 + m_2$ and record it in Data and Calculations Table 2.

GRAPHS

1. On a single sheet of graph paper plot the data for the v versus t for each of the five cases. Also show on the graph the straight line obtained by each fit to the data. Use different symbols to label the five different sets of data.
2. Make a graph of the data for the applied force m_2g versus the acceleration a. Also show on the graph the straight line obtained by the fit to the data.

Name Section Date

LABORATORY 8 Newton's Second Law on the Air Table

PRE-LABORATORY ASSIGNMENT

1. A net force of 25.0 N acts on a 6.50-kg object. What is the acceleration of the object? Show your work.

2. In Figures 8-1 and 8-2, what are the two forces acting on mass m_1 that must be considered in determining the acceleration of the mass? Why don't the other forces have to be considered?

3. Suppose that a 0.450-kg puck (m_1) is attached to a 0.0400-kg mass (m_2) as shown in Figure 8-1. A constant frictional force $f = 0.100$ N acts on the puck. Solve Equation 4 for the acceleration a. Show your work.

4. What is the weight (in N) of a mass of 0.050 kg? Show your work.

The following data for coordinate position x versus time t were taken for a puck that is uniformly accelerated.

t (s)	0.000	0.100	0.200	0.300	0.400	0.500	0.600
x (m)	0.000	0.0102	0.0399	0.0896	0.1603	0.2498	0.3607

Δx (m)						
v (m/s)						
t (s)	0.050	0.150	0.250	0.350	0.450	0.550

5. Calculate the displacement Δx during each time interval and record each of them in the table above. Calculate the instantaneous velocity v at the center of each time interval and record each of them in the table above. Show sample calculations.

6. For the data in the table above perform a linear least squares fit with v the vertical axis and t the horizontal axis. Record the slope of this fit as the acceleration a of the puck. Record the intercept of the fit as the initial velocity v_o. (*This is the calculation that you will do five times in the laboratory.*)

$a =$ ________________ m/s²

$v_o =$ ________________ m/s

Name　　　　Section　　　　Date

Lab Partners

8 LABORATORY 8 *Newton's Second Law on the Air Table*

LABORATORY REPORT

Data and Calculations Table 1

$\Delta t = 0.100$ s	$m_p =$ kg	$m_1 + m_2 =$ kg

t (s)	0.000	0.100	0.200	0.300	0.400	0.500	0.600
x (m)							

m_2 .0100 kg							
	Δx (m)						
	v (m/s)						
	t (s)	0.050	0.150	0.250	0.350	0.450	0.550

t (s)	0.000	0.100	0.200	0.300	0.400	0.500	0.600
x (m)							

m_2 .0200 kg							
	Δx (m)						
	v (m/s)						
	t (s)	0.050	0.150	0.250	0.350	0.450	0.550

t (s)	0.000	0.100	0.200	0.300	0.400	0.500	0.600
x (m)							

m_2 .0300 kg							
	Δx (m)						
	v (m/s)						
	t (s)	0.050	0.150	0.250	0.350	0.450	0.550

t (s)		0.000	0.100	0.200	0.300	0.400	0.500	0.600
x (m)								
m_2 .0400 kg	Δx (m)							
	v (m/s)							
	t (s)	0.050	0.150	0.250	0.350	0.450	0.550	

t (s)		0.000	0.100	0.200	0.300	0.400	0.500	0.600
x (m)								
m_2 .0500 kg	Δx (m)							
	v (m/s)							
	t (s)	0.050	0.150	0.250	0.350	0.450	0.550	

Data and Calculations Table 2

Force $= m_2 g$ (N)					
Acceleration (m/s^2)					
Initial velocity (m/s)					
Correlation Coefficient					

$f =$ N	$(m_1 + m_2)_{exp} =$ kg	$r =$	% error $=$

SAMPLE CALCULATIONS

1. $\Delta x = x_f - x_i =$
2. $v = \Delta x / \Delta t =$
3. $m_1 + m_2 = m_p + 0.0500 =$
4. % error $= (E - K)/K \times 100\% =$

QUESTIONS

1. Each of your five graphs of v versus t contains six data points. According to statistical theory for six data points there is only 0.1% probability (1 chance in 1000) that uncorrelated data would give $r \geq 0.974$. State how your values of r compare to 0.974, and make the best statement you can about the validity of the correlation of these data.

2. Your graph of applied force versus acceleration contains five data points, and statistical theory indicates that for five points there is a 1% probability that $r \geq 0.959$ would be obtained for uncorrelated data. From your value of r make the best statement you can about the validity of the correlation of your data.

3. What is the ratio of the frictional force f to the applied force m_2g for each case? For which of the cases (if any) is f less than 10% of m_2g? Based on this, what is your conclusion about the importance of friction in your data?

4. Summarize the concepts investigated in this laboratory. What was to be proved, and how confident are you that your data confirm the theory?

Newton's Second Law on Atwood's Machine

OBJECTIVES

- ❑ Investigate the acceleration produced by a series of different forces applied to a fixed mass, and demonstrate that the acceleration is proportional to the applied force.
- ❑ Demonstrate that the constant of proportionality between the acceleration and the applied force is the mass to which the force is applied.
- ❑ Determine the frictional force that acts on the system.

EQUIPMENT LIST

- Atwood's machine pulley (very low-friction ball-bearing pulley)
- Calibrated slotted masses, slotted mass holders, laboratory balance
- Laboratory timer (capable of measuring to 0.01 s), strong thin nylon string, and meter stick

THEORY

The relationship between the *net* force F exerted on a mass m and the acceleration a of the mass is **Newton's Second Law.**

$$F = ma \qquad \text{(Eq. 1)}$$

The system shown in Figure 9-1 is called an Atwood's machine. It consists of two masses at the ends of a string passing over a pulley. Also shown in the figure is a free-body diagram of the forces. For $m_2 > m_1$, Equation 1 applied to each mass gives

$$T - m_1 g = m_1 a \quad \textit{and} \quad m_2 g - T = m_2 a \qquad \text{(Eq. 2)}$$

where T is the tension in the string, and a is the magnitude of the acceleration of either mass. Combining Equations 2 leads to

$$(m_2 - m_1)g = (m_1 + m_2)a \qquad \text{(Eq. 3)}$$

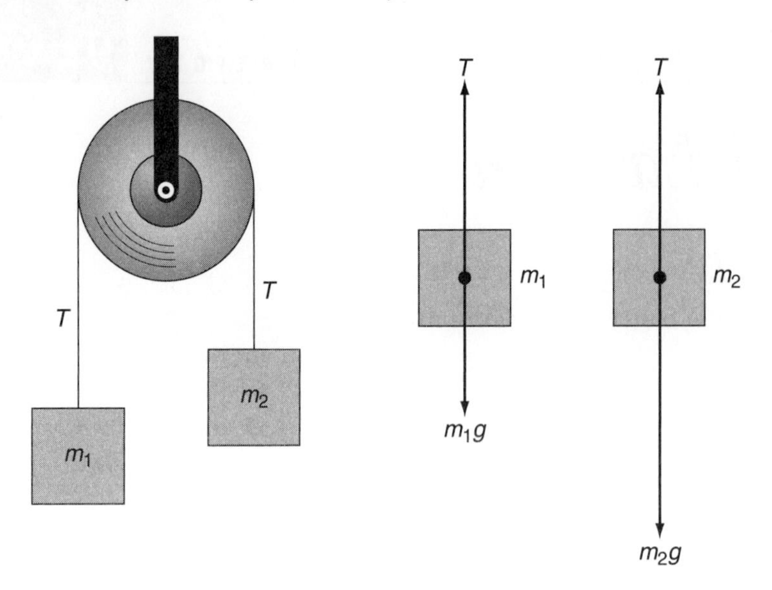

Figure 9-1 Atwood's machine and free-body diagram of the forces on each mass.

Equation 3 states that a force $(m_2 - m_1)g$ equal to the difference in the weight of the two masses acts on the sum of the masses $(m_1 + m_2)$ to produce an acceleration a of the system. There will be a frictional force f in the system that opposes the applied force $(m_2 - m_1)g$. Including the frictional force but moving it to the other side of the equation gives

$$(m_2 - m_1)g = (m_1 + m_2)a + f \qquad \text{(Eq. 4)}$$

An Atwood's machine is shown in Figure 9-2 where $m_2 > m_1$, the mass m_1 is initially on the floor, and m_2 is released from rest at distance x above the floor at $t = 0$. Successive positions of the two masses are shown in Figure 9-2 at later times until the final picture shows m_2 as it strikes the floor at some time t after its release. The relationship between the distance x, the acceleration a of the system, and the time t is

$$x = \frac{at^2}{2} \qquad \text{(Eq. 5)}$$

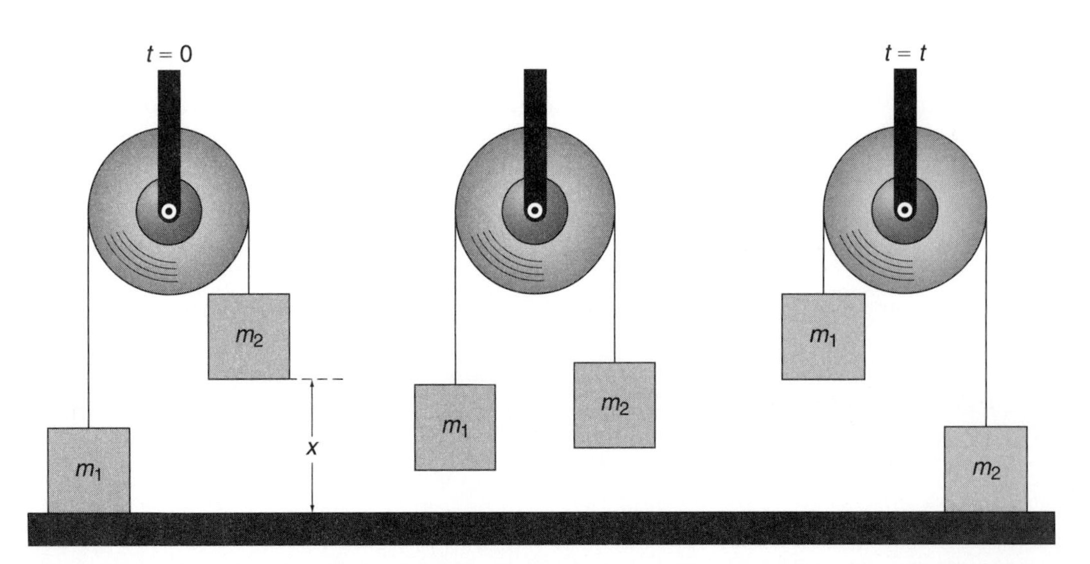

Figure 9-2 Atwood's machine as mass m_2 falls a distance x in time t.

Solving Equation 5 for a in terms of the measured quantities x and t gives

$$a = \frac{2x}{t^2} \quad \text{(Eq. 6)}$$

This laboratory will measure the acceleration for the Atwood's machine for several different values of the applied force $(m_2 - m_1)g$ using a fixed total mass $(m_1 + m_2)$. Because the pulley is not massless, some portion of its mass should be included in the total mass. You will be challenged to discover what fraction of the pulley's mass should be included when you analyze the data that you will take in the laboratory.

EXPERIMENTAL PROCEDURE

1. Place 1.0000 kg of slotted masses (including a 0.0500 kg mass holder) on each pan of a laboratory balance. Include five 0.0020 kg masses on one of the pans. If all the masses are accurate, the scale should be balanced. If the scales are not exactly balanced, add 0.0010 kg slotted masses to whichever side is needed to produce as perfect a balance as can be obtained. The purpose of this procedure is to produce two masses that are as nearly equal as possible. Record in the Data Table the sum of all mass on the balance as $(m_1 + m_2)$. Record in the Data Table the mass of the pulley as m_p.
2. Place these two collections of slotted masses on mass holders at each end of a string over the pulley. Place the group of masses that contain the collection of 0.0020 kg masses on the left side (m_1) of the pulley and the other masses on the right (m_2).
3. Transfer a 0.0060 kg mass from the left side (m_1) to the right side (m_2) while holding the system fixed. This produces a mass difference $(m_2 - m_1)$ of 0.0120 kg.
4. While one partner holds mass m_1 on the floor, another partner should measure the distance x from the bottom of mass m_2 to the floor as shown in Figure 9-2. The height of the pulley above the floor should be chosen as high as feasible but at least 1.000 m. Record this distance x in the Data Table. *Use this same distance for all measurements.*
5. Release mass m_1 and simultaneously start the timer. Stop the timer when mass m_2 strikes the floor. Repeat this measurement four more times for a total of five trials with a mass difference $(m_2 - m_1)$ of 0.0120 kg. Record all times in the Data Table.
6. Repeat Steps 4 and 5 using mass differences $(m_2 - m_1)$ of 0.0160, 0.0200, 0.0240, 0.0280, and 0.0320 kg by transferring a mass of 0.0020 kg each time. Make a total of five trials for each mass difference and record the measured times in the Data Table.

CALCULATIONS

1. Calculate and record the forces $(m_2 - m_1)g$ using $g = 9.800\,\text{m/s}^2$.
2. Calculate the mean time $\bar{t}$ and the standard error α_t for the five measurements of time at each of the mass differences $(m_2 - m_1)$. Record those values in the Calculations Table.
3. Use Equation 6 to calculate the acceleration a from x and $\bar{t}$ for each value of applied force. Record these values of a in the Calculations Table.
4. Perform a linear least squares fit with the applied force $(m_2 - m_1)g$ as the vertical axis and the acceleration a as the horizontal axis. Record in the Calculations Table the slope of the fit as $(m_1+m_2)_{\text{exp}}$, the intercept of the fit as f, and the value of the correlation coefficient r.

GRAPHS

1. Make a graph of the data with the applied force as the vertical axis and the acceleration as the horizontal axis. Also show on the graph the straight line obtained by the least squares fit to the data.

Name | Section | Date

9 LABORATORY 9 *Newton's Second Law on Atwood's Machine*

PRE-LABORATORY ASSIGNMENT

1. A net force of 3.50 N acts on a 2.75 kg object. What is the acceleration of the object? Show your work.

2. Describe the basic concept of the Atwood's machine. What is the net applied force? What is the mass to which this net force is applied? Show your work.

3. An Atwood's machine consists of a 1.060 kg mass and a 1.000 kg mass connected by a string over a massless and frictionless pulley. Use Equation 3 to find the acceleration of the system. Assume that g is 9.80 m/s^2. Show your work.

4. Suppose that the system in Question 3 has a frictional force of 0.056 N. Use Equation 4 to determine the acceleration of the system. Show your work.

The following data were taken with an Atwood's machine for which the total mass $m_1 + m_2$ is kept constant. For each of the values of mass difference $(m_2 - m_1)$ shown in the table, the time for the system to move $x = 1.000$ m was determined.

$(m_2 - m_1)$ (kg)	0.010	0.020	0.030	0.040	0.050
t (s)	8.30	5.06	3.97	3.37	2.98
a (m/s^2)					
$(m_2 - m_1)g$ (N)					

5. From the data above for x and time t, use Equation 6 to calculate the acceleration for each of the applied forces and record them in the table above. Show the calculation for the 0.010 kg mass difference as a sample calculation.

6. From the mass differences $(m_2 - m_1)$ calculate the applied forces $(m_2 - m_1)g$ and record them in the table above. Use a value of 9.80 m/s^2 for g. Show the calculation for the 0.010 kg mass difference as a sample calculation.

7. Perform a linear least squares fit with the applied force as the vertical axis and the acceleration as the horizontal axis. The slope of the fit is equal to the total mass $(m_1 + m_2)_{exp}$ and the intercept is the frictional force f. Record those and the value of the correlation coefficient r. (*This is the calculation that will be performed for the data of the laboratory.*)

$(m_1 + m_2)_{exp}$ = __________kg f = __________N r = __________

Name | Section | Date

Lab Partners

9 LABORATORY 9 *Newton's Second Law on Atwood's Machine*

LABORATORY REPORT

Data Table

$(m_1 + m_2) =$ kg		$m_p =$ kg		$x =$ m	
$(m_2 - m_1)$ (kg)	t_1 (s)	t_2 (s)	t_3 (s)	t_4 (s)	t_5 (s)
0.012					
0.016					
0.020					
0.024					
0.028					
0.032					

Calculations Table

$(m_2 - m_1)g$ (N)	$\bar{t}$ (s)	α_t (s)	a (m/s^2)
$(m_1+m_2)_{exp} =$ kg	$f =$ N	$r =$	

SAMPLE CALCULATIONS

1. % Error = (E – K)/K × 100% =

QUESTIONS

1. According to statistical theory for six data points there is only 1% probability (1 chance in 100) that a value of $r \geq 0.917$ and a 0.1% probability (1 chance in 1000) that a value of $r \geq 0.974$ would be obtained for data that are uncorrelated. Based on this idea, what does your value of r indicate for the level of correlation of your data?

2. Divide the frictional force by the applied force, $(m_2 - m_1)g$, for each applied force and express it as a percentage in the space below. Friction would not be important if these percentages were a few percent, would be small if they were about 10%, and would be very important if they were 25% or greater. Make the best possible statement about the importance of friction for your data.

3. Your value of $(m_1 + m_2)_{exp}$ should be greater than your recorded value of $(m_1 + m_2)$ because of the effect of the pulley. Perform the calculations needed to determine what fraction of the pulley's mass appears to be included in your value of $(m_1 + m_2)_{exp}$. If you were to express that fraction as a whole number fraction, which of the following would best fit your data? (½ ⅓ ¼ ⅕)

4. The laboratory instructed you to transfer mass from one side of the pulley to the other side. Why was this procedure used instead of just adding mass to one side of the pulley to produce a larger force?

5. What concept is this laboratory designed to investigate? Describe the extent to which your data and analysis validate the concept.

Torques and Rotational Equilibrium of a Rigid Body

OBJECTIVES

- ❑ Apply the conditions for equilibrium of a rigid body to a meter stick pivoted on a knife edge.
- ❑ Determine the center of gravity of the meter stick, mass of the meter stick, and mass of an unknown object by applying known torques.
- ❑ For a given applied force needed to produce equilibrium, compare the theoretically predicted location of the force to an experimentally determined location.

EQUIPMENT LIST

- Meter stick with adjustable knife-edge clamp and support stand
- Laboratory balance and calibrated hooked masses
- Thin nylon thread and unknown mass with hook

THEORY

If a force F acts on a rigid body that is pivoted about some axis, the body tends to rotate about that axis. The tendency of a force to cause a body to rotate about some axis is measured by a quantity called **torque** τ. It is defined by

$$\tau = Fd_{\perp} \qquad \text{(Eq. 1)}$$

with F the magnitude of the force, and $d_{\perp}$ the lever arm of the force. The units of torque are N–m. Torque caused by a given force must be defined relative to a specific axis of rotation. Figure 10-1 shows two forces F_1 and F_2 acting on an arbitrarily shaped body.

The axis of rotation is along a line through O perpendicular to the page. The direction of the line of action of each force is shown as a dotted line extended in either direction along the force vector. The lever arm for each force is shown as the perpendicular distance from O to the line of action of the force. In this case there are two torques τ_1 and τ_2 acting on the body given by

$$\tau_1 = F_1 d_1 \quad \text{and} \quad \tau_2 = F_2 d_2 \qquad \text{(Eq. 2)}$$

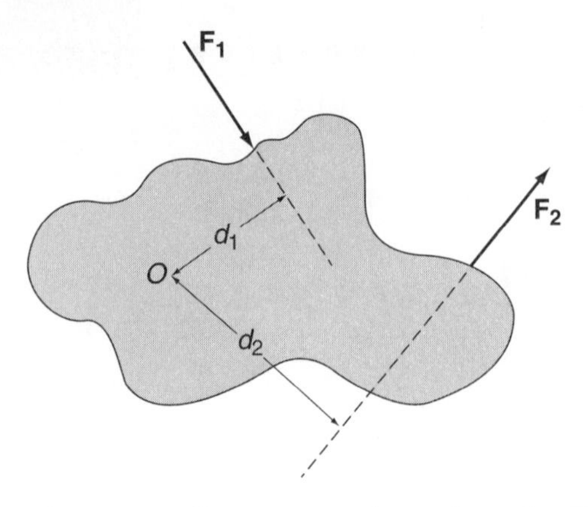

Figure 10-1 Lever arms about the point O for two forces acting on a body.

Torques tend either to rotate the body clockwise or counterclockwise about the axis. The convention in this laboratory will be to consider counterclockwise torques positive and clockwise torques negative. That convention gives net torque due to F_1 and F_2 about an axis through O as

$$\tau_{net} = F_2 d_2 - F_1 d_1 \tag{Eq. 3}$$

The net torque τ_{net} will be either counterclockwise, clockwise, or zero depending upon the magnitudes of $F_2 d_2$ and $F_1 d_1$. A meter stick will be the rigid body to which forces will be applied to produce mechanical equilibrium. The two conditions that must be satisfied for complete equilibrium of a rigid body are:

(1) Translational equilibrium is achieved if the vector sum of all the forces acting on the body is zero.

(2) Rotational equilibrium is achieved if the magnitude of $\sum \tau_{ccw}$ (sum of counterclockwise torques) is equal to the magnitude of $\sum \tau_{cw}$ (sum of clockwise torques).

The **center of gravity** of a body is defined as that point through which the sum of all the torques due to all the differential elements of mass of the body is zero. If the gravitational field is uniform throughout the body, the center of gravity and the **center of mass** are the same point. A uniform and symmetric meter stick has its center of gravity at the 0.500 m (50.0 cm) mark. Any meter stick will probably be close to uniform and symmetric, and its center of gravity will be close to the 0.500 m mark.

A meter stick with a knife-edge clamp on a support stand is shown in Figure 10-2. The mass of the meter stick is m_o, and three other masses m_1, m_2, and m_3 are shown hung from the meter stick. The masses produce forces where they are placed equal to the weight of the masses $m_1 g$, $m_2 g$, and $m_3 g$. The support exerts a force F_S directed upward at the point of the support. The weight of the meter stick $m_o g$ is exerted at the center of gravity of the meter stick x_g.

The meter stick in Figure 10-2 is in equilibrium. Forces in the upward direction are positive, and forces in the downward direction are negative. Take torques about an axis perpendicular to the page

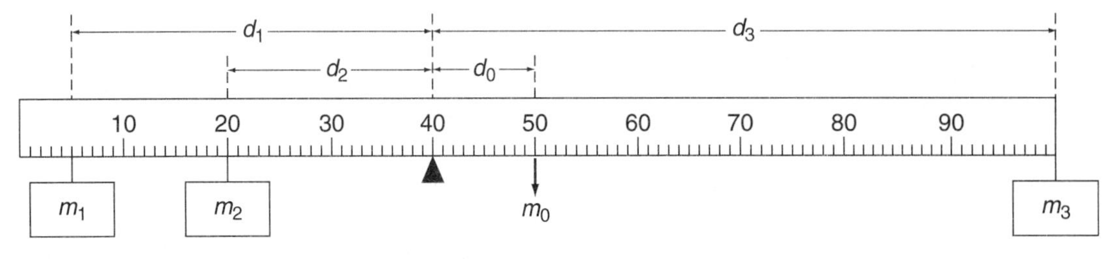

Figure 10-2 Meter stick balanced at point not the center of gravity. There are torques from the three applied masses and from the meter stick mass at the center of gravity.

through the point of support x_o. The lever arm for each mass is $d_i = |x_i - x_o|$ with x_i the position of the ith mass. The two conditions for equilibrium give

$$\sum F = 0 \quad \text{leads to} \quad F_S - m_1g - m_2g - m_3g - m_og = 0 \qquad \text{(Eq. 4)}$$

$$\sum \tau_{ccw} - \sum \tau_{cw} = 0 \quad \text{leads to} \quad m_1gd_1 + m_2gd_2 - m_ogd_o - m_3gd_3 = 0 \qquad \text{(Eq. 5)}$$

The following is a numerical example of the arrangement in Figure 10-2. The mass of the meter stick is $m_o = 0.120$ kg, and masses $m_1 = 0.150$ kg and $m_2 = 0.200$ kg are placed as in the figure. The point of support is at the 0.400 m mark. What value of mass m_3 must be placed at the 1.000 m mark to put the system in equilibrium, and what is the resulting support force F_S? The solution is given by:

$$\sum \tau_{ccw} - \sum \tau_{cw} = 0 \text{ is } (0.150)g(0.350) + (0.200)g(0.200) - (0.120)g(0.100) - m_3g(0.600) = 0$$

which reduces to $\quad 0.0525 + 0.0400 - 0.0120 = m_3\,(0.600)$

Solving the above equation gives $\quad m_3 = \dfrac{0.0805}{0.600} = 0.134\text{ kg}$

$$\sum F = 0 \text{ gives} \quad F_S = (m_1g + m_2g + m_3g + m_og) = (0.150 + 0.200 + 0.134 + 0.120)(9.80) = 5.92\text{ N}$$

In the numerical example, the value of m_3 was determined for torques about an axis through the point of support. *When the conditions of complete equilibrium have been met for some specific axis, the sum of torques about any axis is then ensured to be zero.* To confirm this for the numerical example given above, take torques about the axis perpendicular to the page through the left end of the meter stick. The masses m_1, m_2, m_3, and m_o exert clockwise torques, and the force F_S exerts the only counterclockwise torque. Calculating those values gives

$$\tau_{cw} = [(0.150)(0.050) + (0.200)(0.200) + (0.120)(0.500) + (0.134)(1.00)][9.80] = 2.37\text{ Nm}$$

$$\tau_{ccw} = [(F_S)0.400] = (5.92)(0.400) = 2.37\text{ Nm}$$

In this laboratory, attention will be directed to satisfying the conditions of rotational equilibrium. The support force F_S will always act through the support position. If the support position is chosen as the axis for torques, F_S will not contribute to the torque because it will have a zero lever arm. When the rotational equilibrium conditions are met, the value of the support force F_S will ensure translational equilibrium as well.

EXPERIMENTAL PROCEDURE

Part 1. Torque due to Two Known Forces

1. Remove the knife-edge clamp from the meter stick. Use the laboratory balance to determine the mass of the meter stick. Record it in the Meter Stick Data Table.
2. Place the knife-edge clamp on the meter stick and place it on the support. Adjust the position of the clamp until the best balance is achieved. Record the position of the knife-edge clamp as x_g in the Meter Stick Data Table.
3. With the meter stick supported at x_g, place a mass $m_1 = 0.100$ kg at the 0.100 m mark. Determine and record in Data and Calculations Table 1 the position x_2 at which a mass $m_2 = 0.200$ kg balances the meter stick. Use a small loop of nylon thread to hang the hooked masses at a given position. It may prove helpful to use a very small piece of tape to hold the thread at the desired position.

4. Calculate the lever arm for each force $d_i = |x_g - x_i|$where x_i is the position of the ith mass. With the support at the position x_g, the meter stick mass has zero lever arm and contributes no torque. Record the values of d_1 and d_2 in Data and Calculations Table 1.
5. Calculate and record in Data and Calculations Table 1 the value of the torques. The only counter-clockwise torque is due to m_1 and $\sum \tau_{ccw} = m_1 g d_1$. The only clockwise torque is due to m_2 and $\sum \tau_{cw} = m_2 g d_2$. Use a value of 9.80 m/s^2 for g for these and all other calculations.
6. Calculate the percentage difference between $\sum \tau_{ccw}$ and $\sum \tau_{cw}$ and record it in Data and Calculations Table 1.

Part 2. Torque due to Three Known Forces

1. Support the meter stick at x_g. Place $m_1 = 0.100$ kg at 0.100 m, and $m_2 = 0.200$ kg at 0.750 m. Determine the position x_3 at which $m_3 = 0.050$ kg balances the system. Record the value of x_3 in Data and Calculations Table 2.
2. The meter stick mass m_o makes no contribution to the torque. Calculate the lever arm for each of the masses and record the values in Data and Calculations Table 2. ($d_i = |x_g - x_i|$)
3. Calculate the values of $\sum \tau_{ccw}$ and $\sum \tau_{cw}$ and record them in Data and Calculations Table 2.
4. Calculate the percentage difference between $\sum \tau_{ccw}$ and $\sum \tau_{cw}$ and record it in Data and Calculations Table 2.

Part 3. Determination of the Meter Stick Mass by Torques

1. Place a mass $m_1 = 0.200$ kg at the 0.100 m position. Loosen the knife-edge clamp and slide the meter stick in the clamp until the torque exerted by $m_1 g$ is balanced by the torque of the meter stick weight acting at x_g. When the best balance is achieved, tighten the clamp. The position at which the meter stick is supported is x_o. Record x_o in Data and Calculations Table 3.
2. The values of the lever arms are given by $d_1 = |x_1 - x_o|$ and $d_o = |x_g - x_o|$. Calculate and record the values of d_1 and d_o in Data and Calculations Table 3.
3. For these conditions, $\sum \tau_{ccw} = m_1 g d_1$ and $\sum \tau_{cw} = m_o g d_o$ where m_o stands for the assumed unknown mass of the meter stick. Equating the two torques gives $m_o = m_1(d_1/d_o)$. Calculate and record in Data and Calculations Table 3 this value as $(m_o)_{exp}$.
4. Calculate and record in Data and Calculations Table 3 the percentage error in $(m_o)_{exp}$ compared to the meter stick mass determined by the laboratory balance.

Part 4. Comparison of Experimental and Theoretical Determinations of the Location of an Applied Force

1. Adjust the knife-edge clamp to the 0.400 m mark. Place $m_1 = 0.050$ kg at 0.050 m, $m_2 = 0.300$ kg at 0.300 m, and $m_3 = 0.200$ kg at 0.700 m as shown in Figure 10-3. *With the meter stick supported at the 0.400 m mark,* determine the position at which mass $m_4 = 0.100$ kg balances the meter stick. Record this

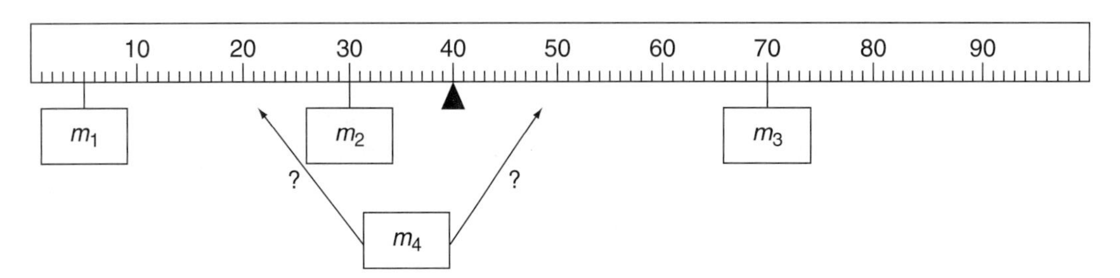

Figure 10-3 The location of mass m_4 needed to place the system in equilibrium is to be determined both experimentally and theoretically, and then compared.

position as x_4 in Data and Calculations Table 4 and use this value of x_4 to calculate the lever arm d_4 and record it in Data and Calculations Table 4 as $(d_4)_{exp}$.

2. In the space provided in Data and Calculations Table 4, write the equation for the rotational equilibrium with counterclockwise torques positive and clockwise torques negative. Use m_1, m_2, m_3, and m_4 as the symbols for the appropriate mass, and d_1, d_2, d_3, and d_4 as the symbols for the lever arms. Include the contribution from the meter stick mass m_o acting at the center of gravity of the meter stick x_g.
3. In the equation, treat the lever arm d_4 of mass m_4 as unknown and all the other quantities as known. Solve the equation to obtain a value for d_4 and record that result in Data and Calculations Table 4 as $(d_4)_{theo}$.
4. Calculate the percentage error in $(d_4)_{exp}$ compared to $(d_4)_{theo}$ and record it in Data and Calculations Table 4.

Part 5. Determination of an Unknown Mass by Torques

1. Use an experimental arrangement with the meter stick that is similar to those that we have used thus far to devise a method to determine the mass of an unknown mass. Describe carefully the procedure that is followed. Use at least one known mass and state its value and location on the meter stick. Write an equation that describes the equilibrium of the system treating the mass as unknown. Include a sketch of the experimental arrangement showing the position of all masses known and unknown. Construct your own Data and Calculations Table 5 listing all the relevant quantities.
2. Use the laboratory balance to determine the value of the unknown mass. Calculate the percentage error in your experimental value of the unknown mass compared to that obtained using the laboratory balance.

Name Section Date

10 LABORATORY 10 *Torques and Rotational Equilibrium of a Rigid Body*

PRE-LABORATORY ASSIGNMENT

1. State a definition of torque and give an equation for torque. Define the terms in the equation.

2. What are the conditions for equilibrium of a rigid body? State in words and equation form and define the terms of the equations.

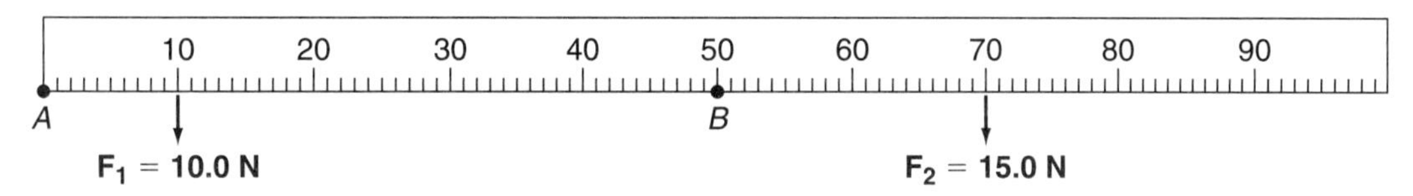

Figure 10-4 Meter stick with two forces F_1 and F_2 acting at points shown.

3. For the meter stick shown in Figure 10-4, the force F_1 10.0 N acts at 10.0 cm. What is the magnitude of the torque due to F_1 about an axis through point A perpendicular to the page? Is it clockwise, or is it counterclockwise? Show your work and give correct units.

4. In Figure 10-4 the force $F_2 = 15.0\,\text{N}$ acts at the point 70.0 cm. What is the magnitude of the torque due to F_2 about an axis through point B and perpendicular to the page? Is the torque clockwise, or is it counterclockwise? Show your work and give correct units.

5. For the meter stick in Figure 10-4, what is the magnitude of the *net* torque due to both forces F_1 and F_2 about an axis perpendicular to the page through point A? Is it clockwise or counterclockwise? Show your work.

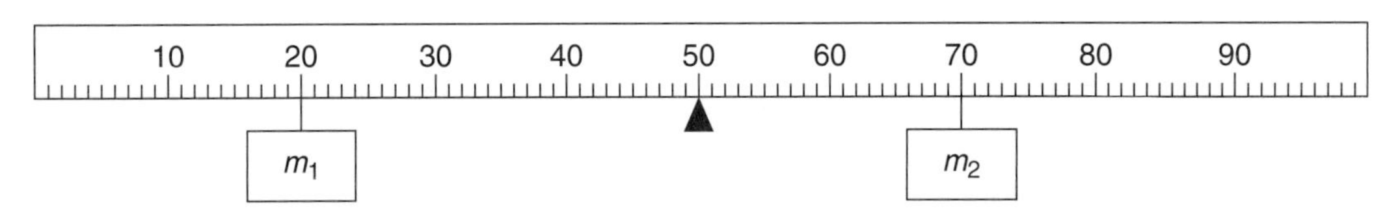

Figure 10-5 Meter stick with forces applied by hanging two masses m_1 and m_2.

6. In Figure 10-5 if mass $m_1 = 0.100\,\text{kg}$ acts at 20.0 cm, what is the value of mass m_2 that must be placed at the position 70.0 cm shown to put the system in equilibrium? Write the equation for $\sum \tau_{ccw} = \sum \tau_{cw}$ with the mass m_2 as unknown and solve for m_2. Assume that the meter stick is uniform and symmetric. Show your work.

Name | Section | Date

Lab Partners

10 LABORATORY 10 *Torques and Rotational Equilibrium of a Rigid Body*

LABORATORY REPORT

Meter Stick Data Table

$m_o =$ kg	$x_g =$ m

Data and Calculations Table 1

Mass (kg)	Position (m)	Lever arm (m)	Torque (N–m)	% Difference
$m_1 = 0.100$	$x_1 = 0.100$	$d_1 =$	$\tau_{ccw} =$	
$m_2 = 0.200$	$x_2 =$	$d_2 =$	$\tau_{cw} =$	

Data and Calculations Table 2

Mass (kg)	Position (m)	Lever arm (m)	Torque (N–m)	% Difference
$m_1 = 0.100$	$x_1 = 0.100$	$d_1 =$	$\tau_{ccw} =$	
$m_2 = 0.200$	$x_2 = 0.750$	$d_2 =$		
$m_3 = 0.050$	$x_3 =$	$d_3 =$	$\tau_{cw} =$	

Data and Calculations Table 3

Support Position $x_o =$ m				
Mass (kg)	Position (m)	Lever arm (m)	$(m_o)_{exp} =$ kg	% Error
$m_1 = 0.200$	$x_1 = 0.100$	$d_1 =$		
$m_o =$	x_g	$d_o =$		

Data and Calculations Table 4

Support Position $x_o = 0.400$ m			
Mass (kg)	*Position* (m)	*Lever Arm* (m)	*Equation for the Torque*
$m_1 = 0.050$	$x_1 = 0.050$	$d_1 =$	
$m_2 = 0.300$	$x_2 = 0.300$	$d_2 =$	
$m_3 = 0.200$	$x_3 = 0.700$	$d_3 =$	Solving equation for d_4 gives $(d_4)_{theo} =$ m
$m_4 = 0.100$	$x_4 =$	$(d_4)_{exp} =$	
$m_o =$	$x_g =$	$d_o =$	

Data and Calculations Table 5

SAMPLE CALCULATIONS

1. $d_i = |x_i - x_o| =$

2. $\sum \tau_{ccw} =$

3. $\sum \tau_{cw} =$

4. % Diff torques $= |\tau_{ccw} - \tau_{cw}| / (0.5(\tau_{ccw} + \tau_{cw})) \times 100\% =$

5. $(m_o)_{exp} = m_1(d_1/d_o) =$

6. % Error for $(m_o)_{exp} = |E - K|/K \times 100\% =$

7. $(d_4)_{theo} =$

8. % Error for $(d_4)_{theo} = |E - K|/K \times 100\% =$

9. Calculation for m_{exp} in Part 5 =

10. % Error for $m_{exp} = |E - K|/K \times 100\% =$

QUESTIONS

1. Consider the percentage difference between the $\sum \tau_{ccw}$ and the $\sum \tau_{cw}$ for the first two parts of the laboratory when known forces are balanced. A difference of 0.5% or less is excellent, a difference of 1.0% or less is good, and a difference of 2% or less is acceptable. Based on these criteria, describe your results for the first two parts of the laboratory and defend your statement.

2. Using the same criteria as in Question 1 for the percentage differences, describe your results for the determination of mass of the meter stick in Part 3 of the laboratory and for the determination of the lever arm of the mass m_4 in Part 4 of the laboratory.

3. In all of the experimental arrangements the mass of the knife-edge clamp is ignored. Is this an approximation because its mass is small, or is there some reason it makes no contribution to the torque? State your reasoning clearly.

4. Suppose an experimental arrangement like the one in Part 2 has mass $m_1 = 0.200\,\text{kg}$ at the 0.100-m mark and a mass $m_2 = 0.100\,\text{kg}$ at the 0.750-m mark. Can the system be put in equilibrium by a 0.050-kg mass? If it can be done, state where it would be placed. If it cannot be done, state why not.

5. In Part 1 of the laboratory, what is the value of the force F_s with which the support pushes upward on the meter stick?

6. For the equilibrium conditions established in Part 4 of the laboratory, calculate the counterclockwise and clockwise torques about an axis perpendicular to the page through a point at the left end of the meter stick. Calculate the percentage difference between the net counterclockwise torque and the net clockwise torque.

Conservation of Energy on the Air Table

OBJECTIVES

- ❑ Determine the value of the spring constant k for two springs.
- ❑ Investigate how the speed, kinetic energy, and total spring potential energy of a puck on a horizontal air table vary with time.
- ❑ Evaluate the extent to which the total mechanical energy (sum of kinetic energy plus spring potential energy) is constant as a function of position of the puck.

EQUIPMENT LIST

- Air table, sparktimer, air pump, foot switch, carbon paper, and recording paper
- Level, two springs, two spring clamps, and a double hook for the puck
- Meter stick, pulley, string, calibrated masses, and laboratory balance

THEORY

The force F exerted by a spring as it is elongated from its natural length x_o to some greater length x is given by

$$F = -k(x - x_o) \qquad \text{(Eq. 1)}$$

where k is a constant called the **spring constant.** The spring constant is defined as the force per unit elongation and has units of N/m. The value of k will be different for each spring depending upon the spring material properties and how tightly the spring is wound. The value of k for a spring is determined by measuring the elongation caused by a known force.

The force exerted by a spring is an example of a class of forces known as conservative forces. A **conservative force** is one for which a **potential energy** function can be defined. For a spring force the potential energy U is given by

$$U = \tfrac{1}{2} k(x - x_o)^2 \qquad \text{(Eq. 2)}$$

A mass M attached to the end of a spring will change speed V as the force of the spring varies. At any instant when the mass M has speed V, it has a **kinetic energy** K given by

$$K = \tfrac{1}{2} MV^2 \qquad \text{(Eq. 3)}$$

The sum of the potential and kinetic energies is equal to the **total energy** E. In an isolated system where only conservative forces are present, the total energy E is a constant throughout the motion. This is the principle of **conservation of mechanical energy.** In equation form the total energy E is given by

$$E = \tfrac{1}{2} MV^2 + \tfrac{1}{2} k(x - x_o)^2 \qquad \text{(Eq. 4)}$$

In this laboratory a puck on an air table will move under the influence of two springs arranged so that they pull in opposite directions on the puck. We will determine the spring constants k_1 and k_2 of the two springs. We will determine the position of the puck as a function of time with sparktimer measurements. At each position of the puck along its recorded path, the speed of the puck and the elongation of each spring will be determined. From these measurements we will determine the kinetic energy K, the spring potential energy U, and the total energy E at each point. A graph of the total energy E as a function of position will show the extent to which the total energy remains constant.

EXPERIMENTAL PROCEDURE

Spring Constants

1. Place a sheet of carbon paper and a sheet of recording paper on the table with the recording paper on top. Label one of the springs as spring #1, and use it in the arrangement described in Figure 11-1. Place enough mass m on the end of the string to elongate the spring slightly and place it under tension. A value for m of 0.020 kg should be about the right choice for the initial value. Record the value of m in Data and Calculations Table 1. With the puck at rest and the spring under tension, mark the position of the puck with a short burst of the sparktimer.
2. To find a convenient range for the mass needed, apply enough mass to extend the spring about 0.300 m. This will be the maximum mass needed with other trials at intermediate values. For six values of mass in the range 0.020 kg to the mass required to extend the spring 0.300 m, record the position of the puck by a short burst of the sparktimer. Record the six values of m used in Data and Calculations Table 1.

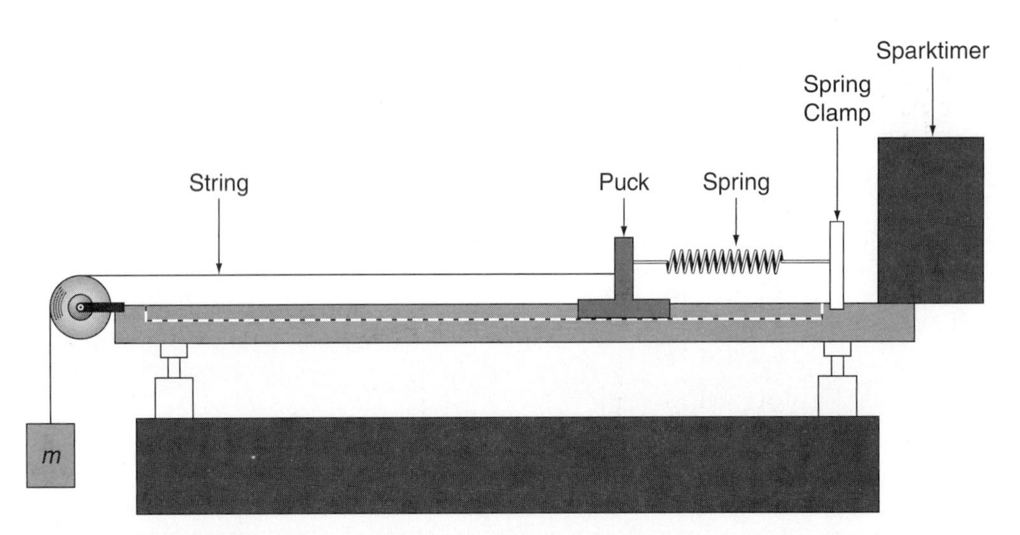

Figure 11-1 Air table arrangement for measuring the spring constant k.

3. Remove the recording paper from the table and draw a straight line through the points obtained from the sparktimer bursts. Arbitrarily choose an origin and measure the distance from the origin to each point. Record these values as x in Data and Calculations Table 1.
4. Label the second spring as spring #2 and repeat Steps 1 through 3 for the second spring.

Energy Conservation

1. Place the double hook on the puck and measure the mass of the puck plus the hook using the laboratory balance. Record this value as M_p in Data and Calculations Table 2.
2. Measure the unstretched length of each of the two springs and record them in Data and Calculations Table 2 as L_{o1} and L_{o2}.
3. Level the air table and on the table place a sheet of carbon paper and a sheet of recording paper with the recording paper on top. Attach the leads to both of the pucks and place one of the pucks in one corner of the table, where it will remain.
4. Place the two spring clamps on opposite sides of the table. Attach the two springs to opposite ends of the double hook on the puck. Attach the other end of each spring to one of the spring clamps.
5. Release the puck so it *moves in an elliptical trajectory* as shown in Figure 11-2. A second condition is that *the two springs must both be under stretched tension at all times during the motion*. Launch the puck for several trial runs to be sure that both conditions are satisfied.
6. On the recording paper mark the position of the two spring clamps and indicate which spring is attached to which clamp. A small piece of paper must be taped to the recording paper to include the location of the spring clamps.
7. Set the sparktimer to 20.0 Hz and record the trajectory of the puck. Note the location of the initial position of the puck.

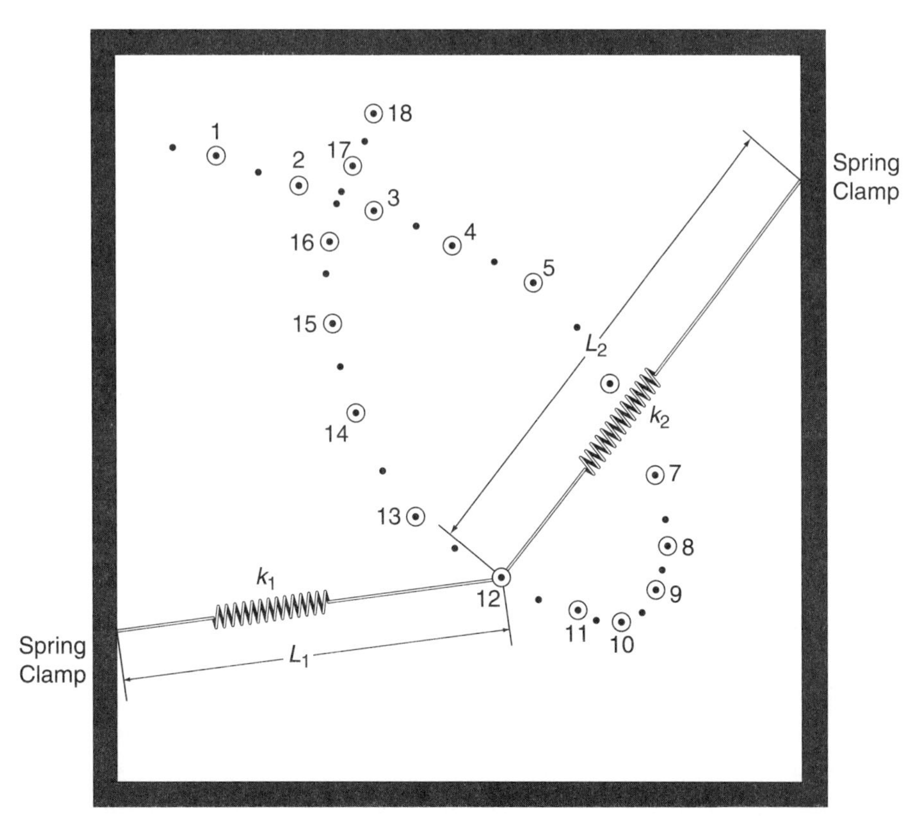

Figure 11-2 Sparktimer record produced by the elliptical motion of puck on an air table under the influence of two springs clamped on opposite sides of the table.

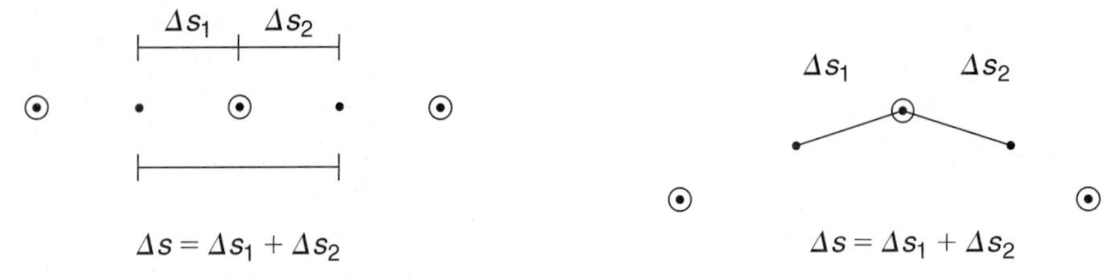

Figure 11-3 Illustrating calculation of Δs when path is sharply curved at the right, as opposed to the case when the path is essentially straight, as shown at the left.

8. Choose a point near the beginning of the puck's motion (but not the very first point). Circle that point and label it as 1. Continue in the direction of the puck's motion, circle every other point, and number them consecutively up to at least 15 (Figure 11-2).
9. For each circled point measure the distance from the point to each spring clamp. These are the lengths L_1 and L_2 of the springs at that point as illustrated for the point labeled 12 in Figure 11-2. Record them in Data and Calculations Table 2.
10. At each *numbered point* measure the distance between the *preceding uncircled point* and the *next uncircled point* as illustrated in Figure 11-3. This gives the average displacement of the puck during the time interval $\Delta t = 0.100$ s including 0.050 s before and 0.050 s after the point. For each circled point record this distance as Δs in Data and Calculations Table 2. If the path has sharp curvature, measure Δs as the distance from uncircled to circled to uncircled points as shown in the right-hand side of Figure 11-3.

CALCULATIONS

Spring Constants

1. For each applied mass m calculate the force mg using $g = 9.80$ m/s^2 and record it in Data and Calculations Table 1.
2. Perform a linear least squares fit to the data for each spring with the force as the vertical axis and x as the horizontal axis. Record the slope of those fits as the spring constants k_1 and k_2 in Data and Calculations Table 1 and Data and Calculations Table 2. Record the values of the correlation coefficient r for each of the fits in Data and Calculations Table 1.

Energy Conservation

1. Input the values of M_p, k_1, k_2, L_{o1}, L_{o2}, L_1, L_2, and Δs into the Excel spreadsheet template Lab 11 Calculations. This template is available at academic.cengage.com/physics/loyd. The spreadsheet will calculate $V = \Delta s/\Delta t$ and $K = ½\, M_p\, V^2$ for each point. It will also calculate $U = ½ k_1 |L_1 - L_{o1}|^2 + ½ k_2 |L_2 - L_{o2}|^2$ and $E = K + U$ for each point. Record all the values calculated by the spreadsheet in Data and Calculations Table 2.

GRAPHS

1. Use a single sheet of linear graph paper to make a graph that has energy units on the vertical scale and position numbers (1 through 15) on the horizontal scale. Use different symbols for each and graph the kinetic energy K, the total spring potential energy U, and the total mechanical energy E as a function of position number. Do not suppress the energy scale. Show the whole range of energy from zero to slightly greater than E.

Name | Section | Date

LABORATORY 11 *Conservation of Energy on the Air Table*

PRE-LABORATORY ASSIGNMENT

1. What is the definition, both in words and equation form, of the spring constant k? What are the units of k?

2. What is the equation for the spring potential energy of a spring? Define all terms used in the equation.

3. What is the principle of conservation of mechanical energy applied to a mass m attached to a spring of spring constant k? State it in words and equation form and define all terms used.

4. What kinds of forces conserve mechanical energy? Is mechanical energy conserved if frictional forces are present?

5. A spring has a spring constant $k = 2.75$ N/m. It has an unstretched length of 0.100 m. What is its spring potential energy when the spring is stretched to a length of 0.170 m? Show your work.

6. A spring is stretched by applying mass m over a pulley to the spring as shown in Figure 11-1. The position of the end of the spring as a function of the applied mass is determined, and the data given below are obtained. Find the spring constant k by performing a linear least squares fit with F as the vertical axis and x as the horizontal axis. Record the slope as k.

m (kg)	0.020	0.040	0.060	0.080	0.100	0.120
x (m)	0.000	0.049	0.100	0.151	0.201	0.249
F (N)						

$k =$ ______________ N/m Intcp = ______________ N $r =$ ______________

7. A 2.00 kg mass moves under the force of a spring with a spring constant of $k = 5.65$ N/m. At one instant of time the mass has a speed of 4.75 m/s when the spring has a displacement from equilibrium of 1.557 m. What is the speed of the mass at a later time when the spring displacement from equilibrium is 0.857 m? (*Hint—Use the conservation of mechanical energy.*) Show your work.

Name | Section | Date

Lab Partners

11 LABORATORY 11 *Conservation of Energy on the Air Table*

LABORATORY REPORT

Data and Calculations Table 1

Spring 1 $k_1 =$ N/m			
Point	m (kg)	mg (N)	x (m)
1			
2			
3			
4			
5			
6			
$r =$			

Spring 2 $k_2 =$ N/m			
Point	m (kg)	mg (N)	x (m)
1			
2			
3			
4			
5			
6			
$r =$			

Data and Calculations Table 2

$M_p =$ kg	$k_1 =$ N/m	$k_2 =$ N/m	$L_{o1} =$ m	$L_{o2} =$ m			
Point	L_1 (m)	L_2 (m)	Δs (m)	V (m/s)	K (J)	U (J)	E (J)
1							
2							
3							
4							
5							
6							
7							
8							
9							
10							
11							
12							
13							
14							
15							

SAMPLE CALCULATIONS

1. $F = mg =$
2. Velocity $= V = \Delta s / \Delta t =$
3. $K = ½ M_p V^2 =$
4. $U = ½ k_1 |L_1 - L_{o1}|^2 + ½ k_2 |L_2 - L_{o2}|^2 =$
5. $E = K + U =$

QUESTIONS

1. Calculate the mean and standard error for the 15 values for the total energy E. Divide the standard error by the mean, and express the ratio as a percentage. Use this calculation as evidence for your statement of how constant are your values of E.

2. Although your values of E should be fairly constant, they normally show a slight decrease from beginning to end of the data caused by friction. Calculate the decrease in E from the first to last point. Also calculate the total path length of the puck by summing the values of Δs. Divide the loss in energy by the total path length, which is the force of friction. What is the friction for your data? Show your work.

3. In the determination of the spring constant k the origin from which the position x was measured was chosen arbitrarily. Would choosing another point for the origin significantly change the value obtained for k? State clearly the reasoning for your answer.

4. Summarize in your own words the physical theory that this laboratory is supposed to demonstrate. Do your results support the theory? Consider the variations in the values of U, K, and E to support your answer.

Conservation of Spring and Gravitational Potential Energy

OBJECTIVES

- ❑ Determine the value of the spring constant k for a spring.
- ❑ Investigate the change in gravitational energy $\Delta U_g = mg(x_f - x_i)$ and the change in spring potential energy $\Delta U_k = (\frac{1}{2})k(x_f^2 - x_i^2)$ for a mass suspended from the spring.
- ❑ Evaluate the extent to which the changes in energy are equal as the mass oscillates.

EQUIPMENT LIST

- Spring in the form of a truncated cone made of spring brass with a spring constant of about 10 N/m. (Available from Central Scientific Co. If another spring is used, appropriate adjustments should be made in the masses and distances used.)
- Set of calibrated hooked masses
- Table clamp, right angle clamps, support rods, meter stick

THEORY

When a spring is stretched or compressed a distance x from its equilibrium length, the spring exerts a restoring force F. The equation relating the force F and the displacement x is

$$F = -kx \qquad \text{(Eq. 1)}$$

and k is a constant called the **spring constant** with units N/m. The negative sign in Equation 1 indicates that the restoring force direction is opposite the displacement.

When a spring is compressed or stretched by x it has stored energy called **spring potential energy** given by $U_k = \frac{1}{2}kx^2$. When the spring is stretched from a displacement of x_1 to a displacement of x_2 the change in spring energy is equal to the work done on the spring (Figure 12-1).

$$\text{Work} = \Delta U_k = \tfrac{1}{2}\, k(x_2^2 - x_1^2) \qquad \text{(Eq. 2)}$$

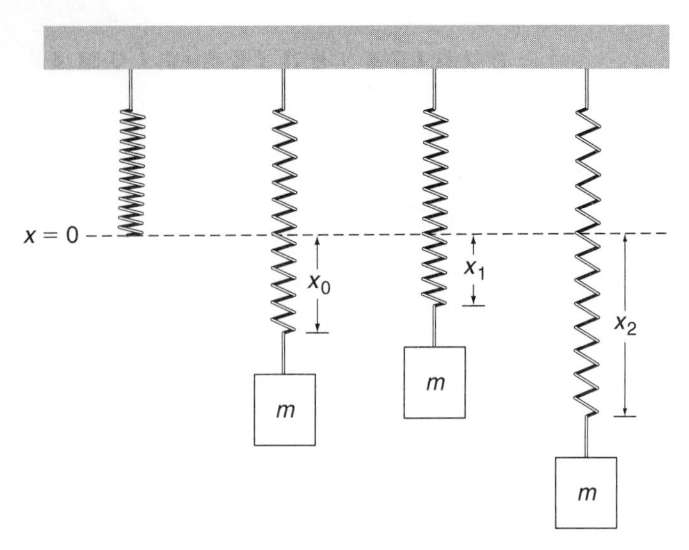

Figure 12-1 Positions of a mass m on a spring in the earth's gravitational field.

A spring with spring constant k is supported at the top by a rigid support and allowed to hang vertically. The vertical position at which the lower end of the spring hangs is the zero of the coordinate x. A hooked mass m is placed by hand on the end of the spring, and the mass is slowly lowered by hand. The mass will extend the spring by an amount x_o when the hand is removed as shown in Figure 12-1.

The mass is raised and supported by hand with the lower end of the spring at position x_1 above position x_o as shown in Figure 12-1. The mass is now released and allowed to fall under the influence of the spring and the earth's gravitational force. The mass will fall to its lowest point with displacement x_2 and then rebound and oscillate.

The **gravitational potential energy** relative to any horizontal plane is mgx where x is the distance above the plane. The **total mechanical energy** of the system is the sum of kinetic energy, spring potential energy, and gravitational potential energy. Consider the total mechanical energy at x_1 and x_2. At each of these points the kinetic energy is zero, and the total mechanical energy is the sum of the spring potential energy and the gravitational potential energy. The center of mass of m is distance d below the lower end of the spring, and the gravitational potential energy zero is the same as the equilibrium point of the spring. The equation for the sum of spring energy and gravitational energy is

$$\tfrac{1}{2} kx_1^2 - mg(x_1 + d) = \tfrac{1}{2} kx_2^2 - mg(x_2 + d) \qquad \text{(Eq. 3)}$$

$$\text{or} \quad \tfrac{1}{2} kx_1^2 - mgx_1 = \tfrac{1}{2} kx_2^2 - mgx_2 \qquad \text{(Eq. 4)}$$

Both gravitational potential energy terms are negative because the mass is below the reference point for both positions. Equation 4 can be rewritten as

$$mg(x_2 - x_1) = \tfrac{1}{2} k(x_2^2 - x_1^2) \qquad \text{(Eq. 5)}$$

Equation 5 states that the change in gravitational energy between points 1 and 2 is equal to the change in spring potential energy between those points because the kinetic energy is zero at both points 1 and 2. This laboratory will consist of a series of measurements that will test the validity of Equation 5.

EXPERIMENTAL PROCEDURE

Spring Constant

1. Use appropriate clamps and rods to provide a horizontal rod sticking out beyond the table as shown in Figure 12-2. Hang the spring on the horizontal rod, and attach it to the rod with a piece of tape.

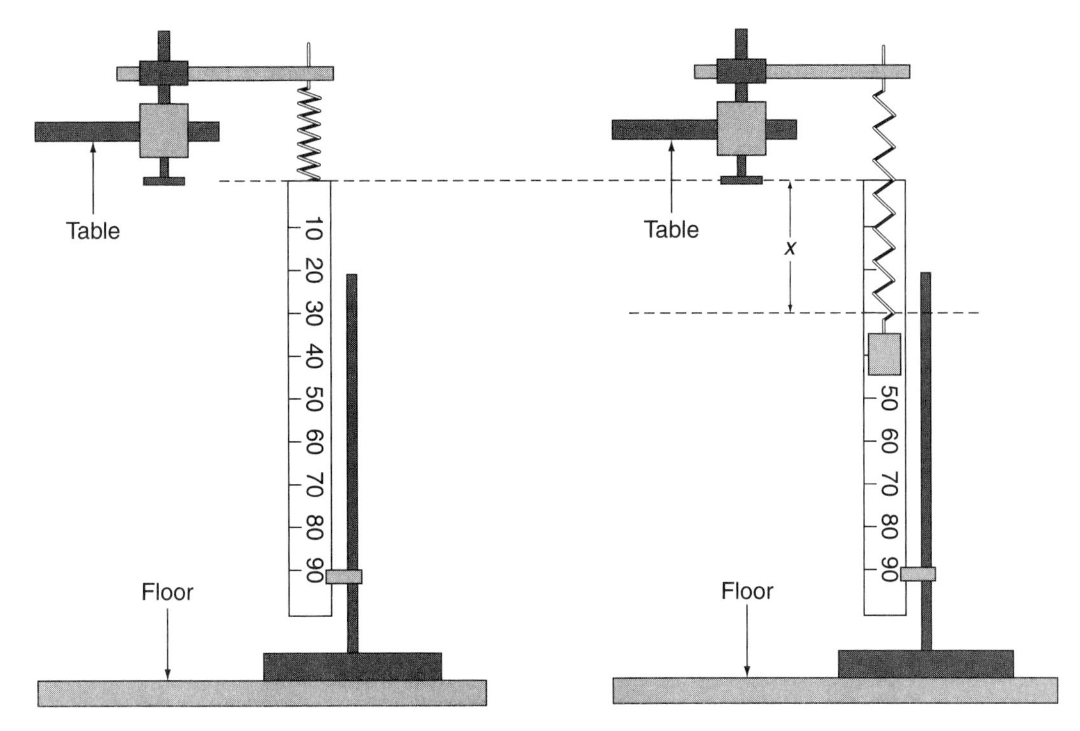

Figure 12-2 Spring supported by table clamp and meter stick aligned with the lower end of the spring. Mass placed on the end of the spring caused displacement x of the spring.

Arrange a clamp for the meter stick so that it can be supported from the floor as shown. Adjust the height of the meter stick until the zero mark of the meter stick is aligned with the bottom of the hanging spring as shown.

2. Place a hooked mass m of 0.1000 kg on the end of the spring. Slowly lower the mass m until it hangs at rest in equilibrium when released. Carefully read the position of the lower end of the spring on the meter stick scale. Record the value of the mass m and the value of the displacement x in Data Table 1.
3. Repeat Step 2, placing in succession 0.2000, 0.3000, 0.4000, and 0.5000 kg on the spring and measuring the displacement x of the spring. Record all values of m and x in Data Table 1. Record x to the nearest 0.1 mm.

Energy Conservation

1. Check that the lower end of the spring is still precisely at the zero mark. Adjust the meter stick if necessary. Hang a 0.5000 kg mass on the end of the spring and support it with your hand with the lower end of the spring precisely at the 0.2500 m mark. Record 0.2500 as x_1, and record the value of the mass in Data Table 2. Release the mass and mark the lowest point of the lower end of the spring. Release the mass several times until you have accurately located the lowest point of the motion. It may be easier to note the lowest position of the mass itself, and then hold the mass at that position to determine the position of the lower end of the spring. Record the distance as x_2 in Data Table 2.
2. Repeat Step 1 for x_1 values of 0.3000, 0.3500, and 0.4000 m. Measure the value of x_2 for each of these values of x_1 and record the values of x_1 and x_2 in Data Table 2.
3. Check that the lower end of the spring is still precisely at the zero mark. Adjust the meter stick if necessary. Use a mass of 0.5000 kg and pull the mass down by hand until the lower end of the spring is precisely at the 0.7500 m mark. Record 0.7500 m as x_2 in Data Table 3. Release the mass and determine how high it rises. The position of the lower end of the spring when the mass is at its highest point is x_1. Again release the mass several times to accurately determine the value of x_1. Record the value of x_1 and the value of the mass in Data Table 3.

4. Repeat Step 3 for x_2 values of 0.7000, 0.6500, and 0.6000 m. Measure the value of x_1 for each of these values of x_2 and record the values of x_1 and x_2 in Data Table 3.

CALCULATIONS

Spring Constant

1. Calculate the force mg for each mass and record the values in Calculations Table 1. Use the value of $9.800\,\text{m/s}^2$ for g.
2. Perform a linear least squares fit with mg as the vertical axis and x as the horizontal axis. Record the slope in Calculations Table 1 as the spring constant k and record r the correlation coefficient.

Energy Conservation

1. For each of the four measurements of the falling mass in Data Table 2, calculate the change in the gravitational potential energy ΔU_g where $\Delta U_g = mg(x_2 - x_1)$. Calculate the change in spring potential energy ΔU_k where $\Delta U_k = (1/2)k(x_2^2 - x_1^2)$. Record the results in Calculations Table 2.
2. Calculate the percentage differences between ΔU_g and ΔU_k for each case of Step 1 and record them in Calculations Table 2.
3. For each of the four measurements of the rising mass in Data Table 3, calculate the change in gravitational potential energy ΔU_g and the change in spring potential energy ΔU_k. Record the results in Calculations Table 3.
4. Calculate the percentage differences between ΔU_g and ΔU_k for each case of Step 3 and record them in Calculations Table 3.

GRAPHS

1. Graph the data from Calculations Table 1 for force mg versus displacement x with mg as the vertical axis and x as the horizontal axis. Also show on the graph the straight line obtained from the fit to the data.

Name Section Date

12 LABORATORY 12 *Conservation of Spring and Gravitational Potential Energy*

PRE-LABORATORY ASSIGNMENT

1. A spring has a spring constant of $k = 7.50$ N/m. If the spring is displaced 0.550 m from its equilibrium position, what is the force that the spring exerts? Assume for this and for all other questions in the pre-laboratory that $g = 9.80\ \text{m/s}^2$. Show your work.

2. A spring of spring constant $k = 8.25$ N/m is displaced from equilibrium by a distance of 0.150 m. What is the stored energy in the form of spring potential energy? Show your work.

3. A spring of spring constant $k = 12.5$ N/m is hung vertically. A 0.500 kg mass is then suspended from the spring. What is the displacement of the end of the spring due to the weight of the 0.500 kg mass? Show your work.

4. A mass of 0.400 kg is raised by a vertical distance of 0.450 m in the earth's gravitational field. What is the change in its gravitational potential energy? Show your work.

5. A spring of spring constant $k = 8.75$ N/m is hung vertically from a rigid support. A mass of 0.500 kg is placed on the end of the spring and supported by hand at a point so that the displacement of the spring is 0.250 m. The mass is suddenly released and allowed to fall. At the lowest position of the mass what is the displacement of the spring from its equilibrium position? (*Hint*—Apply Equation 5 with $x_1 = 0.250$ m and x_2 the unknown. This will lead to a quadratic equation with one of the solutions the unknown x_2, and the other solution the original 0.250 m displacement.) Show your work.

6. The laboratory is based on the assumption that at the two points of the motion being considered, the mass is at rest. What kind of energy does not need to be included under these experimental conditions?

Name | Section | Date

Lab Partners

12 LABORATORY 12 *Conservation of Spring and Gravitational Potential Energy*

LABORATORY REPORT

Data Table 1

m (kg)	x (m)

Calculations Table 1

mg (N)	k (N/m)	r

Data Table 2

x_1 (m)	x_2 (m)	m (kg)

Calculations Table 2

$mg\,(x_2 - x_1)$ (J)	$1/2k\,(x_2^2 - x_1^2)$ (J)	% *Diff*

Data Table 3

x_1 (m)	x_2 (m)	m (kg)

Calculations Table 3

$mg\ (x_2 - x_1)$ (J)	$1/2k\ (x_2^2 - x_1^2)$ (J)	% Diff

SAMPLE CALCULATIONS

1. $mg =$
2. $mg(x_2 - x_1) =$
3. $1/2k(x_2^2 - x_1^2) =$
4. $\%\ \text{Diff} = 2(E_1 - E_2)/(E_1 + E_2) \times 100\% =$

QUESTIONS

1. For five data points, statistical theory states that there is only 0.1% probability that a value of $r \geq 0.992$ would be obtained for uncorrelated data. Based on your value of r, make the best statement you can about the extent to which your data indicate that the force and displacement are linear.

2. Describe the extent to which your data indicate that mechanical energy is conserved in this laboratory. Consider the percentage differences in the energy changes in Data Table 2 and Data Table 3 in your answer.

3. Examine your data in Data Table 2 and in Data Table 3. For the data with the smallest percentage difference, compare the total energy at each point. Calculate the sum $U_k + U_g$ at x_1 as $\frac{1}{2}kx_1^2 - mgx_1$. Calculate that sum at x_2 as $\frac{1}{2}kx_2^2 - mgx_2$. Do you expect them to agree reasonably well? Explain why they should or should not be the same.

4. Consider the same data as used in Question 3. Calculate the value of x halfway between x_1 and x_2. Calculate $U_k + U_g = \frac{1}{2}kx^2 - mgx$ for that point. Do you expect them to agree with the energy calculated in Question 3? If they agree reasonably well, explain why they do. If they do not agree, explain why they do not agree.

The Ballistic Pendulum and Projectile Motion

OBJECTIVES

- ❑ Investigate how the initial velocity of a ball fired into a ballistic pendulum is related to the initial velocity with which the pendulum plus ball moves after the collision.
- ❑ Investigate the kinetic energy loss in the collision of the ball with the pendulum.
- ❑ Determine the initial velocity of the ball by firing it as a projectile and compare it with the velocity determined by the collision.

EQUIPMENT LIST

- Ballistic pendulum apparatus with projectile ball
- Laboratory balance and calibrated masses
- Meter stick, plain paper, carbon paper, and masking tape

THEORY

Ballistic Pendulum

The principle of **conservation of momentum** states that the total momentum of a system of particles remains constant if there are no external forces acting on the system. Collision processes are good examples of this concept. In this laboratory we will use a ballistic pendulum to measure the velocity of a ball projected by a spring gun. Figure 13-1 shows a ball of mass m moving initially in the horizontal direction with speed v_{xo} that then strikes a pendulum designed to catch the ball. The pendulum of mass M catches the ball and swings about pivot point O to some maximum height y_2 above its original height y_1. The system of ball plus pendulum rises a vertical distance of $y_2 - y_1$ as a result of the process.

Momentum is conserved because the only forces acting on the ball and the pendulum in the direction of motion are the forces of the collision. The two particles stick together after the collision and move with the same velocity V. A collision where particles stick together is called a **completely inelastic collision.** The equation for conservation of momentum is

$$mv_{xo} = (m + M)V \qquad \text{(Eq. 1)}$$

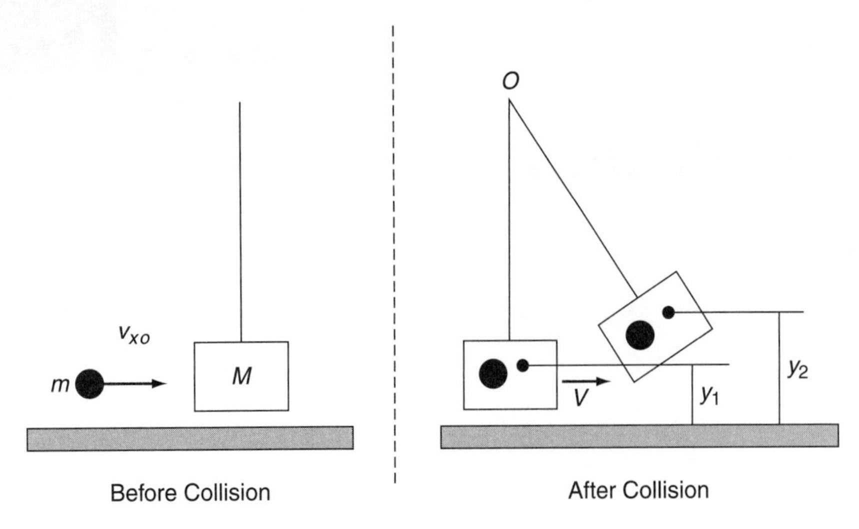

Figure 13-1 Ballistic pendulum of mass M before and after collision with ball of mass m.

The collision does not conserve **mechanical energy,** but mechanical energy is conserved as the ball plus pendulum swings up along the arc. The **kinetic energy** immediately after the collision is converted into **gravitational potential energy.** In equation form

$$\tfrac{1}{2}(m+M)V^2 = (m+M)g(y_2 - y_1) \qquad \text{(Eq. 2)}$$

Solving for V gives

$$V = \sqrt{2g(y_2 - y_1)} \qquad \text{(Eq. 3)}$$

Solving Equation 1 for the initial velocity of the ball gives

$$v_{xo} = \left(\frac{m+M}{m}\right)V \qquad \text{(Eq. 4)}$$

A measurement of $y_2 - y_1$ used in Equation 3 gives a value of V to be used in Equation 4 to determine v_{xo}.

Projectile Motion

If the pendulum is raised and the pawl is placed in one of the notches on the track, the ball can now travel a horizontal distance X while it falls a vertical distance Y as shown in Figure 13-2.

The original velocity of the ball is completely in the x direction with no y component. The acceleration due to gravity in the y direction is the only acceleration of the ball. The horizontal displacement X and the vertical displacement Y as a function of time t after the ball is launched are

$$X = v_{xo}\,t \qquad \text{(Eq. 5)}$$

$$Y = \tfrac{1}{2}gt^2 \qquad \text{(Eq. 6)}$$

Equation 6 has been written with positive displacements down in the same direction as g. Combining Equations 5 and 6 to eliminate time t gives v_{xo} in terms of X and Y as

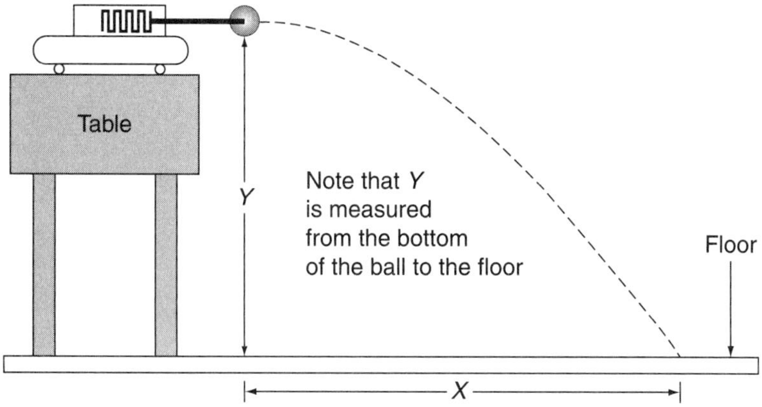

Table

Y

Note that Y is measured from the bottom of the ball to the floor

Floor

X

Figure 13-2 Motion of the ball moving horizontal distance X while free falling height Y.

$$v_{xo} = \frac{X}{\sqrt{2Y/g}} \qquad \text{(Eq. 7)}$$

Equation 7 can be used to determine the initial velocity v_{xo} by firing the projectile from a known height Y and measuring the value of X that results.

The velocity in the y direction is initially zero. As the projectile falls under the influence of gravity, it acquires a velocity in the y direction given by

$$V_y = gt = g\sqrt{2Y/g} = \sqrt{2gY} \qquad \text{(Eq. 8)}$$

EXPERIMENTAL PROCEDURE

Ballistic Pendulum

1. Slide the projectile ball (which has a hole in it) onto the rod of the spring gun (Figure 13-3). When the ball is in the pendulum, be careful when removing it. The spring that catches the ball in the pendulum can be easily broken. With the ball on the rod, cock the gun by pushing against the ball until the latch

Figure 13-3 Ballistic pendulum apparatus. (Photo courtesy of Central Scientific Co., Inc.)

catches. *Be very careful not to get your hand caught in the spring gun mechanism.* Fire the gun several times to see how it operates. There are two common problems. If the ball does not catch in the pendulum bob, the spring in the bob should be adjusted or replaced. If the pawl that is designed to catch on the notched track does not engage, the pendulum suspension should be adjusted by means of the screws at the suspension points.

2. A sharp curved point on the side of the pendulum (or on some models a dot) marks the center of mass of the pendulum-ball system. Let the pendulum bob hang vertically and measure the distance y_1, the center of mass point above the base of the gun (Figure 13-1). Record the value of y_1 in Data Table 1.
3. Fire the ball into the stationary pendulum while it hangs freely at rest. The pendulum will catch the ball, swing up, and then lodge in the notched track. Record in Data Table 1 the position number, p, at which the pawl on the pendulum catches on the track. Measure the distance y_2 of the center of mass point above the base of the apparatus (Figure 13-1) and record it in Data Table 1. Repeat this procedure four more times for a total of five trials, recording the position p and measuring the distance y_2 for each trial.
4. Loosen the screws holding the pendulum in its support and remove the pendulum consisting of the rod and the bob. Determine the mass of the pendulum (bob and rod) using a laboratory balance. Record the pendulum mass as M in Data Table 1. Determine the projectile ball mass and record it in Data Table 1 as m.

Projectile Motion

In the following procedure, be extremely careful not to fire the ball when anyone is in a position to be struck by the ball. Serious injury could result.

1. Raise the pendulum and secure it so that the ball can be fired under it.
2. Place the apparatus near the front edge of the laboratory table so that the ball will strike the floor before it strikes a wall or any other object. The gun must be fired each time from the same position relative to the table. It may be necessary to clamp the apparatus to the table. Place a piece of heavy cardboard or some other object in a position to catch the ball after it strikes the floor but before it strikes a wall. *Do not allow the ball to strike a wall because it will likely damage it.* Make several test firings to locate the approximate place where the ball will land on the floor.
3. Place a sheet of white paper on the floor approximately centered where test firings have landed. Place a piece of carbon paper over the white paper so that the ball striking the carbon paper will leave a dot on the white paper. Tape both of the papers to the floor.
4. Place the ball on the rod of the spring gun. The vertical distance Y that the ball will fall is the distance from the *bottom* of the ball to the floor as shown in Figure 13-2. Measure this distance to the nearest 0.1 mm and record it in Data Table 2 as Y.
5. Fire the ball five times onto the same sheet of paper. Place the ball on the rod and measure the horizontal distance X to the nearest 0.1 mm from the center of the ball to the center of each dot on the paper. Record these five values of X in Data Table 2.

CALCULATIONS

Ballistic Pendulum

1. Calculate the distance $y_2 - y_1$ that the ball plus pendulum rises for each trial. Record these values and all other calculations in this section in Calculations Table 1.
2. From Equation 3 calculate the velocity V for each of the five trials.
3. From Equation 4 calculate the initial speed v_{xo} for the five trials.
4. Calculate the mean $\bar{v}_{xo}$ and the standard error α_v of the five values of v_{xo}.

Projectile Motion

1. Use Equation 7 to calculate the value of v_{xo} for each of the five values of X. Use a value of $g = 9.80\,\text{m/s}^2$. Record these values in Calculations Table 2.
2. Calculate the mean $\overline{v}_{xo}$ and the standard error α_v of the five values of v_{xo}. Record them in Calculations Table 2.

Name Section Date

13 LABORATORY 13 *The Ballistic Pendulum and Projectile Motion*

PRE-LABORATORY ASSIGNMENT

1. What are the conditions under which the total momentum of a system of particles is conserved?

2. What kind of collision conserves kinetic energy?

3. What kind of collision does not conserve kinetic energy? What kind of collision results in the maximum loss of kinetic energy?

4. A ball of mass 0.075 kg is fired horizontally into a ballistic pendulum as shown in Figure 13-1. The pendulum mass is 0.350 kg. The ball is caught in the pendulum, and the center of mass of the system rises a vertical distance of 0.145 m in the earth's gravitational field. What was the original speed of the ball? Assume that $g = 9.80\,\text{m/s}^2$. Show your work.

5. How much kinetic energy was lost in the collision of Question 4? Show your work.

6. A projectile is fired in the earth's gravitational field with a horizontal velocity of $v = 9.00\,\text{m/s}$. How far does it go in the horizontal direction in 0.550 s? Show your work.

7. How far does the projectile of Question 6 fall in the vertical direction in 0.550 s? Show your work.

8. A projectile is launched in the horizontal direction. It travels 2.050 m horizontally while it falls 0.450 m vertically, and it then strikes the floor. How long is the projectile in the air? Show your work.

9. What was the original velocity of the projectile described in Question 8? Show your work.

Name ______ Section ______ Date ______

Lab Partners ______

13 LABORATORY 13 The Ballistic Pendulum and Projectile Motion

LABORATORY REPORT

Data Table 1

Trial	p	y_2 (m)
1		
2		
3		
4		
5		
$m =$ kg	$M =$ kg	$y_1 =$ m

Calculations Table 1

$y_2 - y_1$ (m)	V (m/s)	v_{xo} (m/s)
$\bar{v}_{xo} =$ m/s	$\alpha_v =$ m/s	

Data Table 2

Trial	X (m)
1	
2	
3	
4	
5	
$Y =$ m	

Calculations Table 2

v_{xo} (m/s)	
$\bar{v}_{xo} =$ m/s	$\alpha_v =$ m/s

SAMPLE CALCULATIONS

1. $y_2 - y_1 =$

2. $V = \sqrt{2g(y_2 - y_1)} =$

3. (Ballistic pendulum) $v_{xo} = \left(\frac{m + M}{m}\right)V$

4. (Projectile motion) $= v_{xo} = \frac{X}{\sqrt{2Y/g}}$

QUESTIONS

1. Compare the two different values of $\overline{v}_{xo}$. Calculate the difference and the percentage difference between them. State whether the two measurements agree within the combined standard errors of the two values of $\overline{v}_{xo}$.

2. Can you make any statement about the accuracy of the two values of $\overline{v}_{xo}$? Are either of these values more precise than the other? State clearly the basis for your answer in each case.

3. Calculate the loss in kinetic energy when the ball collides with the pendulum as the difference between $\frac{1}{2} m v_{xo}^2$ (the kinetic energy before) and $\frac{1}{2}(m + M)V^2$ (the kinetic energy immediately after the collision).

4. What is the fractional loss in kinetic energy? Calculate by dividing the loss calculated in Question 3 by the original kinetic energy.

5. Calculate the ratio $M/(m+M)$ for the values of m and M in Data Table 1. Compare this ratio with the ratio calculated in Question 4. Express the fractional loss of kinetic energy in symbol form and use equations from the lab to show it should equal $M/(m+M)$.

6. It was assumed that the ball fired as a projectile moved exactly in the horizontal direction. If it moved at some small angle θ to the horizontal, the correct equation would be $(4.90X^2/v_0^2)\tan^2\theta - X\tan\theta + (Y + 4.90X^2/v_0^2) = 0$ with the initial velocity labeled v_o. Use the value of v_o from the ballistic pendulum measurement and the measured X and Y in the equation and solve for the angle θ. If the ball was fired at angle θ to the horizontal it would account for the difference in the two measured values of v_o. The equation is a quadratic in $\tan\theta$ and Y is negative in the equation. Is the θ you found small enough that it is plausible that the projectile might deviate that much from horizontal?

Conservation of Momentum on the Air Track

OBJECTIVES

- ❑ Determine the total momentum of two gliders on an air track before and after they collide.
- ❑ Evaluate the extent to which the total momentum of the system before the collision is equal to the total momentum of the system after the collision.

EQUIPMENT LIST

- Air track, air blower, three gliders (two approximately equal in mass), rubber band puck launcher, meter stick (if air track does not have marked scale)
- Four laboratory timers, laboratory balance and calibrated masses, masking tape

THEORY

The **momentum** of a mass m moving with a **velocity** $\mathbf{v}$ is

$$\mathbf{p} = m\mathbf{v} \qquad \text{(Eq. 1)}$$

Momentum is a vector quantity because it is the product of a scalar (m) with a vector ($\mathbf{v}$). Forces exerted between particles of a system are called internal forces, and they cannot change the momentum of the system. The total momentum of the system can change only if external forces act on the system.

Momentum is conserved in a collision between two objects because the forces that the objects exert on each other are internal to the system. If $\mathbf{p}_1^{\,i}$ and $\mathbf{p}_2^{\,i}$ stand for the initial momenta of two particles, and $\mathbf{p}_1^{\,f}$ and $\mathbf{p}_2^{\,f}$ stand for their final momenta after the collision, then

$$\mathbf{p}_1^{\,i} + \mathbf{p}_2^{\,i} = \mathbf{p}_1^{\,f} + \mathbf{p}_2^{\,f} \qquad \text{(Eq. 2)}$$

Equation 2 implies that each of the vector components of momentum is conserved. For the linear air track a collision is one-dimensional, and the vectors are specified by writing momenta to the right as positive and momenta to the left as negative. Friction on the gliders of the air track will be an external force, and friction must be negligible if Equation 2 is to be valid in this laboratory.

We will determine the constant velocities of two gliders before the collision and after the collision. A velocity will be determined by measuring the elapsed time for the glider to travel some known distance.

We will arrange the collisions to take place in the center of the air track, with a fixed path length required on either side. A 5-meter air track is ideal for performing this experiment, but a shorter air track can be used. We will investigate three different collisions. We will arrange them so that a glider that has a different velocity before and after the collision moves in opposite directions before and after the collision.

EXPERIMENTAL PROCEDURE

1. Determine the mass of the three gliders m_1, m_2, and m_3. The mass of m_1 should be essentially the same as m_2, and m_3 should be greater than m_1. Record the values of the masses in Data Table Mass and Distance.
2. Use pieces of tape to mark a fixed path length on either side of the center of the track. Leave room in the center of the track for the collision to occur. Leave a distance slightly longer than the sum of the lengths of the gliders. Use as much of the remaining track on either side of the center as possible for the marked paths. Figure 14-1 shows how the marker tape should be placed at positions 1, 2, 3, and 4 to define the distances of the collision region. Place the tape low enough on the air track for the gliders to clear the tape as they move. Measure the distance between 1 and 2 and record it in Data Table Mass and Distance as d_{12}. Measure the distance between 3 and 4 and record it in Data Table Mass and Distance as d_{34}.
3. Collisions I and II involve the collision of a moving glider with a glider at rest. For them, attempt to repeat the collision several times at the same initial velocity. With practice the rubber band glider launcher can launch the glider with the same velocity (Figure 14-2). Launch a glider five or six times, starting a timer as the glider is released, and stopping the timer when the glider has gone the length of the track. With practice the time for the glider to go the length of the track should vary by no more than 0.1 s from the average.
4. **Collision I** (Figure 14-3) This collision will be between gliders m_1 and m_2, which have essentially the same mass, with m_2 initially at rest in the center of the track. Glider m_1 will collide with m_2 and essentially stop, and m_2 will move in the direction m_1 was moving before the collision. Launch glider m_1, start a timer as it passes point 1, and stop the timer when it passes point 2. After the collision, start a second timer when glider m_2 passes point 3 and stop this timer when it passes point 4. Record in Data Table Collision I the time intervals as Δt_{12} and Δt_{34}. Repeat this procedure four more times for a total of five trials. To the extent possible launch the glider m_1 with the same velocity each time.
5. **Collision II** (Figure 14-3) This collision will be between glider m_1 and glider m_3 with glider m_3 initially at rest. Because $m_3 > m_1$, glider m_1 will rebound after the collision and move back past point 2 and then point 1. Launch glider m_1 and start one timer when m_1 passes point 1 and stop the timer when m_1 passes point 2. This is the time Δt_{12}. After the collision m_1 will rebound, and m_3 will move in the original direction of the motion of m_1 . A second timer is used to measure the elapsed time for glider m_3 between points 3 and 4 (time Δt_{34}), and a third timer is used to measure the elapsed time for glider m_1 as it moves back past points 2 and then 1 (time Δt_{21}). Record these three time intervals in Data Table Collision II. Repeat this procedure four more times for a total of five trials. To the extent possible, launch the glider m_1 with the same velocity every time.
6. **Collision III** (Figure 14-3) This collision will be between gliders m_1 and m_3 launched from opposite ends of the air track with velocities directed toward the center of the air track. Launch m_1 and m_3 with

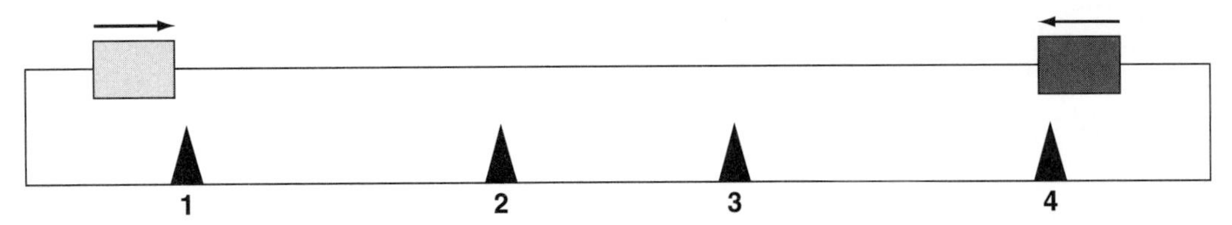

Figure 14-1 Air track with paths 1–2 and 3–4 marked on either side of the center.

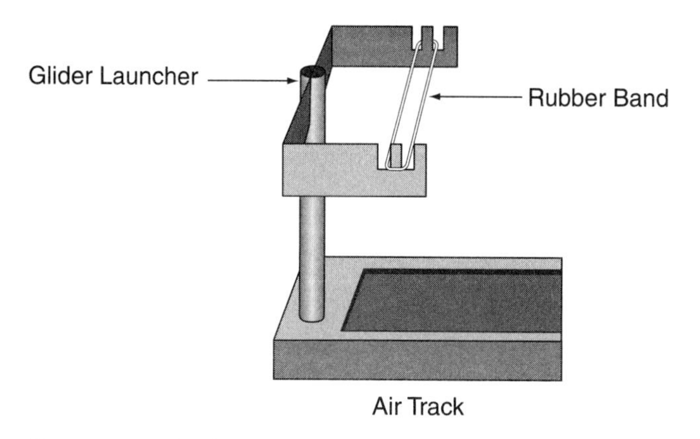

Figure 14-2 Rubber band glider launcher to assure repeated launches at the same velocity.

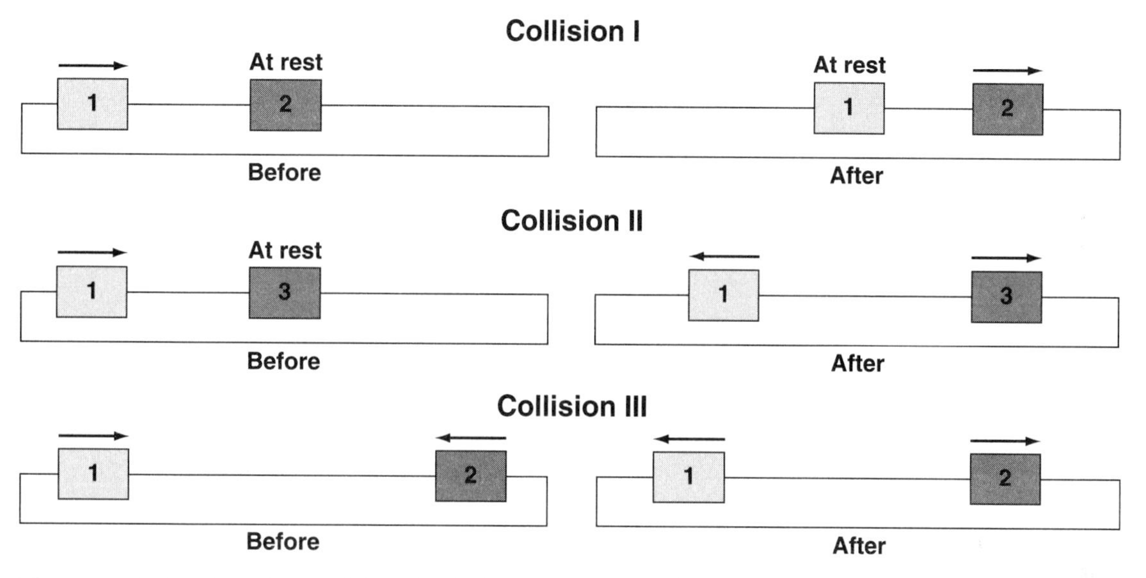

Figure 14-3 Motion of the gliders in Collisions I, II, and III.

the rubber band launchers and give m_1 about 30% greater speed than m_3. Thus m_1 must be launched after m_3 for them to collide at the center of the track. Both gliders will rebound after the collision and reverse direction. Two timers are used to measure the time intervals Δt_{12} and Δt_{21} for m_1 as it moves toward the center and rebounds. Two other timers are used to measure the time intervals Δt_{43} and Δt_{34} for m_3 as it moves toward the center and then rebounds. Record the values of the four time intervals in Data Table Collision III. Repeat the procedure four more times for a total of five trials. To the extent possible launch m_1 at approximately the same speed each time and launch m_3 at approximately the same speed each time.

CALCULATIONS

1. Calculate the velocities v from $v = d/\Delta t$ for each of the measured time intervals and distances. Let velocities to the right be positive and velocities to the left be negative. Record these values and all other calculated quantities in Calculations Tables Collision I, Collision II, and Collision III.
2. Calculate the momentum for each of the gliders before and after the collision using Equation 1. Let momenta to the right be positive and momenta to the left be negative.

3. For Collision I compare the percentage difference of the momentum of the first glider before the collision to the momentum of the second glider after the collision.
4. For Collisions II and III, calculate the total momentum of both gliders before and the total momentum of both gliders after the collision. Then calculate the percentage difference in the total momentum before the collision and the total momentum after the collision.

(Instead of using hand timers the laboratory can be done with photogates to more accurately measure the time intervals with the appropriate changes in procedures.)

Name | Section | Date

14 LABORATORY 14 *Conservation of Momentum on the Air Track*

PRE-LABORATORY ASSIGNMENT

1. What is the definition of momentum?

2. What conditions must be satisfied for momentum to be conserved?

3. Two pieces of tape are placed a distance 1.50 m apart on an air track. A 0.350 kg glider on the air track takes time $\Delta t = 1.30$ s to move between the two pieces of tape. What is the velocity of the glider? What is its momentum? Show your work.

4. A glider of mass $m_1 = 0.350$ kg moves with a velocity of 0.850 m/s to the right on an air track. It collides with a glider of mass $m_2 = 0.350$ kg at rest. Glider m_1 stops, and m_2 moves in the direction that m_1 was traveling. What is the velocity of m_2? Show your work.

5. An air track glider of mass $m_1 = 0.200$ kg moving at 0.750 m/s to the right collides with a glider of mass $m_2 = 0.400$ kg at rest. If m_1 rebounds and moves to the left with a speed of 0.250 m/s, what is the speed and direction of m_2 after the collision? (*Hint*—Momentum is a vector quantity, and direction is indicated by the sign of the momentum.) Show your work.

6. For the collision in Question 5, calculate the original kinetic energy of the system before the collision. Calculate the total kinetic energy after the collision. What happened to the lost energy? Show your work.

7. An air track glider of mass $m_1 = 0.300$ kg moving at a speed of 0.800 m/s to the right collides with a glider of mass $m_2 = 0.300$ kg moving at a speed of 0.400 m/s in the opposite direction. After the collision m_1 rebounds at speed 0.200 m/s to the left. After the collision, what is the speed and direction of m_2? Show your work.

Name ____ *Section* ____ *Date* ____

Lab Partners ____

14 **LABORATORY 14** *Conservation of Momentum on the Air Track*

LABORATORY REPORT

Data Table Mass and Distance

m_1 (kg)	m_2 (kg)	m_3 (kg)	d_{12} (m)	d_{34} (m)

Data Table Collision I

		Trial 1	*Trial 2*	*Trial 3*	*Trial 4*	*Trial 5*
m_1 Before	Δt_{12} (s)					
m_2 After	Δt_{34} (s)					

Calculations Table Collision I

		Trial 1	*Trial 2*	*Trial 3*	*Trial 4*	*Trial 5*
m_1 Before	v_1 (m/s)					
	$\mathbf{p}_1$ (kg–m/s)					
m_2 After	v_2 (m/s)					
	$\mathbf{p}_2$ (kg–m/s)					
% Difference $\mathbf{p}_1$ and $\mathbf{p}_2$						

Data Table Collision II

		Trial 1	Trial 2	Trial 3	Trial 4	Trial 5
m_1 Before	Δt_{12} (s)					
m_1 After	Δt_{21} (s)					
m_2 After	Δt_{34} (s)					

Calculations Table Collision II

		Trial 1	Trial 2	Trial 3	Trial 4	Trial 5
m_1 Before	v_1^i (m/s)					
	$\mathbf{p}_1^i$(kg–m/s)					
m_1 After	v_1^f (m/s)					
	$\mathbf{p}_1^f$(kg–m/s)					
m_3 After	v_3^f (m/s)					
	$\mathbf{p}_3^f$(kg–m/s)					
After = $\mathbf{p}_1^f + \mathbf{p}_3^f$ =						
% Diff						

Data Table Collision III

		Trial 1	Trial 2	Trial 3	Trial 4	Trial 5
m_1 Before	Δt_{12} (s)					
m_3 Before	Δt_{43} (s)					
m_1 After	Δt_{21} (s)					
m_3 After	Δt_{34} (s)					

Calculations Table Collision III

		Trial 1	*Trial 2*	*Trial 3*	*Trial 4*	*Trial 5*
m_1 Before	$v_1{}^{i}$ (m/s)					
	$p_1{}^{i}$(kg–m/s)					
m_3 Before	$v_3{}^{i}$ (m/s)					
	$P_3{}^{i}$(kg–m/s)					
m_1 After	$v_1{}^{f}$ (m/s)					
	$p_1{}^{f}$(kg–m/s)					
m_3 After	$v_3{}^{f}$ (m/s)					
	$p_3{}^{f}$(kg–m/s)					
$p_1{}^{i} + p_3{}^{i} =$						
$p_1{}^{f} + p_3{}^{f} =$						
% Diff						

SAMPLE CALCULATIONS

1. $v_1 = d_{12}/\Delta t_{12} =$
2. $p_1 = (m_1)(v_1) =$
3. Total p (before) $= p_1{}^{i} + p_3{}^{i} =$
4. % Difference =
5. $K_i = ½ m_1(v_1{}^{i})^2 + ½ m_2(v_2{}^{i})^2 =$
6. $\Delta K = K_f - K_i =$
7. % Lost $= \Delta K/K_i \times 100\% =$

QUESTIONS

1. Consider the percentage differences between the total momentum before the collision and the total momentum after the collision for the various trials of Collisions I, II, and III. If they are less than 10% they are good evidence that momentum is conserved, and if they are less than 5% they are very good evidence. To what extent do your data indicate that momentum is conserved?

2. For each of the Collisions I, II, and III, consider the one trial that has the smallest percentage difference and calculate K_i, K_f, ΔK, and the % Lost for that trial. What happens to the lost energy?

Collision I	Trial ______	$K_i =$ ________	$K_f =$ ________	$\Delta K =$ ________	%Lost = ________
Collision II	Trial ______	$K_i =$ ________	$K_f =$ ________	$\Delta K =$ ________	%Lost = ________
Collision III	Trial ______	$K_i =$ ________	$K_f =$ ________	$\Delta K =$ ________	%Lost = ________

3. Is the kinetic energy approximately conserved for any of the collisions that you calculated? If so, state which one or ones and give your evidence.

Conservation of Momentum on the Air Table

OBJECTIVES

- ❑ Determine the velocities and momenta of pucks on an air table before and after several different two-dimensional collisions.
- ❑ Impose a coordinate system on each collision to resolve the momenta into components and evaluate the extent to which momentum is conserved for each component.

EQUIPMENT LIST

- Air table with pucks, sparktimer, air pump, foot switch, lead weights, Velcro collars, and level
- Carbon paper, recording paper, meter stick or ruler, protractor, and square

THEORY

The **momentum** of an object of mass m moving with a **velocity v** is

$$\mathbf{p} = m\mathbf{v} \qquad \text{(Eq. 1)}$$

Momentum is a vector quantity because it is the product of a scalar (m) with a vector (**v**). For a system of particles the total momentum of the system is constant if there are no external forces acting on the system. The forces exerted between particles of the system are called internal forces, and they cannot change the momentum of the system.

Momentum is conserved in collisions between two objects because the forces that the objects exert on each other are internal to the system. If $\mathbf{p}_1{}^{i}$ and $\mathbf{p}_2{}^{i}$ are the initial momenta of two particles, and $\mathbf{p}_1{}^{f}$ and $\mathbf{p}_2{}^{f}$ are their final momenta after the collision, then

$$\mathbf{p}_1{}^{i} + \mathbf{p}_2{}^{i} = \mathbf{p}_1{}^{f} + \mathbf{p}_2{}^{f} \qquad \text{(Eq. 2)}$$

A collision where particles stick together is called a **completely inelastic collision.** Such a collision will result in the maximum possible loss of kinetic energy.

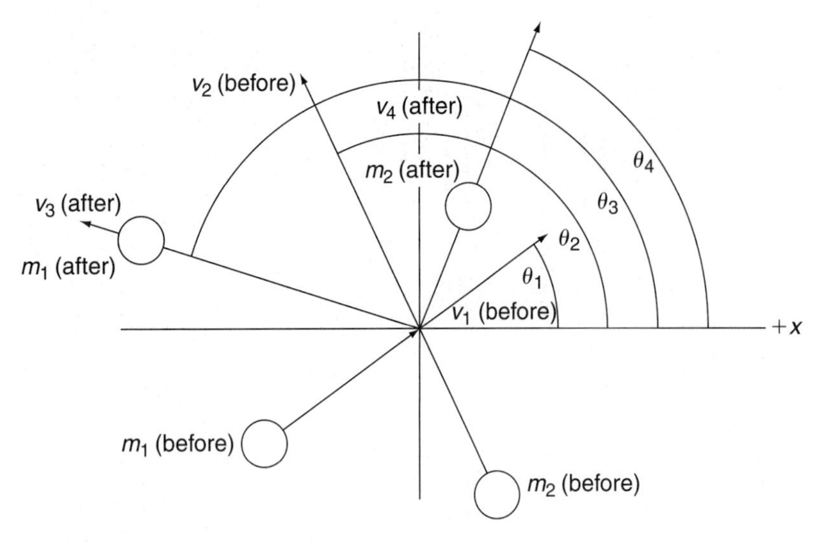

Figure 15-1 Collision between mass m_1 and m_2, both with an initial velocity.

Figure 15-1 shows a mass m_1 with velocity v_1 and a mass m_2 with velocity v_2 that collide at the origin of a coordinate system. After the collision they have velocities v_3 and v_4. For that collision with coordinates as shown, the equations that represent the conservation of momentum are

$$x \text{ direction} \qquad m_1v_1\cos\theta_1 + m_2v_2\cos\theta_2 = m_2v_4\cos\theta_4 + m_1v_3\cos\theta_3 \qquad \text{(Eq. 3)}$$

$$y \text{ direction} \qquad m_1v_1\sin\theta_1 + m_2v_2\sin\theta_2 = m_1v_3\sin\theta_3 + m_2v_4\sin\theta_4 \qquad \text{(Eq. 4)}$$

We will use collisions on the air table to investigate this general case and several simpler special cases. Simplified collisions occur when $m_1 = m_2$, when one of the masses is originally at rest, and when the collision is head-on with all motion along a single line.

EXPERIMENTAL PROCEDURE

1. Unless you have prior experience with the air table (Figure 15-2) read the instruction manual for the air table to become familiar with its operation including the air pump, sparktimer, and foot switch. Be careful not to touch the pucks except by the insulated tubes when the sparktimer is in operation.
2. Level the air table with the three adjustable legs until a puck placed near the center of the table is essentially motionless. *It is especially critical to have the table as level as possible* for these collision events because several of the collisions are to be made with one of the pucks initially at rest.
3. Set the sparktimer to 20.0 Hz. Perform each of the four collisions described below. Start the sparktimer after the pucks are launched, and stop it before the pucks collide with the edge of the table. Friction will cause the experimental results to disagree with the theory, and the effect is larger for low speeds. For best results the pucks should move as fast as reasonable. However, it is possible to damage the glass tops of the tables if the pucks are thrown too fast. *Be very careful not to damage the air table.* Consult your instructor for advice about the speed of the pucks. Use the same sheet of carbon paper for all measurements, but use a new sheet of recording paper for each collision.
4. **Collision I** Use a laboratory balance to determine the mass of two pucks that are approximately the same mass and record in the Data Table. Label one puck #1 and the other #2. Launch puck 1 from near one corner so that it strikes puck 2, which is at rest near the center of the table. The collision should not be head-on. The pucks should move at a large angle relative to each other after the collision.

Figure 15-2 Precision air table. (Photo courtesy of Central Scientific Co., Inc.)

Before removing the recording paper, label each sparktimer track with the puck number and the direction that each puck moved before and after the collision. Label each collision before removing the recording paper from the table. Label this recording as Collision I, and set it aside for future analysis.

5. **Collision II** Use the same pucks as Collision I, but launch them both at the same time from *adjacent corners* of the table so they collide at the center at an angle of about 90°. Label the tracks in the manner described above. Label it as Collision II, and set it aside for future analysis.
6. **Collision III** Use the same procedure as Collision II but use masses that are not equal. Use puck 1 as it is. Place several of the lead weights that fit over the pucks to increase the mass of puck 2. Determine the mass of puck 2 with the lead on it, and record the results in the Data Table. Launch the pucks from adjacent corners so they collide near the center. Label the tracks and recording paper as above, and set the paper aside for future analysis.
7. **Collision IV** Remove the lead from puck 2. Place Velcro collars on each of the pucks. Determine the mass of each puck, and record it in the Data Table. With Velcro collars the pucks will stick together, and the collision will be completely inelastic. Launch the two pucks of approximately equal mass from adjacent corners of the table so they collide near the center. It might be necessary to try this several times. When the pucks stick together they might have a tendency to rotate about a common center of mass as they move together. Instead, it is necessary that after the collision the trace shows the combined pucks making parallel tracks as they move together. That is evidence that no rotation is occurring. It may take several traces to achieve parallel tracks. Label the acceptable recording, and set it aside for future analysis.
8. For each collision determine the displacement Δs along the four tracks that represent the motion of the pucks before and after the collision. Measure the Δs of the puck for four spark intervals or $\Delta t = 0.200$ s. Be sure that the appropriate Δs is associated with each of the pucks both before and after the collision. Record all displacements Δs in the Data Tables.
9. For each of the collisions draw an x coordinate axis on the recording paper. This is an arbitrary choice, but it will save work to choose the $+x$ direction to be the direction of the initial motion of one of the pucks. That puck would then initially have only an x component, with y component zero. Determine the components for the other puck before the collision, and for both the pucks after the collision, by finding the angle their tracks make with the $+x$ axis. This process is illustrated in Figure 15-3. The $+x$ axis has been chosen to be in the direction of the initial motion of puck 1. The other angles are then determined by the angle each puck's motion makes with the $+x$ axis. The x components are given by the cosine and the y components by the sine.

For each of the four collisions draw the $+x$ axis through the points of the initial motion of puck 1 in the direction of its motion. Using a protractor, carefully measure the angle of the other three tracks with the

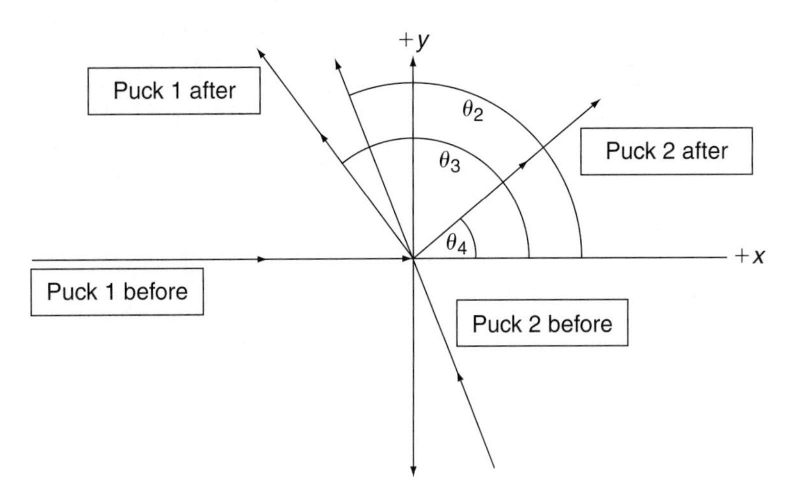

Figure 15-3 Example of how to determine the components of each momentum.

x axis and record the values of those angles in the Data Tables. With this choice of $+x$ axis, θ_1 will be zero for all of the collisions.

CALCULATIONS

1. For the four collisions calculate each velocity as $v = \Delta s/\Delta t$, before and after the collision, from the values of Δs in the Data Tables and $\Delta t = 0.200$ s. Record the values of the velocities as v_1, v_2, v_3, and v_4 in the Calculations Tables.
2. For each of the four collisions calculate the x component and the y component of the momentum for each puck before and after the collision. The equations for the momentum components are given by

$$p_{2x} = m_2 v_2 \cos\theta_2 \quad \text{and} \quad p_{2y} = m_2 v_2 \sin\theta_2 \qquad \text{(Eq. 5)}$$

   with the equations written using v_2 as an example. The extension of Equations 5 for the other velocities should be clear. Calculate the components for each of the collisions and record them in the Calculations Tables. With the angles defined as in Figure 15-3, the sign of each component will be determined by the sign of the sine or cosine function.
3. Calculate the sum of the momentum for each component for each puck and record the results in the Calculations Tables.

Name | Section | Date

15 LABORATORY 15 *Conservation of Momentum on the Air Table*

PRE-LABORATORY ASSIGNMENT

1. What is the definition of momentum?

2. What conditions must be satisfied for momentum to be conserved?

3. A particle of mass $m_1 = 0.350\,\text{kg}$ has speed $v_1 = 0.135\,\text{m/s}$ at a direction of 53.0° above the $+x$ axis as shown in Figure 15-4. What is the magnitude of the particle's momentum? Show your work.

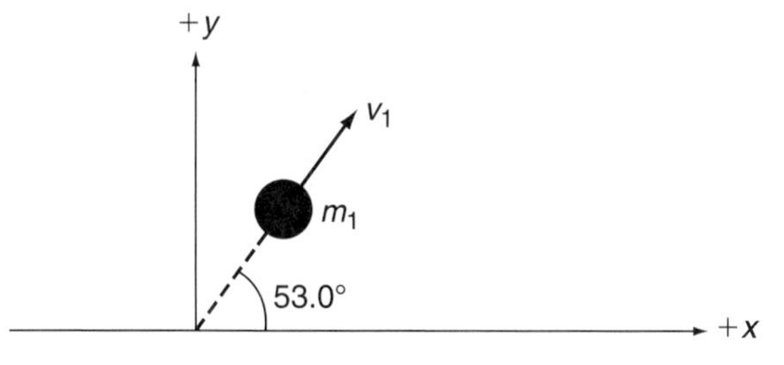

Figure 15-4 Particle of mass m_1 moving at speed v_1 53.0° relative to $+x$ axis.

4. What is the x component, and what is the y component of the momentum of the particle shown in Figure 15-4? Show your work.

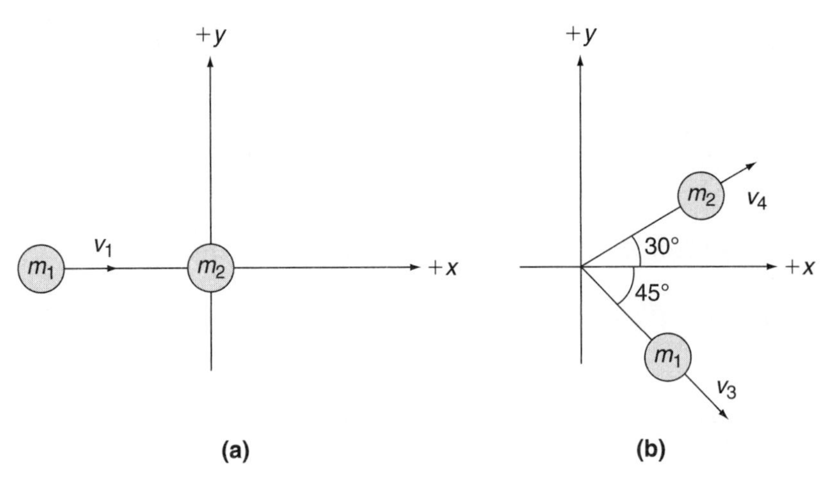

Figure 15-5 Particle of mass m_1 at speed v_1 collides with a particle of mass m_2 at rest.

5. A particle of mass $m_1 = 1.000$ kg moves at speed $v_1 = 0.500$ m/s as shown in Figure 15-5(a). It collides with a particle of mass $m_2 = 2.000$ kg at rest at the origin. What is the total momentum of the system in the x direction before the collision? What is the total momentum of the system in the y direction before the collision? Show your work.

6. Figure 15-5(b) shows that after the collision m_1 moves with speed v_3 at an angle $\theta_3 = 315.0°$ with respect to the x axis, and m_2 moves with a speed v_4 at an angle $\theta_4 = 30.0°$ with respect to the x axis. Write an expression for the total momentum of the system in the x direction and another expression for the total momentum in the y direction after the collision in terms of the symbols m_1, m_2, v_3, v_4, and the angles θ_3 and θ_4.

7. Equate the expression for the x component in Question 6 to the value of the x component in Question 5. Equate the expression for the y component in Question 6 to the value of the y component in Question 5. In the resulting two equations v_3 and v_4 are the only two unknowns. Solve the two equations for v_3 and v_4. Show your work.

Name   Section   Date

Lab Partners

15 LABORATORY 15 Conservation of Momentum on the Air Table

LABORATORY REPORT

(Record all masses in kg, all displacements in m, all velocities in m/s, and all momenta in kg-m/s.)

Data Table Collision I

m_1	m_2	Δs_1	Δs_2	Δs_3	Δs_4	θ_1	θ_2	θ_3	θ_4

Calculations Table Collision I

v_1	v_2	p_{1x}	p_{1y}	p_{2x}	p_{2y}	$p_{1x}+p_{2x}$	$p_{1y}+p_{2y}$
v_3	v_4	p_{3x}	p_{3y}	p_{4x}	p_{4y}	$p_{3x}+p_{4x}$	$p_{3y}+p_{4y}$

Data Table Collision II

m_1	m_2	Δs_1	Δs_2	Δs_3	Δs_4	θ_1	θ_2	θ_3	θ_4

Calculations Table Collision II

v_1	v_2	p_{1x}	p_{1y}	p_{2x}	p_{2y}	$p_{1x}+p_{2x}$	$p_{1y}+p_{2y}$
v_3	v_4	p_{3x}	p_{3y}	p_{4x}	p_{4y}	$p_{3x}+p_{4x}$	$p_{3y}+p_{4y}$

Data Table Collision III

m_1	m_2	Δs_1	Δs_2	Δs_3	Δs_4	θ_1	θ_2	θ_3	θ_4

Calculations Table Collision III

v_1	v_2	p_{1x}	p_{1y}	p_{2x}	p_{2y}	$p_{1x}+p_{2x}$	$p_{1y}+p_{2y}$
v_3	v_4	p_{3x}	p_{3y}	p_{4x}	p_{4y}	$p_{3x}+p_{4x}$	$p_{3y}+p_{4y}$

Data Table Collision IV

m_1	m_2	Δs_1	Δs_2	Δs_3	Δs_4	θ_1	θ_2	θ_3	θ_4

Calculations Table Collision IV

v_1	v_2	p_{1x}	p_{1y}	p_{2x}	p_{2y}	$p_{1x}+p_{2x}$	$p_{1y}+p_{2y}$
v_3	v_4	p_{3x}	p_{3y}	p_{4x}	p_{4y}	$p_{3x}+p_{4x}$	$p_{3y}+p_{4y}$

SAMPLE CALCULATIONS

1. $v_i = \Delta s_i / \Delta t =$
2. $p_i = m_i v_i =$
3. $p_{ix} = p_i \cos \theta_i =$
4. $p_{iy} = p_i \sin \theta_i =$
5. $\%\ \text{Diff}\ x = ((p_{3x} + p_{4x}) - (p_{1x} + p_{2x}))/(½(p_{1x} + p_{2x} + p_{3x} + p_{4x})) \times 100\% =$
6. $p_1(\text{initial}) = \sqrt{(p_{1x} + p_{2x})^2 + (p_{1y} + p_{2y})^2} =$
7. $\%\ \text{Diff}/p(\text{initial}) = ((p_{3x} + p_{4x}) - (p_{1x} + p_{2x}))/(p(\text{initial})) \times 100\% =$
8. $K = ½ \text{mv}^2 =$
9. $\Delta K = K_f - K_i =$
10. $\%\ \text{Loss}\ K = \Delta K / K \times 100\% =$

QUESTIONS

1. In each collision, if momentum is conserved the total x component before the collision $(p_{1x} + p_{2x})$ should equal the total x component after the collision $(p_{3x} + p_{4x})$. Similarly the y component $p_{1y} + p_{2y}$ should equal $p_{3y} + p_{4y}$. Calculate and record the percentage differences between these quantities. The y component of Collision I is zero.

Collision I	x– comp % diff = ____________	
Collision II	x– comp % diff = ____________	y– comp % diff = ____________
Collision III	x– comp % diff = ____________	y– comp % diff = ____________
Collision IV	x– comp % diff = ____________	y– comp % diff = ____________

2. If the percentage differences above are less than 15% it is considered good confirmation of conservation of momentum because frictional forces are likely significant for the data. Do your data show good confirmation of conservation of momentum?

3. For each collision, the magnitude of the total initial momentum is given by the quantity $\sqrt{(p_{1x} + p_{2x})^2 + (p_{1y} + p_{2y})^2}$. Calculate that quantity for each of the four collisions and record the results below.

Collision I	p_{int} = ____________	Collision II	p_{int} = ____________
Collision III	p_{int} = ____________	Collision IV	p_{int} = ____________

4. Divide the difference between each component before and after the collision by the value of p_{int} calculated in Question 3 and express it as a percentage. The smaller the value of this quantity, the better the data are consistent with momentum conservation.

Collision I	x– comp % diff/p_{int} = ____________	y– comp % diff/p_{int} = ____________
Collision II	x– comp % diff/p_{int} = ____________	y– comp % diff/p_{int} = ____________
Collision III	x– comp % diff/p_{int} = ____________	y– comp % diff/p_{int} = ____________
Collision IV	x– comp % diff/p_{int} = ____________	y– comp % diff/p_{int} = ____________

Centripetal Acceleration of an Object in Circular Motion

OBJECTIVES

- ❑ Investigate how the period T of an object that rotates in a circle is related to the mass of the object M, speed v, and radius R of the circle.
- ❑ Determine the centripetal force F as the force required to stretch a spring.

EQUIPMENT LIST

- Hand-operated centripetal force apparatus (The device described is available from Sargent-Welch Scientific Company.)
- Laboratory balance, calibrated slotted masses, mass holder, laboratory timer, and metal ruler

THEORY

An object moving in a circle at constant speed has a velocity vector that is always tangent to the circle. The direction of the velocity is continuously changing. The object is accelerated because acceleration is by definition a change in velocity per unit time. Figure 16-1 shows the velocity vector at points around the circle for an object moving in a circle at constant speed. The lengths of the vectors are the same because the speed is constant, and the direction of the vectors indicates the direction of the velocity at that point. Also shown in Figure 16-1 are the velocity vectors $\mathbf{v}_i$ and $\mathbf{v}_f$ at two times t_i and t_f. In the third part of the figure is the vector difference $\Delta\mathbf{v} = \mathbf{v}_f - \mathbf{v}_i$ indicating that the change in velocity $\Delta\mathbf{v}$ always points toward the center of the circle. The **acceleration a** is

$$\mathbf{a} = \Delta\mathbf{v}/\Delta t \quad \text{(Eq. 1)}$$

The acceleration **a** is in the direction of $\Delta\mathbf{v}$. It points toward the center of the circle and has magnitude

$$a = v^2/R \quad \text{(Eq. 2)}$$

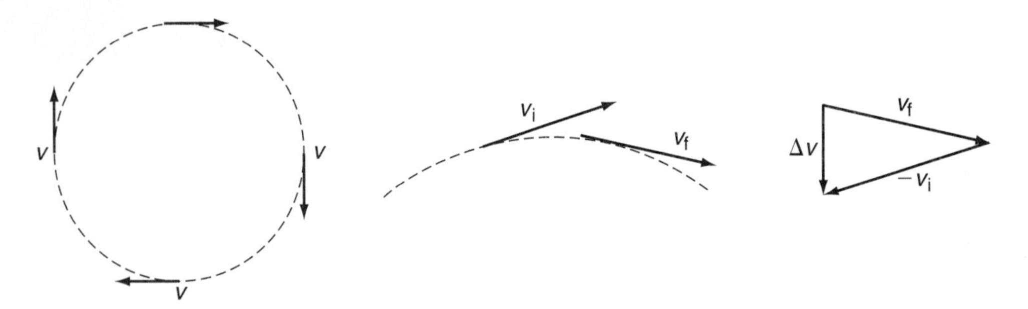

Figure 16-1 Velocity vectors for circular motion at constant speed. Vectors at two times t_i and t close together and the change in velocity **Δv** pointing toward the center of the circle.

By Newton's second law the centripetal force F and the **centripetal acceleration** a are related by $F = Ma$ where M is the mass of the object moving in a circle at speed v. Using Equation 2 gives

$$F = mv^2/R \qquad \text{(Eq. 3)}$$

The time for one complete revolution around the circle is the period T, which is related to the speed v by the expression

$$v = (2\pi R)/T \qquad \text{(Eq. 4)}$$

The centripetal force apparatus has a mass bob with a pointed tip at the bottom suspended from a horizontal rotating bar. The bob has a spring hooked between the side of the bob and the central rotating shaft. The spring provides a horizontal centripetal force when the bob rotates in a horizontal plane. The bob rotates at a fixed radius R from the central rotating shaft when the tip of the bob passes over a pointer located at distance R from the central rotating shaft. For the spring used, mass M will rotate at radius R for only one rotation period T. Figure 16-2(a) shows the system rotating at the period necessary to rotate at radius R. The period T will be measured for a given R and M. Equation 4 allows a determination of v, and using that value in Equation 3 allows determination of F. This will be referred to as F_{theo} for the theoretical value of the force.

The force the spring exerts on the bob when it is rotating at distance R depends on the amount the spring is stretched under those conditions. This force can be measured by determining the force needed to stretch the spring the same amount when the apparatus is not rotating. Figure 16-2(b) shows a string attached to the other side of the bob with slotted masses applied over a pulley. The weight of the total mass needed to stretch the spring until the tip of the bob is aligned with the pointer is the experimental value of the centripetal force F. This will be referred to as F_{exp}.

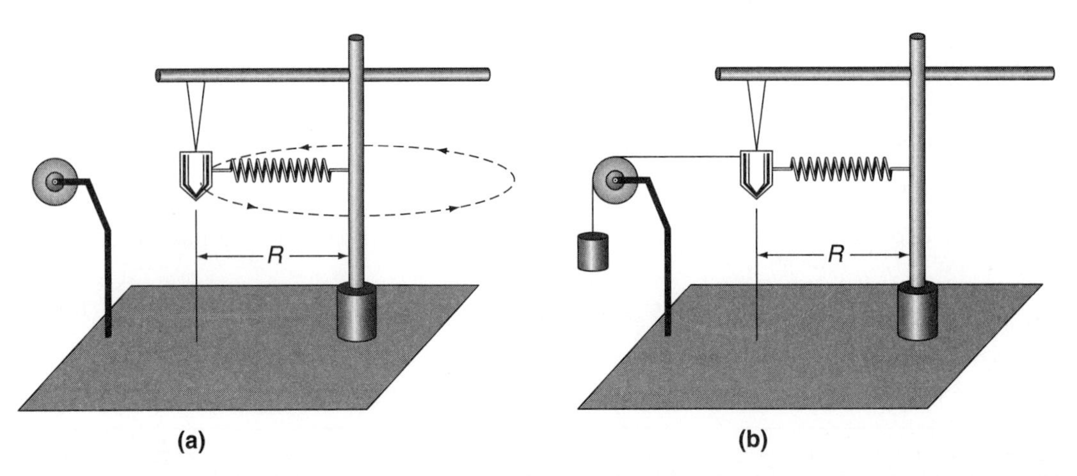

Figure 16-2 (a) Centripetal force apparatus rotating. (b) Determination of the centripetal force by measuring the force needed to stretch the spring under static conditions.

EXPERIMENTAL PROCEDURE

1. Detach the bob from its support strings, remove the spring, and determine its mass. Record this value as m_b in the Data Table.
2. Hang the bob from the cross arm by its support strings (Figure 16-3). Adjust the position of the pointer to its closest position to the rotating shaft for the minimum value of R. Loosen the screw holding the cross arm in the rotating shaft and adjust its position until the tip of the bob is precisely above the tip of the pointer. The tip of the bob should be about 1 mm above the pointer. Measure the distance from the center of the pointer to the center of the rotating shaft. Record this value as R in the Data Table.
3. Attach the spring to the bob and to the rotating shaft. Rotate the system as shown in Figure 16-2(a) by twirling the rotating shaft between the thumb and first finger of your hand. The bob will pass over the pointer at radius R for only one rotation period T. Continue to rotate the apparatus by hand while keeping the rotation speed as constant as possible, and at the same time ensuring that the bob passes over the pointer on each rotation. At this rotation rate measure the time for 25 complete revolutions of the bob and record it in the Data Table as Time 1. Repeat this process two more times, recording the two other measurements of the time for 25 revolutions as Time 2 and Time 3. Record in the Data Table the value of the rotating mass as m_b for this part of the procedure.
4. With the system not rotating, measure directly the centripetal force by attaching a string to the side of the bob opposite the spring. Apply slotted weights over the pulley as shown in Figure 16-2(b) until the tip of the bob is just above the tip of the pointer. Let m_a stand for the total mass needed to stretch the spring by the proper amount. Record in the Data Table the value of m_a needed to stretch the spring to the pointer at position R.
5. Repeat Steps 3 and 4 above using the same R but using two other values of rotating mass. First, add a 0.050 kg slotted mass to the bob. Then remove the 0.050 kg mass and add a 0.100 kg slotted mass. In each case, place the slotted mass with the open end pointed outward, and secure it with the knurled nut on the bob. Record the results of the measurements for rotating mass values of $m_b + 0.050$ kg and $m_b + 0.100$ kg in the Data Table.
6. Perform the measurements with the rotating mass m_{rot} again equal to m_b, but use three new values of R differing by about 1 cm. Each time R is to be changed, remove the spring from the bob, and position the pointer 1 cm further from the rotating shaft. Then adjust the cross arm so that the bob is above the pointer. Perform the measurements of Steps 2–4 for the three values of R and record all results in the Data Table.

Figure 16-3 Centripetal force apparatus. (Photo by Sargent-Welch Scientific Co.)

CALCULATIONS

1. Calculate the mean of the three trials of the time for 25 complete revolutions, and record it in the Calculations Table as $\overline{Time}$. Divide the value of $\overline{Time}$ by 25, and record the result in the Calculations Table as the period T.
2. Use Equation 4 to calculate v from the measured values of R and T, and record the results in the Calculations Table.
3. Use Equation 3 to calculate the theoretical value for the centripetal force from the values of M, v, and R. Record the results in the Calculations Table as F_{theo}.
4. Calculate the experimental value for the centripetal force as $m_a g$ from the values of m_a. Use a value of $9.80\ m/s^2$ for g. Record the results in the Calculations Table as F_{exp}.
5. Calculate the percentage difference between the values of F_{theo} and F_{exp}. Record the results in the Calculations Table.

Name Section Date

16

LABORATORY 16 *Centripetal Acceleration of an Object in Circular Motion*

PRE-LABORATORY ASSIGNMENT

1. If a particle moves in a circle of radius R at constant speed v its acceleration is (a) directed toward the center of the circle (b) equal to v^2/R (c) because the direction of the velocity vector changes continuously (d) all of the above are true.

2. If a particle moves in a circle of radius $R = 1.35$ m at a constant speed of $v = 6.70$ m/s, what is the magnitude and direction of its centripetal acceleration?

3. If the mass of the particle in Question 2 is 0.350 kg, what is the magnitude and direction of the centripetal force on it? Show your work.

4. A 0.500 kg particle moves in a circle of radius $R = 0.150$ m at constant speed. The time for 20 complete revolutions is 31.7 s. What is the period T of the motion? What is the speed of the particle? Show your work.

5. What is the centripetal acceleration of the particle in Question 4? What is the centripetal force on the particle? Show your work.

6. For the apparatus used in this laboratory the centripetal force is the same for a fixed radius R of rotation. Why is that statement true for this apparatus? (*Hint*—What provides the centripetal force on the rotating mass for this apparatus?)

7. A mass of 0.450 kg rotates at constant speed with a period of 1.45 s at a radius R of 0.140 m in the apparatus used in this laboratory. What is the rotation period for a mass of 0.550 kg at the same radius? Show your work.

Name *Section* *Date*

Lab Partners

16 **LABORATORY 16** *Centripetal Acceleration of an Object in Circular Motion*

LABORATORY REPORT

Data Table (m_b = ______ kg)

m_{rot} (kg)	R (m)	*Time 1* (s)	*Time 2* (s)	*Time 3* (s)	m_a (kg)

Calculations Table

m_{rot} (kg)	R (m)	$\overline{Time}$ (s)	T (s)	v (m/s)	F_{theo} (N)	F_{exp} (N)	% *Diff*

SAMPLE CALCULATIONS

1. $\overline{Time} =$
2. $T = \overline{Time}/25 =$
3. $v = (2\pi R)/T =$
4. $F_{theo} = mv^2/R =$
5. $F_{exp} = m_a g =$
6. % Diff =

QUESTIONS

1. Do your results confirm the theoretical relationship for the centripetal acceleration given by $F = Mv^2/R$? Consider the agreement between F_{theo} and F_{exp} to answer this question. Explain your reasoning.

2. Because the centripetal force is provided by a spring for this apparatus, the centripetal force at a given distance R is fixed by the spring constant of the spring. Therefore, Mv^2 should be constant for a given radius R. Calculate the quantity Mv^2 for the four data points taken at the same radius R. Describe the agreement of those values.

3. Equation 3 can be written in the form $v^2 = (1/M)FR$. For a constant value of M this would imply that the quantity v^2 should be proportional to the quantity FR with the reciprocal of the mass as the constant of proportionality. For your data points with the same mass, perform a linear least squares fit with v^2 as the vertical axis and FR as the horizontal axis. Use the values of F_{exp} to calculate FR. Compare the slope of the fit to the reciprocal of the mass. Record the correlation coefficient r.

4. Suppose that a spring with a larger spring constant was used in this same apparatus. If a given mass were rotated at the same radius at which it had been rotated with the original spring, would the new period of rotation using the new spring be greater, or would it be less than the period of rotation using the original spring? Explain your reasoning.

Moment of Inertia and Rotational Motion

OBJECTIVES

- ❑ Investigate the dependence of angular acceleration α of a cylinder on applied torque τ.
- ❑ Determine the moment of inertia I of the cylinder from the slope of applied torque τ versus angular acceleration α and compare the experimental value with a theoretical calculation of the moment of inertia.

EQUIPMENT LIST

- Rotational inertia apparatus (wheel and axle with hub and tripping platform)
- Meter stick, string, mass holder, slotted masses, timer, laboratory balance, vernier calipers, and large calipers

THEORY

For linear motion, Newton's second law $F = ma$ describes the relationship between the applied force F, the mass m of an object, and its acceleration a. Force is the cause of the acceleration, and mass is a measure of the tendency of an object to resist a change in its linear translational motion.

For rotational motion of some object about a fixed axis, an equivalent description for the relationship between the applied **torque** τ, the **moment of inertia** I, and the **angular acceleration** α of the object is given by

$$\tau = I\alpha \qquad \text{(Eq. 1)}$$

Torque τ is the cause of the angular acceleration α, and the moment of inertia I is a measure of the tendency of a body to resist a change in its rotational motion.

The moment of inertia I of a rigid body depends upon the mass of the body and the way in which the mass is distributed relative to the axis of rotation. A thin solid cylinder is shown in Figure 17-1. Three arbitrarily chosen elements of mass m_1, m_2, and m_3 are shown located at distances r_1, r_2, and r_3 from the rotation axis AB. Their contribution to the moment of inertia is $m_1r_1^2 + m_2r_2^2 + m_3r_3^2$. The total moment of inertia is

$$I = \sum_{i=1}^{N} m_i r_i^2 \qquad \text{(Eq. 2)}$$

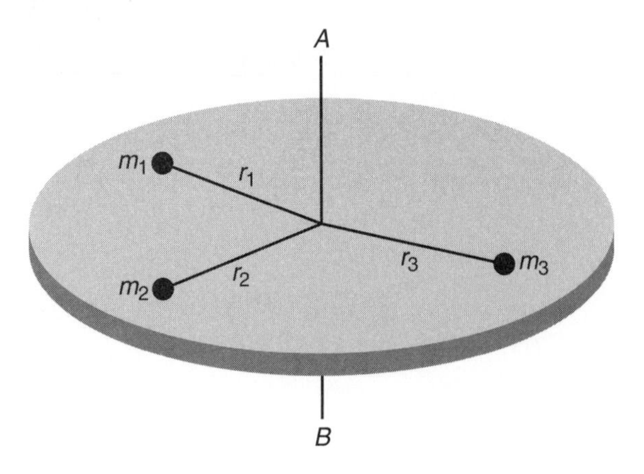

Figure 17-1 Elements of mass contributing to the moment of inertia of a cylinder.

where N is arbitrarily chosen, but must be large enough so that each element of mass approximates a point mass. Equation 2 is the general definition of the moment of inertia about a particular axis for any object. For a solid cylinder of radius R and total mass M, the results are

$$I = \tfrac{1}{2}MR^2 \qquad \text{(Eq. 3)}$$

It is not necessary to know the thickness of the cylinder to calculate the moment of inertia.

In this laboratory, a wheel and axle with a small hub will be caused to rotate by a torque. The torque is produced by the weight of a mass m on a string wrapped around the hub as shown in Figure 17-2. The mass m moves linearly downward as the cylinder rotates about its fixed axis. The two forces acting on the mass m are the weight mg and the tension T in the string. Applying Newton's second law to the linear motion of the mass gives

$$mg - T = ma \quad \text{or} \quad T = m(g - a) \qquad \text{(Eq. 4)}$$

The force producing the torque is the tension in the string T, and the lever arm of the force is the radius of the hub r or

$$\tau = Tr \qquad \text{(Eq. 5)}$$

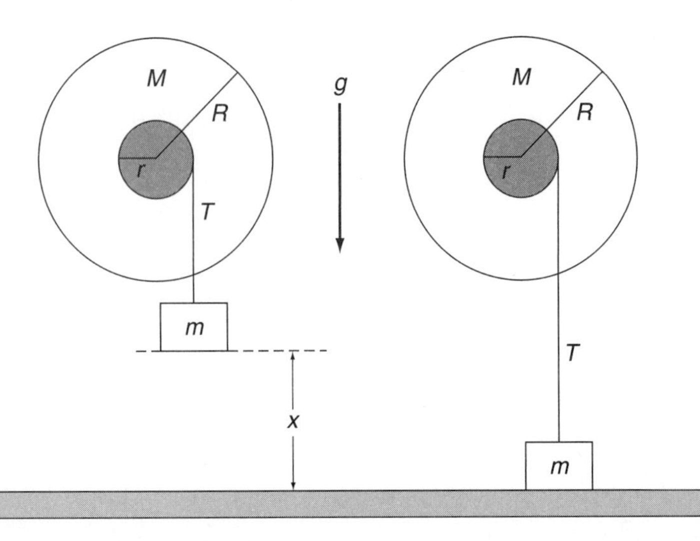

Figure 17-2 Motion of mass m as it moves linearly x and rotates a cylinder.

Figure 17-3 Rotational inertia apparatus. (Photo by Sargent Welch Scientific Co.)

The mass m is released from rest, accelerates for a distance x, and then strikes the floor as shown in Figure 17-2. It moves the distance x in time t. The acceleration a is related to x and t by

$$a = 2x/t^2 \qquad \text{(Eq. 6)}$$

The linear and angular accelerations are related by

$$\alpha = a/r \qquad \text{(Eq. 7)}$$

The relationships described above will be used in this laboratory to determine the angular acceleration α produced by different values of torque τ applied to the moment of inertia I of the wheel. The moment of inertia of the wheel will be determined theoretically by treating the system as a simple cylinder. This is a good approximation because the hub is hollow and made of a lightweight material, and its mass contributes very little to the moment of inertia of the system.

EXPERIMENTAL PROCEDURE

1. Determine the values of the total mass of the wheel M, the radius of the wheel R, and the radius of the small hub r. The mass M may be stamped on the wheel, or else it may be available from your instructor. Record the values of M, R, and r in the Data Table.
2. The wheel and axle system may already be mounted on the wall in a permanent position. If not, clamp the system to the laboratory table with the wheel hanging over the edge of the table and the rotation axis horizontal. Use thick, rigid rods to make the system as vibration-free as possible. The wheel should be mounted above the floor high enough that the mass will travel at least 1 m and preferably more. The longer distance gives longer time intervals, which can be measured with greater precision.
3. The apparatus has a small tripping platform on which the mass holder on the end of the string rests before it is released. Measure the distance from the top of this platform to the floor and record it in the Data Table as x.
4. Make a small loop in each end of a string that is long enough to wrap around the hub several times and still touch the floor. Place one loop on the peg of the wheel hub and wrap the string around the hub. On the loop at the other end of the string place a 0.0500 kg mass holder, and put it on

the platform. Release the tripping platform and simultaneously start a timer. Stop the timer when the mass holder strikes the floor. Make sure that the mass holder is resting on the platform with no slack in the string when it is released. Record the value of the time t in the Data Table. Using this value of m, repeat this process three more times for a total of four trials.

5. Repeat the procedure in Step 4 for a series of different values of the mass m on the string. Add 0.050 kg each time to produce values of m of 0.1000, 0.1500, 0.2000, 0.2500, and 0.3000 kg. Perform four trials with each value of m. Record all the values of m and t in the Data Table.

CALCULATIONS

1. Use the values of the mass of the wheel M and the radius of the wheel R in Equation 3 to calculate a theoretical value of I the moment of inertia. Record this value as I_{theo} in the Calculations Table.
2. Calculate the mean $\bar{t}$ and standard error for the repeated trials of the time for each mass. Record the values of $\bar{t}$ and the values of the standard error in the Calculations Table. (*Note*: To avoid confusion with the angular acceleration, the standard error is noted in that table as Std Err instead of its usual symbol α.)
3. Using the values of $\bar{t}$ in Equation 6, calculate the acceleration a for each value of m. Record the values of a in the Calculations Table.
4. Use the values of acceleration determined above in Equation 7 to calculate the angular acceleration α for each value of m. Note that the units of α are rad/s^2. Record the values of α in the Calculations Table.
5. Use the values of acceleration determined above in Equation 4 to calculate the tension in the string T for each value of m. Record these values of T in the Calculations Table.
6. Use the values of T in Equation 5 to calculate the value of the torque τ for each value of m. Record the values of τ in the Calculations Table.
7. Perform a linear least squares fit with τ as the vertical axis and α as the horizontal axis. Record the slope as I_{exp} the experimental value for the moment of inertia, the intercept as τ_f the frictional torque acting on the wheel, and r the correlation coefficient.

GRAPHS

1. Graph the data for τ versus α with τ as the vertical axis and α as the horizontal axis. Also show on the graph the straight line that was obtained by the fit to the data.

Name　　　　Section　　　　Date

17 LABORATORY 17 *Moment of Inertia and Rotational Motion*

PRE-LABORATORY ASSIGNMENT

1. What equation is the rotational equivalent of Newton's second law? Give the meaning of each symbol and state which rotational quantities are analogous to which linear quantities.

2. What equation defines moment of inertia? Define the terms used in the equation.

3. What is the moment of inertia of a solid cylinder of radius $R = 0.0950$ m, thickness $t = 0.015$ m, and total mass $M = 3.565$ kg? Show your work.

4. A mass hung on a string that is wrapped around an axle on a wheel produces a tension in the string of 5.65 N. The axle has a radius of 0.045 m. The wheel has a mass of 4.000 kg and a radius of 0.125 m. What is the torque produced by the tension on the axle? Show your work.

5. The mass in Question 4 has an acceleration of 0.655 m/s^2. What is the angular acceleration α of the system? Show your work.

6. In the experimental procedure, why is a path length longer than 1 m suggested for the motion of the mass on the string?

7. The following data were taken with a system like the one described in this laboratory. The path length of the falling mass was $x = 1.434$ m, and the radius of the hub around which the string was wrapped was $r = 0.040$ m. For the values of mass m on the string listed, the times to accelerate the distance x are given in the table. Use these data to calculate the values for the acceleration a, the tension T, the torque τ, and the angular acceleration α. Perform a linear least squares fit with τ as the vertical axis and α as the horizontal axis. Record the slope as I the moment of inertia, the intercept as τ_f the frictional torque, and the correlation coefficient r.

m (kg)	t (s)	a (m/s^2)	T (N)	τ (N–m)	α (rad/s^2)
0.050	10.60				
0.100	7.40				
0.150	6.10				
0.200	5.20				
0.250	4.60				
0.300	4.20				

$I =$ ____________ $kg{-}m^2$

$\tau_f =$ ____________ N–m

$r =$ ____________

Name Section Date

Lab Partners

17 LABORATORY 17 *Moment of Inertia and Rotational Motion*

LABORATORY REPORT

Data Table

$M =$ kg	$R =$ m	$r =$ m	$x =$ m	
m (kg)	t_1 (s)	t_2 (s)	t_3 (s)	t_4 (s)

Calculations Table

$\bar{t}$ (s)	*Std Err* (s)	a (m/s^2)	T (N)	τ (N–m)	α (rad/s^2)
$I_{theo} =$ kg–m^2	$I_{exp} =$ kg–m^2	$\tau_f =$ N–m	$r =$		

SAMPLE CALCULATIONS

1. $I_{theo} = ½\, MR^2 =$
2. $a = 2x/(\bar{t})^2 =$
3. $T = m(g - a) =$

QUESTIONS

1. Calculate the percentage error of the experimental value I_{exp} compared to the theoretical value I_{theo}. Comment on the accuracy of your experimental results based on this comparison.

2. For six data points, statistical theory states that there is a 0.1% probability (1 chance in 1000) of obtaining a value of $r \geq 0.974$ for uncorrelated data. What does your value of r imply about the agreement of your data with the theory?

3. Would there be any advantage to using a smaller mass m than was used in the laboratory? What would be the disadvantage of doing that?

4. Would there be any advantage to using a larger mass m than was used in the laboratory? What would be the disadvantage of doing that?

5. What amount of mass placed on the string would produce a torque equal to the value of the frictional torque τ_f?

Archimedes' Principle

OBJECTIVES

- ❑ Apply Archimedes' principle to measurements of specific gravity.
- ❑ Determine the specific gravity of several metal objects that are denser than water, a cork that is less dense than water, and a liquid (alcohol).

EQUIPMENT LIST

- Laboratory balance, calibrated masses, 1000 mL beaker, string, alcohol
- Metal cylinders, and cork or piece of wood (to serve as unknowns)

THEORY

If the mass of an object is distributed uniformly, its **density** ρ_o is defined as the mass m of the object divided by its volume V. In equation form this is

$$\rho_o = m/V \qquad \text{(Eq. 1)}$$

The SI units for density are kg/m^3. Other commonly used units for density are the cgs system units g/cm^3. The **specific gravity** is defined as the ratio of the density of an object to the density of water ρ_w. The equation for the specific gravity is given by

$$\text{Specific gravity} = SG = \rho_o/\rho_w \qquad \text{(Eq. 2)}$$

Specific gravity is a dimensionless quantity because it is the ratio of two densities. The density of water in the cgs system is $1.000\ g/cm^3$, and densities in that system are numerically equal to the specific gravity SG. Water has a density of $1000\ kg/m^3$, and densities in the SI system are equal to $SG \times 10^3$.

Archimedes' principle states that an object placed in a fluid experiences an upward buoyant force equal to the weight of fluid displaced by the object. The principle applies to both liquids and gases. In this laboratory, we will examine the application of the principle to liquids. An object floats if its density is less than the density of the liquid in which it is placed. It sinks in the liquid until it displaces a weight of liquid equal to its own weight. The object is in equilibrium because its weight acts downward and a buoyant force acts upward. An object with density greater than the liquid in which it is placed will sink to the bottom of the liquid. It experiences an upward buoyant force equal to the weight of the displaced fluid, but that is less than the weight of the object, and it sinks.

For objects with a density greater than the density of water, Archimedes' principle allows a simple determination of the specific gravity of the object. Consider an object that is tied to a string and submerged in water with the string attached beneath the arm of a laboratory balance as shown in Figure 18-1. Also in the figure is a free-body diagram of the forces acting on the submerged object. If B stands for the buoyant force on the object, W is the weight of the object, and W_1 is the tension in the string, the following is true at equilibrium

$$B + W_1 = W \qquad \text{(Eq. 3)}$$

The quantity W_1 is the apparent weight read by the laboratory balance in Figure 18-1. By Archimedes' principle, the buoyant force is the weight of displaced water. This can be written as

$$B = m_w g = \rho_w V_w g = \rho_w V_o g \qquad \text{(Eq. 4)}$$

with m_w the mass of the displaced water, and V_w and V_o the volumes of the displaced water and the object. The last step in Equation 4 is true because the object sinks and displaces water equal to its volume, and therefore $V_w = V_o$. Using Equations 3 and 4 with $W = mg$ and $W_1 = m_1 g$ gives

$$W_1 = W - B = mg - \rho_w V_w g = \rho_o V_o g - \rho_w V_o g \qquad \text{(Eq. 5)}$$

$$\frac{W}{W - W_1} = \frac{mg}{mg - m_1 g} = \frac{\rho_o V_o g}{\rho_o V_o g - (\rho_o V_o g - \rho_w V_o g)} = \frac{\rho_o}{\rho_w} \qquad \text{(Eq. 6)}$$

$$SG = \frac{\rho_o}{\rho_w} = \frac{m}{m - m_1} \qquad \text{(Eq. 7)}$$

This laboratory will use Equation 7 to determine the specific gravity of several metals with densities greater than that of water.

To use Archimedes' principle to determine the specific gravity of an object that floats, the object must be submerged by attaching a lead weight to the object. Figure 18-2(a) shows the lead weight in water and the object in air, and the balance reading is W_1. In Figure 18-2(b) both the object and the lead weight are below the water, and the balance reads W_2. By analysis similar to that used to derive Equation 7, it can be shown that the specific gravity of an object that floats in water is given by

$$SG = \frac{W}{W_1 - W_2} = \frac{mg}{m_1 g - m_2 g} = \frac{m}{m_1 - m_2} \qquad \text{(Eq. 8)}$$

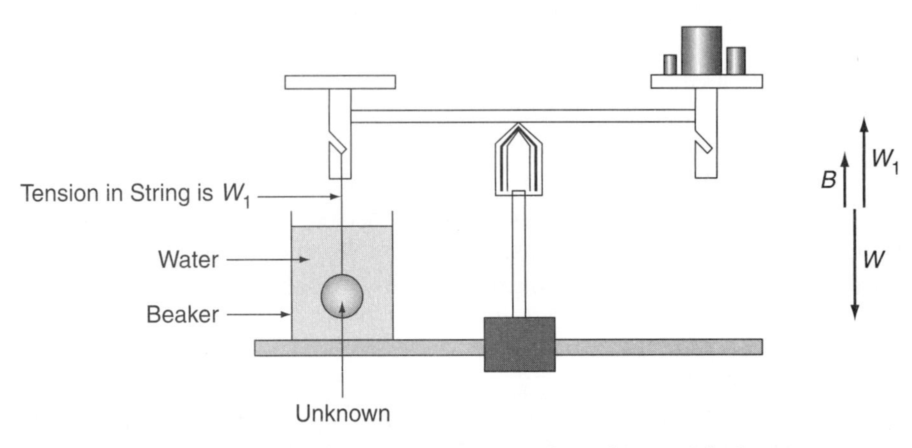

Figure 18-1 Determining the apparent mass of an object with density greater than water.

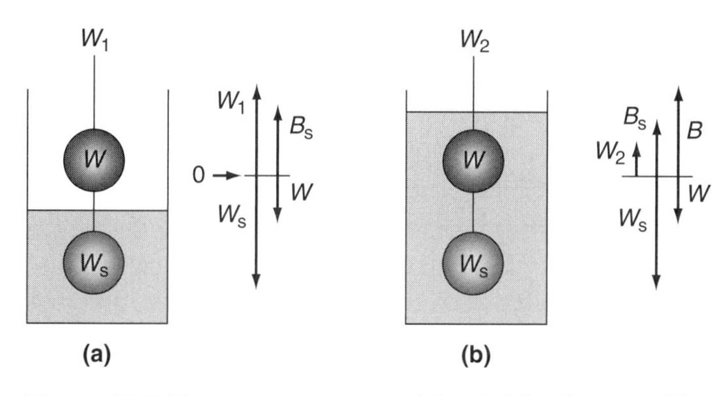

Figure 18-2 Forces acting on an object held submerged by a lead weight.

In Equation 8 m is the mass of the object, m_1 is the apparent mass with the object in air and the lead weight submerged in water, and m_2 is the apparent mass with the object and the lead weight both submerged in water.

The specific gravity of a liquid can be determined by measuring the mass of some object m, the apparent mass of the object submerged in water m_w, and the apparent mass of the object submerged in the liquid m_L. It is assumed that the object sinks in both liquids. By similar analysis as used to derive Equation 7, it can be shown that the specific gravity of the liquid is given by

$$SG(\text{liquid}) = \frac{m - m_L}{m - m_w} \qquad \text{(Eq. 9)}$$

EXPERIMENTAL PROCEDURE

Object Density Greater than Water

1. Use the laboratory balance to determine the mass of each of the two unknown metal cylinders. Record these values as m in Data Table 1. The unknowns should have their metal type stamped on them, or else ask your instructor for the metal type of the unknowns. Record the metal of the unknowns in Calculations Table 1.
2. Place a clamp on the laboratory table and screw a threaded rod in the clamp. Slide the rod into the hole designed for that purpose in the base of the laboratory balance. Adjust the height of the balance above the table to allow the 1000 mL beaker to fit under the balance. Fill the beaker about three-fourths full of water.
3. Tie a piece of light string or thread around one of the unknowns and suspend it from one of the slots in the left arm underneath the balance. Determine the apparent mass with the unknown suspended completely below the surface of the water. Be sure that the unknown is not touching the side of the beaker or is in any way supported by anything other than the string. Record the mass reading of the balance as m_1 in Data Table 1.
4. Repeat Steps 1 through 3 for the second unknown.
5. From Appendix II obtain the known SG of the metal for each of the unknowns and record those values in Data Table 1.

Object Density Less than Water

1. Using the laboratory balance, determine the mass of the cork and record it as m in Data Table 2.
2. Determine the apparent mass with the cork suspended in air and a lead weight tied below submerged in water as shown in Figure 18-2(a). Record that value as m_1 in Data Table 2.

3. Determine the apparent mass with both the cork and the lead weight submerged in water as shown in Figure 18-2(b). Record that value as m_2 in Data Table 2.

Liquid Unknown

1. Use one of the unknown metal cylinders from the original procedure. Its mass m and its apparent mass in water have already been determined. Record those values in Data Table 3.
2. Use a very clean beaker and fill it about three-fourths full of alcohol. Determine the apparent mass of the metal cylinder in alcohol. Record it as m_L in Data Table 3. Return the alcohol to its container when you have finished with it. Be very careful not to have any spark or flame near the alcohol.
3. Repeat Steps 1 and 2 for the other unknown metal.
4. From Appendix II determine the known value of the specific gravity of alcohol and record it in Data Table 3.

CALCULATIONS

Density Greater than Water

1. Use Equation 7 to calculate the *SG* for each of the unknowns. Record the experimental values of the specific gravity in Calculations Table 1.
2. Calculate the percentage error in your value of the *SG* for each of the unknowns compared to the known values. Record them in Calculations Table 1.

Density Less than Water

1. Use Equation 8 to calculate the specific gravity *SG* for the cork and record it in Calculations Table 2.

Liquid Unknown

1. Use Equation 9 to determine the specific gravity of the alcohol for the two trials with the different metals. Record those values in Calculations Table 3.
2. Calculate the percentage error in each of the two measurements of the specific gravity of alcohol. Record the results in Calculations Table 3.

Name

Section

Date

18 LABORATORY 18 *Archimedes' Principle*

PRE-LABORATORY ASSIGNMENT

1. What is the definition of density? What are its units?

2. What is specific gravity? What are its units?

3. State Archimedes' principle.

4. The buoyant force on an object placed in a liquid is (a) always equal to the volume of the liquid displaced (b) always equal to the weight of the object (c) always equal to the weight of the liquid displaced (d) always less than the volume of the liquid displaced.

5. An object that sinks in water displaces a volume of water (a) equal to the object's weight (b) equal to the object's volume (c) less than the object's volume (d) greater than the object's weight.

6. An object that sinks in water has a mass in air of 0.0675 kg. Its apparent mass when submerged in water, as in Figure 18-1, is 0.0424 kg. What is the specific gravity *SG* of the object? Considering the densities given in Appendix II, of what material is the object probably made? Show your work.

7. A piece of wood that floats on water has a mass of 0.0175 kg. A lead weight is tied to the wood, and the apparent mass with the wood in air and the lead weight submerged in water is 0.0765 kg. The apparent mass with both the wood and the lead weight both submerged in water is 0.0452 kg. What is the specific gravity of the wood? Show your work.

8. An object has a mass in air of 0.0832 kg, apparent mass in water of 0.0673 kg, and apparent mass in another liquid of 0.0718 kg. What is the specific gravity of the other liquid? Show your work.

Name

Section

Date

Lab Partners

18 LABORATORY 18 *Archimedes' Principle*

LABORATORY REPORT

Data Table 1

Metal	*Known SG*	m (kg)	m_1 (kg)

Calculations Table 1

$SG = \frac{m}{m - m_1}$	*% Error*

Data Table 2

m (kg)	m_1 (kg)	m_2 (kg)

Calculations Table 2

$SG = \frac{m}{m_1 - m_2}$

Data Table 3

Metal	m (kg)	m_w (kg)	m_L (kg)
Known SG of Alcohol =			

Calculations Table 3

$SG = \frac{m - m_L}{m - m_w}$	*% Error*

SAMPLE CALCULATIONS

1. $SG = m/(m - m_1) =$
2. $SG = m/(m_1 - m_2) =$
3. $SG = (m - m_L)/(m - m_w)$
4. $\%\ \text{Error} = |E - K|/K \times (100\%) =$

QUESTIONS

1. Do your data indicate that Archimedes' principle is valid? State clearly the evidence for your answer.

2. An object with a specific gravity of 0.900 is placed in a liquid with a specific gravity of 0.900. Describe if the object will sink, float, or behave in some other way.

3. An object with a specific gravity of 0.850 is placed in water. What fraction of the object is below the surface of the water?

4. Consider a swimmer who swims first in a freshwater lake, and then swims in the ocean. Is the buoyant force on her the same in both cases? If the buoyant force is different for the two cases, state which one is greater and why.

The Pendulum—Approximate Simple Harmonic Motion

OBJECTIVES

- ❑ Investigate the dependence of the period T of a pendulum on the length L and the mass M of the bob.
- ❑ Demonstrate that the period T of a pendulum depends slightly on the angular amplitude of the oscillation for large angles, but that the dependence is negligible for small angular amplitude of oscillation.
- ❑ Determine an experimental value of the acceleration due to gravity g by comparing the measured period of a pendulum with the theoretical prediction.

EQUIPMENT LIST

- Pendulum clamp, string, and calibrated hooked masses, laboratory timer
- Protractor and meter stick

THEORY

A mass M moving in one dimension is said to exhibit **simple harmonic motion** if its displacement x from some equilibrium position is described by a single sine or cosine function. This happens when the particle is subjected to a force F directly proportional to the magnitude of the **displacement** and directed toward the equilibrium position. In equation form this is

$$F = -kx \qquad \text{(Eq. 1)}$$

The **period** T of the motion is the time for one complete oscillation, and it is determined by the mass M and the constant k. The equation that describes the dependence of T on M and k is

$$T = 2\pi\sqrt{\frac{M}{k}} \qquad \text{(Eq. 2)}$$

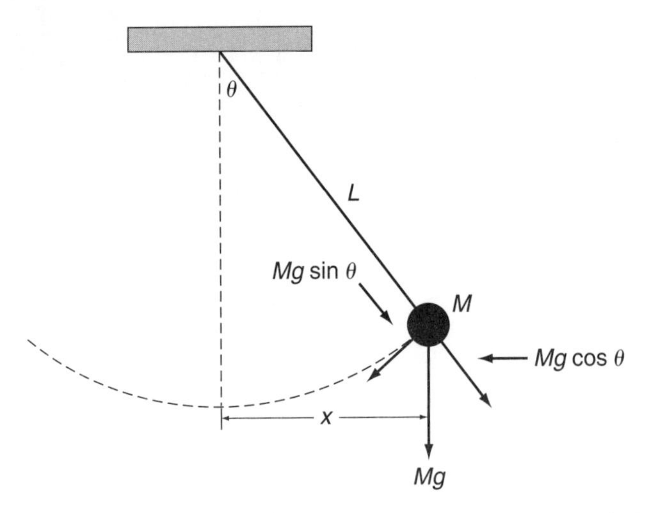

Figure 19-1 Force components acting on the mass bob of a simple pendulum.

A pendulum does not exactly satisfy the conditions for simple harmonic motion, but it approximates them under certain conditions. An ideal pendulum is a point mass M on one end of a massless string with the other end fixed as shown in Figure 19-1. The motion of the system takes place in a vertical plane when the mass M is released from an initial angle θ with respect to the vertical.

The downward weight of the pendulum can be resolved into two components as shown in Figure 19-1. The component $Mg\cos\theta$ equals the magnitude of the tension N in the string. The component $Mg\sin\theta$ acts tangent to the arc along which the mass M moves. This component provides the force that drives the system. In equation form the force F along the direction of motion is

$$F = -Mg\sin\theta \qquad \text{(Eq. 3)}$$

For small values of the initial angle θ, we can use the approximation $\sin\theta \approx \tan\theta \approx x/L$ in Equation 3, which gives

$$F = -\frac{Mg}{L}x \qquad \text{(Eq. 4)}$$

Although Equation 4 is an approximation, it is of the form of Equation 1 with $k = Mg/L$. Using that value of k in Equation 2 gives

$$T = 2\pi\sqrt{\frac{M}{Mg/L}} = 2\pi\sqrt{\frac{L}{g}} \qquad \text{(Eq. 5)}$$

Equation 5 predicts that the period T of a simple pendulum is independent of the mass M and the angular amplitude θ and depends only on the length L of the pendulum.

The exact solution to the period of a simple pendulum without making the small angle approximation leads to an infinite series of terms, with each successive term becoming smaller. Equation 6 gives the first three terms in the series. They are sufficient to determine the very slight dependence of the period T on the angular amplitude of the motion.

$$T = 2\pi\sqrt{\frac{L}{g}}\,[1 + 1/4\sin^2(\theta/2) + 9/64\sin^4(\theta/2) + \cdots] \qquad \text{(Eq. 6)}$$

For an ideal pendulum with no friction, the motion repeats indefinitely with no reduction in the amplitude as time goes on. For a real pendulum there will always be some friction, and the amplitude of the motion decreases slowly with time. However, for small initial amplitudes, the change in the period as the amplitude decreases is negligible. This fact is the basis for the pendulum clock. Pendulum clocks, in one form or another, have been used for more than 300 years. For more than 100 years, clockmakers have built extremely accurate clocks by successfully employing devices to compensate for small changes in the length of the pendulum caused by temperature variations.

EXPERIMENTAL PROCEDURE

Length

1. The dependence of T on the length of the pendulum will be determined with a fixed mass and fixed angular amplitude. Place a 0.2000 kg hooked calibrated mass on a string with a loop in one end. Adjust the position at which the other end of the string is clamped in the pendulum clamp until the distance from the point of support to the center of mass of the hooked mass is 1.0000 m. The length L of each pendulum is from the point of support to the center of mass of the bob. The center of mass of the hooked masses will usually not be in the center because the hooked masses are not solid at the bottom. Estimate how much this tends to raise the position of the center of mass and mark the estimated center of mass on each hooked mass.
2. Displace the pendulum 5.0° from the vertical and release it. Measure the time Δt for 10 complete periods of motion and record that value in Data Table 1. It is best to set the pendulum in motion, and then begin the timer as it reaches the maximum displacement, counting 10 round trips back to that position. Repeat this process two more times for a total of three trials with this same length. The pendulum should move in a plane as it swings. If the mass moves in an elliptical path, it will lead to error.
3. Repeat the procedure of Step 2, using the same mass and an angle of 5.0° for pendulum lengths of 0.8000, 0.6000, 0.5000, 0.3000, 0.2000, and 0.1000 meters. Do three trials at each length. The length of the pendulum is from the point of support to the center of mass of the hooked mass.

Mass

1. The dependence of T on M will be determined with the length L and amplitude θ held constant. Place a 0.0500 kg mass on the end of the string and adjust the point of support of the string until the pendulum length is 1.0000 m. Displace the mass 5.0° and release it. Measure the time Δt for 10 complete periods of the motion and record it in Data Table 2. Repeat the procedure two more times for a total of three trials.
2. Keep the length constant at $L = 1.0000$ m and repeat the procedure above for M of 0.1000, 0.2000, and 0.5000 kg. Because L is from the point of support to the center of mass, you will need to make slight adjustments in the string length to keep L constant for the different masses.

Amplitude

1. The dependence of T on amplitude of the motion will be determined with L and M constant. Construct a pendulum with $L = 1.000$ m and $M = 0.200$ kg. Measure the time Δt for 10 complete periods with amplitude 5.0°. Repeat two more times for a total of three trials at this amplitude. Record all results in Data Table 3.
2. Repeat the procedure above for amplitudes of 10.0°, 20.0°, 30.0°, and 45.0°. Do three trials for each amplitude and record the results in Data Table 3.

CALCULATIONS

Length

1. Calculate the mean $\overline{\Delta t}$ and standard error α_t of the three trials for each of the lengths. Record those results in Calculations Table 1.

2. Calculate the period T from $T = \overline{\Delta t}/10$ and record it in Calculations Table 1.
3. According to Equation 5 the period T should be proportional to $\sqrt{L}$. For each of the values of L calculate $\sqrt{L}$ and record the results in Calculations Table 1. Perform a linear least squares fit with T as the vertical axis and $\sqrt{L}$ as the horizontal axis. By Equation 5 the slope of this fit should equal $2\pi/\sqrt{g}$. Equate the slope determined from the fit to $2\pi/\sqrt{g}$, treating g as unknown. Solve this equation for g and record that value as g_{exp} in Calculations Table 1. Also record the value of the correlation coefficient r for the fit.

Mass

1. Calculate the mean $\overline{\Delta t}$ and standard error α_t of the three trials for each of the masses. Record those results in Calculations Table 2.
2. Calculate the period T from $T = \overline{\Delta t}/10$ and record in Calculations Table 2.

Amplitude

1. Calculate the mean $\overline{\Delta t}$ and standard error α_t for the three trials at each amplitude. Record the results in Calculations Table 3.
2. Calculate $T = \overline{\Delta t}/10$ and record the results in Calculations Table 3 as T_{exp}.
3. Equation 6 is the theoretical prediction for how the period T should depend on the amplitude. Use $L = 1.000$ m and $M = 0.200$ kg in Equation 6 to calculate the T predicted for the values of θ. Record them in Calculations Table 3 as T_{theo}.
4. For the experimental values of the period T_{exp} calculate the ratio of the period at the other angles to the period at $\theta = 5.0°$. Call this ratio $(T_{exp}(\theta)/T_{exp}(5.0°))$. Record these values in Calculations Table 3.
5. For the theoretical values of the period T_{theo} calculate the ratio of the period at the other angles to the period at $\theta = 5.0°$. Call this ratio $(T_{theo}(\theta)/T_{theo}(5.0°))$. Record these values in Calculations Table 3.

GRAPHS

1. Consider the data for the dependence of the period T on the length L. Graph the period T as the vertical axis and $\sqrt{L}$ as the horizontal axis. Also show on the graph the straight line obtained by the linear fit to the data.
2. Consider the data for the dependence of the period T on the mass M. Graph the period T as the vertical axis and the mass M as the horizontal axis.

Name Section Date

19 LABORATORY 19

The Pendulum—Approximate Simple Harmonic Motion

PRE-LABORATORY ASSIGNMENT

1. What is the requirement for a force to produce simple harmonic motion?

2. A particle of mass $M = 1.35$ kg is subject to a force $F = -0.850\ x$, where x is the displacement of the particle from equilibrium. The units of force F are Newtons, and the units of x are meters. What is the period T of its motion? Show your work.

3. A simple pendulum of length $L = 0.800$ m has a mass $M = 0.250$ kg. What is the tension in the string when it is at an angle $\theta = 12.5°$? Show your work.

4. In Question 3, what is the component of the weight of M that is directed along the arc of the motion of M? Show your work.

5. What is the period T of the motion of the pendulum in Question 3? Assume that the period is independent of θ. Show your work.

6. What would be the period T of the pendulum in Question 3 if L was unchanged but $M = 0.500$ kg? Assume that the period is independent of θ. Show your work.

7. Determine the period T of the pendulum in Question 3 if everything else stays the same, but $\theta = 45.0°$. Do not assume that the period is independent of θ. Show your work.

Name | Section | Date

Lab Partners

19 **LABORATORY 19** *The Pendulum—Approximate Simple Harmonic Motion*

LABORATORY REPORT

Data and Calculations Table 1

L (m)	Δt_1 (s)	Δt_2 (s)	Δt_3 (s)	$\overline{\Delta t}$ (s)	α_t (s)	T (s)	$\sqrt{L}$ ($\sqrt{\text{m}}$)
1.0000							
0.8000							
0.6000							
0.5000							
0.3000							
0.2000							
0.1000							
Slope =		g_{exp} =		m/s^2	r =		

Data and Calculations Table 2

M (kg)	Δt_1 (s)	Δt_2 (s)	Δt_3 (s)	$\overline{\Delta t}$ (s)	α_t (s)	T (s)
0.0500						
0.1000						
0.2000						
0.5000						

Data and Calculations Table 3

θ	Δt_1 (s)	Δt_2 (s)	Δt_3 (s)	$\overline{\Delta t}$ (s)	α_t (s)	T_{exp} (s)	T_{theo} (s)	$\frac{T_{exp}(\theta)}{T_{exp}(5^\circ)}$	$\frac{T_{theo}(\theta)}{T_{theo}(5^\circ)}$
5.0°									
10.0°									
20.0°									
30.0°									
45.0°									

SAMPLE CALCULATIONS

1. $T_{exp} = \overline{\Delta t}/10 =$
2. $\sqrt{L} =$
3. $g_{exp} = 4\pi^2/(\text{slope})^2 =$
4. $T_{theo} =$

QUESTIONS

1. In general, what is the precision of the measurements of T? Answer this question by considering what percentage is α_t of $\overline{\Delta t}$ for the measurements as a whole.

2. Do your data confirm the expected dependence of the period T on the length L of a pendulum? Consider the correlation coefficient r for the least squares fit in your answer.

3. Comment on the accuracy of your experimental value for the acceleration due to gravity g.

4. What does the theory predict for the shape of the graph of period T versus M? Do your data confirm this expectation? Calculate the mean and standard error of the periods for the four masses and comment on how this relates to mass independence of T.

5. Do your measured values for the period T as a function of the amplitude θ confirm the theoretical predictions? State clearly what is expected and what your data show.

6. The values of T were determined by measuring the time for 10 periods. Why is the time for more than one period measured? If there is an advantage to measuring for 10 periods, why not measure for 1000 periods?

Simple Harmonic Motion—Mass on a Spring

OBJECTIVES

- ❑ Directly determine the spring constant k of a spring by measuring the elongation versus applied force.
- ❑ Determine the spring constant k from measurements of the period T of oscillation for different values of mass.
- ❑ Investigate the dependence of the period T of oscillation of a mass on a spring on the value of the mass and on the amplitude of the motion.

EQUIPMENT LIST

- Spring, masking tape, laboratory timer, meter stick, table clamps, and rods
- Right-angle clamps, laboratory balance, and calibrated hooked masses

THEORY

A mass that experiences a restoring force proportional to its displacement from an equilibrium position is said to obey **Hooke's law.** In equation form this relationship can be expressed as

$$F = -ky \qquad \text{(Eq. 1)}$$

where k is a constant with dimensions of N/m. The negative sign indicates that the force is in the opposite direction of the displacement. If a spring exerts the force, the constant k is the **spring constant.**

A force described by Equation 1 will produce an oscillatory motion called **simple harmonic motion** because it can be described by a single sine or cosine function of time. A mass displaced from its equilibrium position by some value A, and then released, will oscillate about the equilibrium position. Its **displacement** y from the equilibrium position will range between $y = A$ and $y = -A$ with A called the **amplitude** of the motion. For the initial conditions described above, the displacement y as a function of time t is given by

$$y = A\cos(\omega t + \phi) \qquad \text{(Eq. 2)}$$

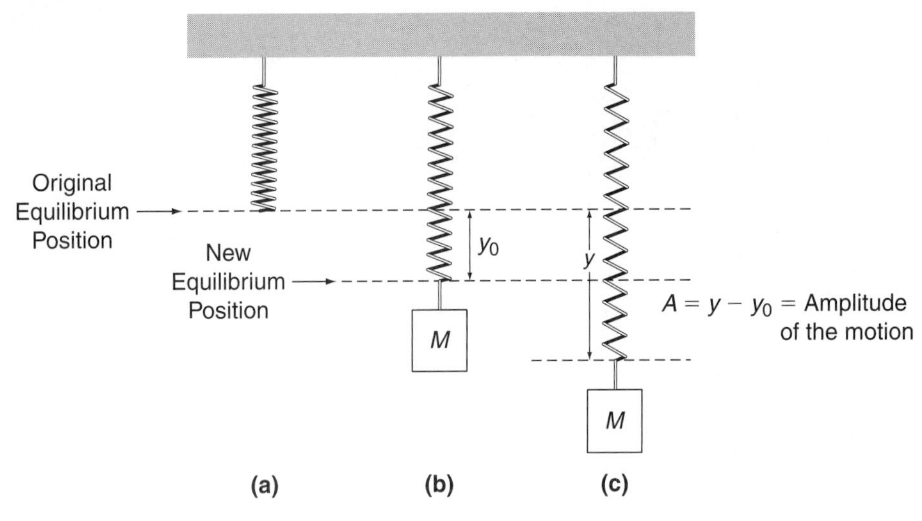

Figure 20-1 New equilibrium position with mass M placed on a spring.

with angular frequency ω related to the frequency f and the period T by

$$\omega = \sqrt{\frac{k}{M}} \qquad \omega = 2\pi f \qquad T = 1/f \qquad T = 2\pi\sqrt{\frac{M}{k}} \qquad \text{(Eq. 3)}$$

A mass M placed on the end of a spring hangs vertically as shown in Figure 20-1. The original equilibrium position of the lower end of the spring is shown in Figure 20-1(a). The position of the lower end of the spring when the mass is applied, shown in Figure 20-1(b), can be considered as the new equilibrium position. In Figure 20-1(c) the mass is pulled down to a displacement A from this new equilibrium position. When released, the mass will oscillate with amplitude A and period T given above.

Equation 3 for the period is strictly true only if the spring is massless. For real springs with finite mass, a fraction of the spring mass m_s must be included along with the mass M. If C stands for the fraction of the spring mass to be included, the period is

$$T = 2\pi\sqrt{\frac{M + Cm_s}{k}} \qquad \text{(Eq. 4)}$$

You will be challenged to discover what fraction C of the spring mass should be included from your analysis of the data that you will take in the laboratory.

EXPERIMENTAL PROCEDURE

Spring Constant

1. Attach the table clamp to the edge of the laboratory table and screw a threaded rod into the clamp vertically as shown in Figure 20-2. Place a right-angle clamp on the vertical rod and extend a horizontal rod from the right-angle clamp. Hang the spring on the horizontal rod and attach it to the horizontal rod with a piece of tape. Screw a threaded vertical rod into a support stand, which rests on the floor. Place a right-angle clamp on the vertical rod and place a meter stick in the clamp so that the meter stick stands vertically. Adjust the height of the clamp on the vertical rod until the zero mark of the meter stick is aligned with the bottom of the hanging spring as shown in Figure 20-2.
2. Place a hooked mass M of 0.050 kg on the end of the spring. Slowly lower the mass M until it hangs at rest in equilibrium when released. Carefully read the position of *the lower end of the spring* on the meter

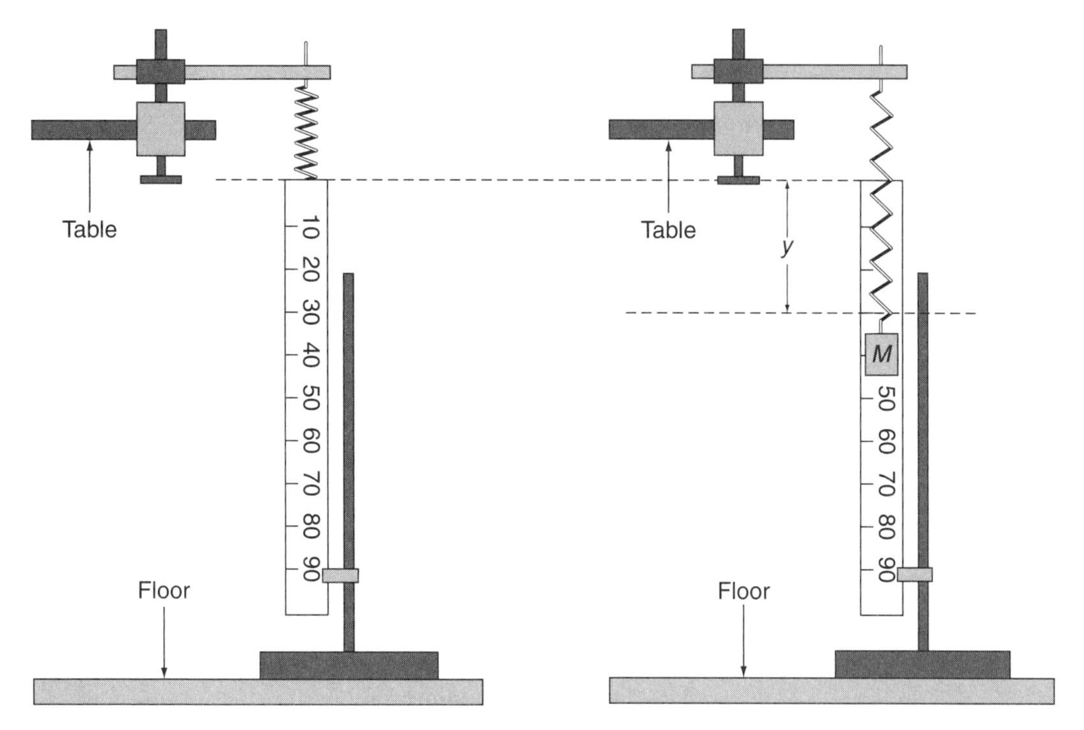

Figure 20-2 Arrangement to measure displacement of spring caused by mass M.

stick scale. Record the value of the mass M and the value of the displacement x in Data and Calculations Table 1.

3. Repeat Step 2, placing in succession 0.100, 0.200, 0.300, 0.400, and 0.500 kg on the spring and measuring the displacement y of the spring. Record all values of M and y in Data and Calculations Table 1.

Amplitude Variation

1. We will investigate dependence of the period T on the amplitude A for a fixed mass of 0.500 kg. Place the mass on the end of the spring and slowly lower the mass until it hangs at rest when released. Record this position of the lower end of the spring as y_o.
2. Displace the mass downward to $y = y_o + 0.0200$ m as shown in Figure 20-1, which will produce $A = 0.0200$ m. Release the mass, and let it oscillate. Measure the time for 10 complete periods and record it in Data Table 2 as Δt. Repeat the procedure two more times for a total of three trials at this amplitude.
3. Repeat Step 2 above for A of 0.0400, 0.0600, 0.0800, 0.1000, and 0.1200 m. Make three trials for each amplitude and measure the time for 10 periods for each trial. Record all results in Data Table 2.

Mass Variation

1. Place a hooked mass of 0.050 kg on the spring and let it hang at rest. Displace the mass 0.0500 m below the equilibrium ($A = 0.0500$ m), release it, and let the system oscillate. Measure the time for 10 periods of the motion and record it in Data Table 3 as Δt. Repeat the procedure two more times for a total of three trials with this mass.
2. Repeat the procedure of Step 1 with the same A for values of the mass M of 0.100, 0.200, 0.300, 0.400, and 0.500 kg. Perform three trials of the time for 10 periods for each mass and record the results in Data Table 3.
3. Determine the mass of the spring m_s and record it in Data Table 3.

CALCULATIONS

Spring Constant

1. Calculate the force Mg for each mass and record the values in Data and Calculations Table 1. Use the value of $9.80\,\text{m/s}^2$ for g.
2. Perform a linear least squares fit to the data with Mg as the vertical axis and y as the horizontal axis. Record in Data and Calculations Table 1 the slope of the fit as the spring constant k and the correlation coefficient r.

Amplitude Variation

1. Calculate the mean $\overline{\Delta t}$ and standard error α_t of the three trials for each amplitude. Record the results in Calculations Table 2.
2. Calculate the period T from $T = \overline{\Delta t}/10$. Record the results in Calculations Table 2.

Mass Variation

1. Calculate the mean $\overline{\Delta t}$ and standard error α_t for the three trials for each mass. Record the results in Calculations Table 3.
2. Calculate the period T from $T = \overline{\Delta t}/10$. Record the results in Calculations Table 3.
3. If both sides of Equation 4 are squared the result is

$$T^2 = \frac{4\pi^2}{k}(M + Cm_s) \qquad \text{(Eq. 5)}$$

4. Equation 5 states that T^2 is proportional to M with $4\pi^2/k$ as the slope and $4\pi^2 Cm_s/k$ as the intercept. Calculate and record the values of T^2 in Calculations Table 3. Perform a linear least squares fit with T^2 as the vertical axis and M as the horizontal axis. Record the values of the slope, intercept, and r in Calculations Table 3.
5. Equate the value of the slope determined in Step 4 to $4\pi^2/k$ and solve for the value of k in the resulting equation. Record this value of k in Calculations Table 3.
6. Calculate the percentage difference between the value of k determined in Step 5 and the value of k determined earlier and record it in Calculations Table 3.
7. Equate the value of the intercept determined in Step 4 to $4\pi^2 Cm_s/k$ and solve for the value of C in the resulting equation. In the equation, use the value of k determined in Step 5. Record the value C in Calculations Table 3.

GRAPHS

1. Graph the data from Calculations Table 1 for force Mg versus displacement y with Mg as the vertical axis and y as the horizontal axis. Show on the graph the straight line obtained from the fit to the data.
2. Graph the data from Calculations Table 2 for the period T versus the amplitude A with T as the vertical axis and A as the horizontal axis.
3. Graph T^2 versus M with T^2 as the vertical axis and M as the horizontal axis. Also show on the graph the straight line obtained from the linear least squares fit to the data.

Name | Section | Date

20 LABORATORY 20 *Simple Harmonic Motion—Mass on a Spring*

PRE-LABORATORY ASSIGNMENT

1. Describe in words and give an equation for the kind of force that produces simple harmonic motion.

2. Other than the type of force that produces it, what characterizes simple harmonic motion?

3. A spring has a spring constant $k = 8.75$ N/m. If the spring is displaced 0.150 m from its equilibrium position, what is the force that the spring exerts? Show your work.

4. A spring of constant $k = 11.75$ N/m is hung vertically. A 0.500 kg mass is suspended from the spring. What is the displacement of the end of the spring due to the weight of the 0.500 kg mass? Show your work.

5. A spring with a mass on the end of it hangs in equilibrium a distance of 0.4200 m above the floor. The mass is pulled down a distance 0.0600 m below the original position, released, and allowed to oscillate. How high above the floor is the mass at the highest point in its oscillation? Show your work.

6. A massless spring has a spring constant of $k = 7.85$ N/m. A mass $M = 0.425$ kg is placed on the spring, and it is allowed to oscillate. What is the period T of oscillation? Show your work.

7. A massless spring of $k = 6.45$ N/m has a mass $M = 0.300$ kg on the end of the spring. The mass is pulled down 0.0500 m and released. What is the period T of the oscillation? What is the period T if the mass is pulled down 0.1000 m and released? State clearly the reasoning for your answer.

Name | Section | Date

Lab Partners

LABORATORY 20 Simple Harmonic Motion—Mass on a Spring

LABORATORY REPORT

Data and Calculations Table 1

M (kg)	Mg (N)	y (m)	k (N/m)	r
0.050				
0.100				
0.200				
0.300				
0.400				
0.500				

Data and Calculations Table 2 $y_o =$ ______ m

A (m)	Δt_1 (s)	Δt_2 (s)	Δt_3 (s)	$\overline{\Delta t}$ (s)	α_t (s)	T (s)
0.0200						
0.0400						
0.0600						
0.0800						
0.1000						
0.1200						

Data and Calculations Table 3 $m_s =$ ______________ kg

M (kg)	Δt_1 (s)	Δt_2 (s)	Δt_3 (s)	$\overline{\Delta t}$ (s)	α_t (s)	T (s)	T^2 (s^2)
0.050							
0.100							
0.200							
0.300							
0.400							
0.500							
Slope =			Intercept =			$r =$	
$k =$		N/m	$C =$			% Diff =	

SAMPLE CALCULATIONS

1. $Mg =$
2. $T = \overline{\Delta t}/10$
3. $T^2 =$
4. $k = 4\pi^2/(\text{Slope})$
5. $C = k(\text{Intercept})/(4\pi^2 m_s) =$

QUESTIONS

1. Do the data for the displacement of the spring y versus the applied force Mg indicate that the spring constant is constant for this range of forces? State clearly the evidence for your answer.

2. How is the period T expected to depend upon the amplitude A? State how your data do or do not confirm this expectation.

3. Consider the value you obtained for C. If you express that fraction as a whole number fraction, which of the following would best fit your data? (½ ⅓ ¼ ⅕)

4. Calculate T predicted by Equation 3 for $M = 0.050$ kg. Calculate T predicted by Equation 4 with the same M and your value of C. What is the percentage difference between these two values of T? Do the same calculations for $M = 0.500$ kg. For which case are the percentage differences greater and why are they greater?

5. The determination of T was done by measuring for 10 periods. Why was the time for more than one period measured? If there is an advantage to measuring for 10 periods, why not measure for 1000 periods?

Standing Waves on a String

OBJECTIVES

- ❏ Demonstrate formation of standing waves on a string from the interference of traveling waves traveling in opposite directions.
- ❏ Determine the tension T in the string required to produce standing waves.
- ❏ Investigate the relationship between tension T and the wavelength λ of the wave.
- ❏ Determine an experimental value for the frequency f of the wave and compare it to the known value of 120 Hz.

EQUIPMENT LIST

- String vibrator (60 Hz AC), string, clamps, pulley, support rods, meter stick
- Mass holder, slotted masses
- Laboratory balance capable of measuring to 0.00001 kg (one for class)

THEORY

Waves are one means by which energy can be transported. Waves on a string are an example of transverse waves. These are waves in which individual particles of the medium (in this case the string) move perpendicular or transverse to energy moving along the string. In Figure 21-1 a string tied to a vibrator at one end passes over a pulley, and the weight of masses on the other end provides **tension** T in the string. The vibrator moves up and down at a **frequency** f, which causes a wave of that same frequency to propagate down the string. The vibrator used in this laboratory is driven by an electromagnet at a frequency of 60 Hz, but because the electromagnet attracts the steel blade twice in each cycle, the vibrator frequency is 120 Hz.

The point at which the string passes over the pulley is a fixed point, and the wave is reflected from that point. Thus the string is a medium in which two waves of the same speed, frequency, and wavelength travel in the opposite direction. These two waves will interfere with each other to produce a standing wave when the proper relationship exists between the string length L and the wavelength λ of the wave. When a standing wave is produced, its characteristic features are the existence of nodes and antinodes at points along the string. A node N is a point for which there is no displacement of the string from its equilibrium position. An antinode A is a point on the string for which the amplitude of vibration is a maximum at all times. To form a standing wave a node must occur at each end of the string, and an antinode must occur between each node. The distance between nodes is $\lambda/2$ or one-half of a wavelength of the wave. In terms of the string length L, a standing wave is possible when

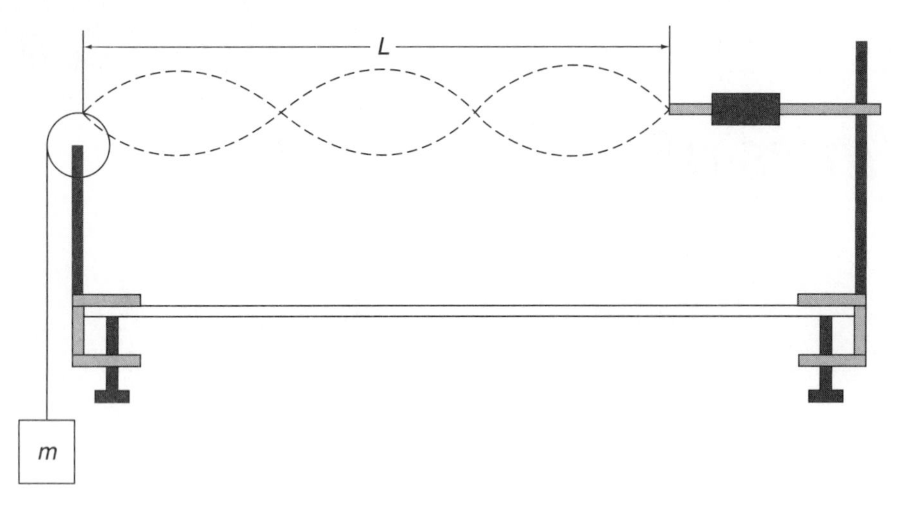

Figure 21-1 Experimental arrangement of vibrator, pulley, and masses.

$$L = n\,(\lambda/2) \qquad \text{where } n = 1, 2, 3, 4, \ldots.. \tag{Eq. 1}$$

Figure 21-2 shows the first four standing waves that are possible. From the figure it is clear that n is the number of segments of half wavelengths in each standing wave. Solving Equation 1 for the possible values of the wavelength λ gives

$$\lambda = 2L/n \qquad \text{where } n = 1, 2, 3, 4, \ldots... \tag{Eq. 2}$$

Because the frequency is fixed at 120 Hz, each different standing wave will have a different wave speed V. The wave speed is determined by the string tension T and the string mass per unit length ρ. The relationship between these quantities is given by

$$V = \sqrt{\frac{T}{\rho}} \tag{Eq. 3}$$

The frequency, wavelength, and speed are related by $V = f\lambda$. Using that and the speed V from Equation 3 gives

$$\lambda = \frac{1}{f}\sqrt{\frac{T}{\rho}} \tag{Eq. 4}$$

The experimental arrangement differs slightly from the ideal situation because the node at the vibrator end of the string is moving up and down rather than being fixed in space. This effect means that the wavelength is somewhat difficult to define. For example, in the case $n = 2$ the node will not be exactly in the center of the string, and each of the segments will be slightly different in length. The effect decreases

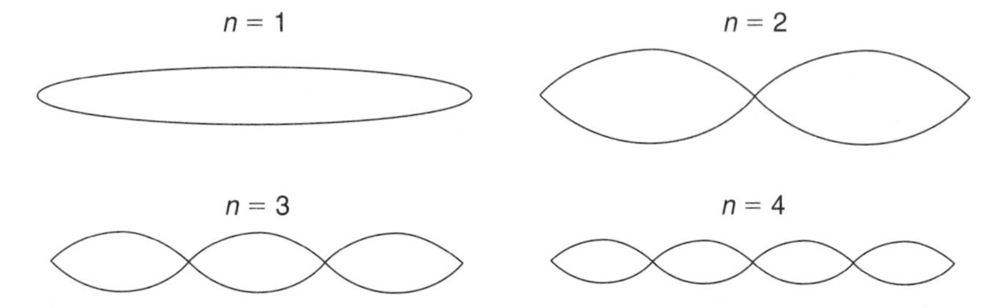

Figure 21-2 First four standing waves for waves in a string.

with each addition to the number of segments. One way to account for this effect is to determine the wavelength using only those segments that do not include the vibrator. Because it is somewhat difficult to locate the nodes precisely, there is usually less error involved if we assume that each wavelength is the total length of the string divided by the number of segments. That is the assumption we will make in this laboratory.

EXPERIMENTAL PROCEDURE

1. The mass per unit length of the string should be determined by the class as a whole. Carefully measure a 1.0000 m length of the type string to be used. Measure the mass of the string m_s to the nearest 0.00001 kg. From these data find ρ and record it in the Data Table.
2. Clamp the vibrator and the pulley at opposite ends of the laboratory table approximately 1.5 m apart if the table will allow that. Tie one end of a piece of string to the vibrator and make a loop in the other end that hangs over the pulley as shown in Figure 21-1. Place a 0.0500 kg mass holder on the loop.
3. Measure the length of the string from the tip of the vibrator to the point where the string touches the pulley and record this as string length L in the Data Table.
4. Plug in the vibrator and place on the mass holder the mass required to produce a standing wave that corresponds to $n=3$ in Figure 21-2. Do not attempt the modes with $n=1$ and $n=2$ because they require such high string tension that the string is likely to break. To determine the mass needed to produce a particular standing wave, it will help to pull down slightly or lift up slightly on the mass holder to determine if mass needs to be added or subtracted. Determine to the nearest 0.001 kg the mass needed to produce the largest possible amplitude of vibration with $n=3$. Record the value of the mass M in the Data Table for $n=3$.
5. Remove mass and determine to the nearest 0.001 kg the mass needed to produce the largest amplitude with $n=4$ and record that mass value in the Data Table.
6. Continue the process of removing mass to produce standing waves for the cases of $n=5, 6, 7, 8,$ and 9. In each case n refers to the number of segments into which the string length is divided. For each case determine the mass needed (to the nearest 0.001 kg) to produce the largest amplitude. Record all values of mass in the Data Table.

CALCULATIONS

1. Use the value of L in Equation 2 to calculate the wavelength λ for each of the standing waves $n=3$ through $n=9$ and record the results in the Calculations Table.
2. From the measured values of M calculate the tension $T=Mg$ with $g=9.80\ \text{m/s}^2$. Record the values of T in the Calculations Table.
3. Calculate the values of $\sqrt{T}$ for each of the values of T and record them in the Calculations Table. Note that when taking the square root, generally an extra significant figure is allowed in the value of the square root. For example, if a measured tension T were 7.83, then the recorded value of $\sqrt{T}$ should be 2.798.
4. According to Equation 4, λ should be proportional to $\sqrt{T}$ with $1/f\sqrt{\rho}$ as the constant of proportionality. Perform a linear least squares fit with λ as the vertical axis and $\sqrt{T}$ as the horizontal axis. Record the slope of the fit in the Calculations Table. Equate the slope to $1/f\sqrt{\rho}$ treating f as an unknown. Use the known value of ρ to solve the resulting equation for f. Record that value of the frequency as f_{exp} in the Calculations Table. Also record the value of r the correlation coefficient in the Calculations Table.
5. Calculate and record the percentage error of f_{exp} compared to the known value of the frequency $f=120\ \text{Hz}$.

Name Section Date

LABORATORY 21 *Standing Waves on a String*

PRE-LABORATORY ASSIGNMENT

1. What is the name given to a point on a vibrating string at which the displacement is always zero? What is the name given to a point at which the displacement is always a maximum?

2. What are the conditions (with respect to the points of zero amplitude and maximum amplitude) that must hold to produce a standing wave on a vibrating string?

3. How is the length of the string L related to the wavelength λ for standing waves?

4. What is the longest possible wavelength λ for a standing wave in terms of the string length L? Show your work.

5. An arrangement like that shown in Figure 21-1 has a frequency of $f = 120$ Hz. The length of the string from the vibrator to the point where the string touches the top of the pulley is 1.200 m. What is the wavelength λ of the standing wave corresponding to the third resonant mode of the system? Show your work.

6. What is the velocity of the wave for the same system described in Question 5, but for the case of the fifth resonant mode? Show your work.

7. Suppose that the system described in Question 5 has string with a mass density ρ equal to 2.95×10^{-4} kg/m. What is the tension T in the string for the second resonant mode? Show your work.

Name | Section | Date

Lab Partners

21

LABORATORY 21 *Standing Waves on a String*

LABORATORY REPORT

Data Table

n	M (kg)
3	
4	
5	
6	
7	
8	
9	
$\rho =$ kg/m	$L =$ m

Calculations Table

$T = Mg$ (N)	λ (m)	$\sqrt{T}$ ($\sqrt{N}$)
Slope =	$r =$	
$f_{exp} =$ Hz	% Error =	

SAMPLE CALCULATIONS

1. $\lambda = (2L)/n =$
2. $T = Mg =$
3. $\sqrt{T} =$
4. $f_{exp} = 1/(\sqrt{\rho}\ \ \text{slope}) =$
5. % Error $= |E - K|/K\,(100\%) =$
6. (Question 3) $f = \dfrac{1}{\lambda}\sqrt{\dfrac{T}{\rho}} =$

QUESTIONS

1. What is the accuracy of your experimental value for the frequency? State clearly the basis for your answer.

2. For the way in which the data were analyzed, the precision of the measurement would be related to the uncertainty of the measured slope. Instead, use each measurement of T and λ to calculate an independent value for the frequency f. Calculate the seven values of the frequency using the equation $f = (1/\lambda)\sqrt{T/\rho}$. Calculate the mean and standard error of those seven values of the frequency and use that to comment on the precision of the measurements of f.

3. Calculate the velocity for the $n = 2$ and for the $n = 8$ standing wave.

4. Calculate the tension T that would be required to produce the $n = 1$ standing wave in your string. Test a piece of the string with that much force and describe the results. In particular, does that much force break the string?

5. Suppose that the string stretched significantly as the tension was increased. How would that affect the value of ρ for the string? How would that affect the results, and would it cause an error in the direction of your observed experimental error?

Speed of Sound—Resonance Tube

OBJECTIVES

- ❑ Determine the effective length of a closed tube at which resonance occurs for several tuning forks.
- ❑ Determine the wavelength of the standing wave from the effective length of the resonance tube for each tuning fork.
- ❑ Determine the speed of sound from the measured wavelengths and known tuning fork frequencies and compare with the accepted value.

EQUIPMENT LIST

- Resonance tubes (with length scale marked on the tube)
- Tuning forks (range 500 to 1040 Hz) and rubber hammer
- Thermometer (one for the class)

THEORY

Traveling waves of **speed** V, **frequency** f, and **wavelength** λ are described by

$$V = f\lambda \qquad \text{(Eq. 1)}$$

We can determine the speed of a traveling wave for known frequency and wavelength from Equation 1. It is difficult to measure the properties of a traveling wave directly. When two waves of exactly the same speed, frequency, and wavelength travel in opposite directions in the same region, they produce **standing waves.** These standing waves can be measured easily.

This laboratory uses a device called a resonance tube to produce standing waves from the sound waves emitted from a tuning fork. The can shown in Figure 22-1 contains water, and the level of the water in the tube can be varied as the can is moved up and down. The water acts as the closed end of the tube, and changing the water level changes the effective length of the resonance tube.

A tuning fork, clamped just above the open end of the tube, is struck with a rubber hammer. Sound waves travel down the tube and are reflected when they strike the water. Standing waves are produced by these traveling waves going in both directions inside the tube. The waves reflected from the

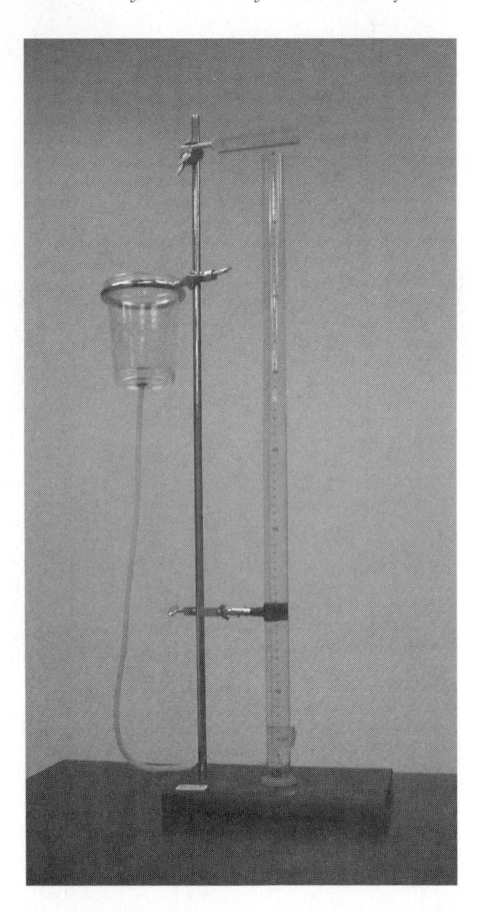

Figure 22-1 Resonance tube apparatus.

closed end of the tube undergo a phase change of 180°, and are completely out of phase with the incident waves. Therefore, the combined amplitude of the incident and reflected waves must be zero at the closed end of the tube. A point in space with wave amplitude zero at all times is called a node N. From similar considerations of the relative phase between the incident and reflected waves, at the open end of the tube the wave amplitude must be a maximum at all times. Such a point is called an antinode A. The speed of sound is fixed, and, for a given tuning fork, the frequency is fixed. Therefore, the resonance conditions can be satisfied for only certain specific lengths of the tube.

Figure 22-2 illustrates the necessary relationship between the length of the tube and the wavelength of the wave for the first four resonances of the tube. Sound waves are a type of wave known as longitudinal. The amplitude of a sound wave is determined by pressure variations in the air along the direction of wave motion. The sound waves in the figure are pictured as if they were transverse waves for ease of representation. The resonances are pictured from left to right as they are encountered when the level of the water in the tube is lowered, increasing the effective length of the tube. The distances L_1, L_2, L_3, and L_4 refer to the distance from the top of the tube to the water level for the first four resonances. The locations of the nodes N and antinodes A are shown for each of these resonances. In the first resonance there is a single node and antinode. Each successive resonance adds an additional node and antinode. The distance between a node and the next antinode is one-fourth wavelength ($1/4\,\lambda$). The distance between nodes is one-half wavelength ($1/2\,\lambda$).

The location of several of the resonances for each tuning fork will be determined experimentally. If the situation were ideal, the following relationships would be implied by Figure 22-2 for the first four resonances shown.

$$L_1 = 1/4\,\lambda \qquad L_2 = 3/4\,\lambda \qquad L_3 = 5/4\,\lambda \qquad L_4 = 7/4\,\lambda \qquad \text{(Eq. 2)}$$

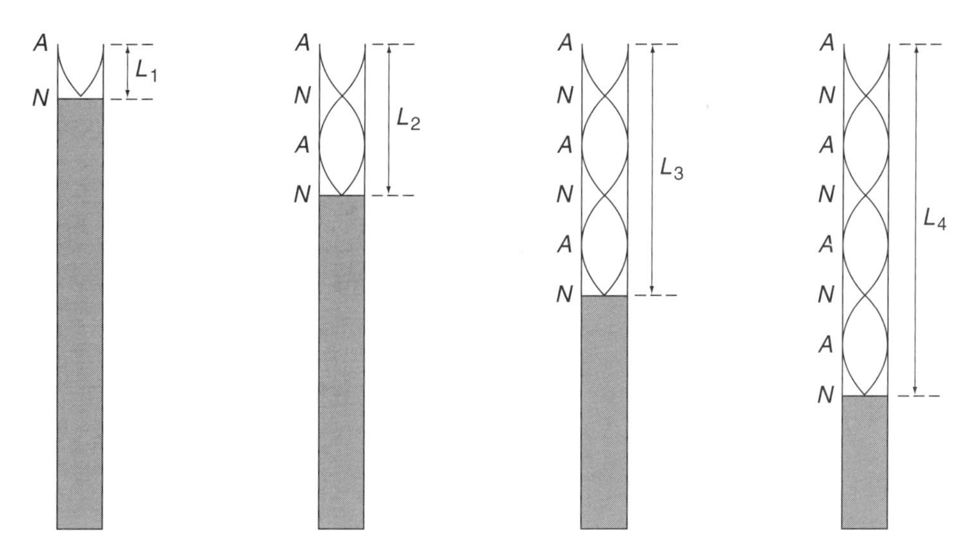

Figure 22-2 Nodes and antinodes of first four resonances of a tube closed at one end.

Examine Figure 22-2 carefully to be sure that you understand how the relationships given in Equations 2 are implied by the figure.

The relationships given in Equations 2 are not valid for a real resonance tube because the point at which the upper antinode actually occurs is just outside the end of the tube. The exact location depends upon the diameter of the tube. Equations 2 are not directly useful to determine the wavelength λ of the wave.

The end effect is the same for each of the resonances and will cancel if differences between the locations of the individual resonances are considered. Considering the differences between adjacent resonances gives the following

$$L_2 - L_1 = L_3 - L_2 = L_4 - L_3 = \lambda/2 \qquad \text{(Eq. 3)}$$

Equations 3 determine the wavelength, and the frequency of the tuning fork is known. Equation 1 then allows determination of the speed of sound.

If Equations 3 are used and the results are then averaged, it would amount to taking the sum of twice the three differences and then dividing by three. In that process, all but the first and last resonance positions cancel from the calculation. In effect, one might as well have not measured the middle two resonances. There is nothing incorrect about such a procedure, but it loses some of the information contained in the data. This shows that there is often more than one way to analyze data, but often one technique gives more information than the others.

All the data contribute to the result if each wavelength is computed, not from the adjacent differences, but from the differences between each resonance and the first resonance. The resulting equations for the wavelength are given below. A subscript has been placed on the wavelength, but it is still understood that each of the wavelengths, λ_1, λ_2, and λ_3, refer to the same wavelength calculated from three different sets of resonances. The equations are

$$\lambda_1 = 2(L_2 - L_1) \qquad \lambda_2 = (L_3 - L_1) \qquad \lambda_3 = 2/3(L_4 - L_1) \qquad \text{(Eq. 4)}$$

The speed of sound in air has a slight linear dependence on the air temperature for a limited range of temperature. The speed of sound V_T at a temperature of T° C will be determined from

$$V_T = (331.5 + 0.607\,T)\ \text{m/s} \qquad \text{(Eq. 5)}$$

where T is the temperature in °C.

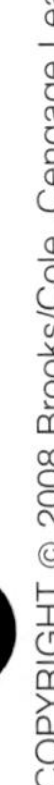

EXPERIMENTAL PROCEDURE

Note carefully that tuning forks should be struck only with the rubber hammer. Take care to ensure that neither the hammer nor a vibrating tuning fork comes into contact with the tube.

1. Measure the room temperature of the air and record it in Data Table 1.
2. Adjust the water level until the can is essentially empty when the tube is almost full. The water level in the tube should come to at least within 0.050 m of the open end of the tube. It may be necessary to remove some water from the can when the water level is near the bottom of the tube.
3. Clamp a tuning fork above the top of the tube, and one partner should strike it repeatedly with the rubber hammer. Keep the fork vibrating continuously with a large amplitude. With the tuning fork vibrating, another partner should slowly lower the water level from the top while listening for a resonance. The sound will be very loud when a resonance is achieved. Try to measure the position of each resonance to the nearest millimeter. Raise and lower the water level several times to produce three trials for the measured position of the first resonance and record the values in Data Table 2. Record the frequency of the tuning fork in Data Table 2.
4. Repeat the procedure in Step 3 to locate as many other resonances as possible. Depending upon the frequency of the tuning fork, either three or four resonances should be attainable. Record in Data Table 2 the location of resonances that are attained.
5. Use a second tuning fork of different frequency and repeat Steps 1 through 4. Record in Data Table 3 the frequency of the tuning fork and the position of as many resonances as are attained.

CALCULATIONS

1. Use Equation 5 to calculate the accepted value of the speed of sound from the measured room temperature. Record it in Data Table 1.
2. Calculate the mean and standard error of the three trials for the location of each of the resonances. Record each of the means and standard errors in Calculations Tables 2 and 3.
3. Use Equations 4 to calculate the wavelengths that are appropriate. If four resonances were found, then all three values of λ can be determined. If only the first three resonances were measured, then only two values of λ can be determined. If this is the case, just leave the Calculations Table blank at the appropriate position. Use the mean values of the lengths to calculate the wavelengths.
4. Calculate the mean and standard error for the number of independent wavelengths measured for each tuning fork. Record those values in the Calculations Tables as $\bar{\lambda}$ and α_{λ}.
5. From the values of $\bar{\lambda}$ and the known values of the tuning fork frequencies, calculate the experimental value for V, the speed of sound.
6. Calculate the percentage error of the experimental values of V compared to the accepted value of the speed of sound in Data Table 1.

Name Section Date

22 **LABORATORY 22** *Speed of Sound—Resonance Tube*

PRE-LABORATORY ASSIGNMENT

1. What is the equation that relates the speed V, the frequency f, and the wavelength λ of a wave?

2. How are standing waves produced?

3. What name is given to a point in space where the wave amplitude is zero at all times?

4. What name is given to a point in space where the wave amplitude is a maximum at all times?

5. What are the conditions that must be satisfied to produce a standing wave in a tube open at one end and closed at the other end?

6. For an ideal resonance tube an antinode occurs at the open end of the tube. What property of real resonance tubes slightly alters the position of this antinode?

7. A student using a tuning fork of frequency 512 Hz observes that the speed of sound is 340 m/s. What is the wavelength of this sound wave? Show your work.

8. A student using a resonance tube determines that three resonances occur at distances of $L_1 = 0.172$ m, $L_2 = 0.529$ m, and $L_3 = 0.884$ m below the open end of the tube. The frequency of the tuning fork used is 480 Hz. What is the average speed of sound from these data? Show your work.

Name | Section | Date

Lab Partners

22 LABORATORY 22 *Speed of Sound—Resonance Tube*

LABORATORY REPORT

Data Table 1

Room Temperature = °C	Speed of sound = m/s

Data Table 2

Frequency Fork One = Hz			
L_1 (m)	L_2 (m)	L_3 (m)	L_4 (m)

Data Table 3

Frequency Fork Two = Hz			
L_1 (m)	L_2 (m)	L_3 (m)	L_4 (m)

Calculations Table 2

$\bar{L}_1 =$ m	$\bar{L}_2 =$ m	$\bar{L}_3 =$ m	$\bar{L}_4 =$ m
$\alpha_{L1} =$ m	$\alpha_{L2} =$ m	$\alpha_{L3} =$ m	$\alpha_{L4} =$ m
$\lambda_1 = 2(L_2 - L_1) =$ m	$\lambda_2 = (L_3 - L_1) =$ m	$\lambda_3 = 2/3(L_4 - L_1) =$ m	
$\bar{\lambda} =$ m	$\alpha_{\lambda} =$ m	$V = f\bar{\lambda} =$ m/s	% Err =

Calculations Table 3

$\bar{L}_1 =$ m	$\bar{L}_2 =$ m	$\bar{L}_3 =$ m	$\bar{L}_4 =$ m
$\alpha_{L1} =$ m	$\alpha_{L2} =$ m	$\alpha_{L3} =$ m	$\alpha_{L4} =$ m
$\lambda_1 = 2(L_2 - L_1) =$ m	$\lambda_2 = (L_3 - L_1) =$ m	$\lambda_3 = 2/3(L_4 - L_1) =$ m	
$\bar{\lambda} =$ m	$\alpha_{\lambda} =$ m	$V = f\bar{\lambda} =$ m/s	% Err =

SAMPLE CALCULATIONS

1. Speed Sound $= 331.5 + 0.607\ T =$
2. $\lambda_1 = 2(L_2 - L_1) =$
3. $\lambda_2 = (L_3 - L_1) =$
4. $\lambda_3 = 2/3(L_4 - L_1) =$
5. $V = f\bar{\lambda} =$
6. % Error $= |E - K|/K\ (100\%) =$

QUESTIONS

1. What is the accuracy of each of your measurements of the speed of sound? State clearly the evidence for your answer.

2. What is the precision of each of your measurements of the speed of sound? State clearly the evidence for your answer.

3. Equations 2 provide a means to determine the end correction for the tube. Using the value of $\bar{\lambda}$ for the first tuning fork, calculate values for L_1 and L_2 from those equations. They should be larger than the measured values of L_1 and L_2 by an amount equal to the end correction. Repeat the calculation for the second tuning fork. Compare these values for the end correction and comment on the consistency of the results.

4. Suppose that the temperature had been 10 °C higher than the value measured for the room temperature. How much would that have changed the measured value of $L_2 - L_1$ for each tuning fork? Would $L_2 - L_1$ be larger or smaller at this higher temperature?

5. Draw a figure showing the fifth resonance in a tube closed at one end. Show also how the length of the tube L_5 is related to the wavelength λ.

Specific Heat of Metals

OBJECTIVES

- ❑ Demonstrate the heat exchange within a calorimeter.
- ❑ Determine the specific heat of two samples of different metals.

EQUIPMENT LIST

- Calorimeter and stirrer, steam generator, Bunsen burner, or electric heating plate
- Metal shot (two different kinds of metal)
- Two thermometers (preferably one 0–100°C and the other 0–50°C) and glycerin

THEORY

Two objects at different temperatures, placed in thermal contact with each other, exchange thermal energy until they are in **thermal equilibrium** at the same temperature. The exchanged thermal energy is known as **heat**. The temperature change that each object experiences is determined by the mass and specific heat of each object. The specific heat c is the amount of heat per unit mass required to change the temperature by one degree. The units used in this laboratory for specific heat are cal/g–C°. The heat or thermal energy change for any object in the system can be written as

$$\text{Heat} = \text{Thermal energy change} = Mc\Delta T \quad \text{(Eq. 1)}$$

where M is mass in grams, c is **specific heat** in cal/g–C°, and ΔT is the temperature change in C°. The units of the thermal energy change or heat will be calories. In an ideal experimental arrangement designed to investigate this concept, all exchanges of heat are among the objects of the system. In real systems, heat losses from the system or gains into the system are held to a minimum. A device that produces a thermally isolated environment to maximize heat exchanges inside the system is called a **calorimeter.** The calorimeter used in this laboratory consists of two metal cups held apart by a plastic ring that produces an insulating air space between the cups. The cups also have an insulating plastic top that contains a hole into which the stirrer is placed, and a rubber stopper with a hole into which a thermometer is placed. A picture of a calorimeter is shown in Figure 23-1.

Metal shot will be heated to the temperature of steam in a steam generator and then placed into a calorimeter containing water near room temperature. The metal shot will lose heat, and that heat will be gained by the water, the calorimeter cup, and the stirrer in the calorimeter. The stirrer is used to mix the

Figure 23-1 Calorimeter. (Photo courtesy of Sargent-Welch Scientific Co.)

system to ensure that all parts of the system quickly come to thermal equilibrium. In equation form, the statement that the heat lost by the metal shot equals the heat gained by the other parts of the system is

$$M_m c_m (T_m - T_e) = M_w c_w (T_e - T_o) + M_{\text{cal}} c_{\text{cal}} (T_e - T_o) + M_s c_s (T_e - T_o) \qquad \text{(Eq. 2)}$$

where the M's are masses and the c's are specific heats. The subscripts are m for metal, w for water, cal for calorimeter, and s for stirrer. The initial temperature of the metal shot is T_m, the original temperature of the water, cup, and stirrer is T_o, and the final equilibrium temperature of the system is T_e.

EXPERIMENTAL PROCEDURE

1. Place about 250 g of one kind of metal shot into the cup that is designed to fit into the steam generator.
2. Fill the steam generator about one-half full of water. Keep the water level below the bottom of the cup that is placed in the generator. The cup should be heated by steam from the water. Use whatever means of heating is provided (Bunsen burner or electric heating plate) to heat the generator.
3. Place a thermometer that reads 0–100°C into the metal shot. If the thermometer is a mercury thermometer, be very careful not to break the bulb as you work it down into the shot. Place paper or some other insulation around the top of the cup to keep the outside air from cooling the top layer of shot.
4. While the shot is heating, determine the mass of the inner calorimeter cup and the stirrer separately and record their mass in the Data Table. Obtain from your instructor the values for c_{cal} and c_s, the specific heats of the calorimeter cup and the stirrer, and record them in the Data Table.
5. Place about 100 g of water in the inner calorimeter cup and determine the combined mass of the cup, stirrer, and water. Determine the mass of the water by subtraction and record it in the Data Table. The initial temperature of the water should ideally be a few degrees below the room temperature. This will tend to compensate for interaction with the air.
6. Assemble the calorimeter, placing the second thermometer in the rubber stopper. Use glycerin on the thermometer so that it may be easily slipped into the hole in the stopper. *If using a glass thermometer, be extremely careful. If the thermometer breaks, you can cut yourself severely.* Stir the water slowly to be sure that all parts of the system are in equilibrium. Record the temperature of the water, calorimeter, and stirrer as T_o in the Data Table. Take this reading for T_o just prior to performing Step 8 below.

7. Take the temperature of the metal shot every few minutes. Move the bulb around carefully in the shot to be sure all parts are in equilibrium. The temperature should rise and then level off at essentially the boiling point of water. This will probably be a few degrees below 100°C depending upon the local elevation above sea level. Record the value (to the nearest 0.1°) of the maximum temperature as T_m in the Data Table.
8. Remove the thermometer from the shot and quickly transfer the shot to the water in the inner calorimeter cup. It is very important not to splash water out of the cup in this process. This can easily happen when removing the insulating plastic cover or when placing the shot in the water.
9. Place the insulating cover back on the calorimeter and slowly stir the water while watching the thermometer in the calorimeter. When the maximum temperature is reached, record its value (to the nearest 0.1°) as the equilibrium temperature T_e.
10. Remove the inner calorimeter cup and determine the mass of the cup, stirrer, water, and metal shot. Record that value in the Data Table. Determine the mass of the shot used by subtraction and record that value in the Data Table.
11. Discard the water but do not lose any of the shot. Place the wet shot on paper towels, spread it out, and allow the shot to dry. Before leaving the laboratory and when the shot is completely dry, return it to its proper container. Be careful not to mix the two kinds of metal shot.
12. From Appendix II, determine the known value of the specific heat for the type of metal of the shot and record it in the Data Table.
13. Repeat all of the above procedure for a second type of metal shot.

CALCULATIONS

1. In Equation 2 all of the variables are known except for the specific heat of the metal shot. Solve Equation 2 for the specific heat of the metal shot and record that value as the experimental value in the Calculations Table.
2. Calculate the percentage error between your experimental value and the known value of the specific heat for each type of metal.

Name | Section | Date

LABORATORY 23 *Specific Heat of Metals*

PRE-LABORATORY ASSIGNMENT

1. What is the definition of specific heat?

2. What is the name for a device that provides a thermally isolated environment in which substances exchange heat?

A heated piece of metal at a temperature T_1 is placed into a calorimeter containing water and a stirrer. The temperature of the calorimeter, water, and stirrer is initially T_2 where $T_1 > T_2$. The system is stirred continuously until it comes to equilibrium at a temperature of T_3. Answer Questions 3–5 concerning what happens.

3. The final equilibrium temperature of the system is such that

(a) $T_3 > T_1 > T_2$ **(b)** $T_1 < T_2 < T_3$ **(c)** $T_1 < T_3 < T_2$ **(d)** $T_1 > T_3 > T_2$

4. The heat lost by the metal ΔQ_m and the heat gained by the calorimeter system ΔQ_c obey which of the following relationships?

(a) $\Delta Q_m > \Delta Q_c$ **(b)** $\Delta Q_m < \Delta Q_c$ **(c)** $\Delta Q_m = \Delta Q_c$

5. What is the purpose of stirring the system continuously?

6. A 350 g piece of metal is at an initial temperature of 22.0°C. It absorbs 1000 cal of thermal energy, and its final temperature is 45.0°C. What is the specific heat of the metal? Show your work.

7. A 250.0 g sample of metal shot is heated to a temperature of 98.0°C. It is placed in 100.0 g of water in a brass calorimeter cup with a brass stirrer. The total mass of the cup and the stirrer is 50.0 g. The initial temperature of the water, stirrer, and calorimeter cup is 20.0°C. The final equilibrium temperature of the system is 30.0°C. What is the specific heat of the metal sample? (The specific heat of brass is 0.092 cal/g–C°.) Show your work.

Name | Section | Date

Lab Partners

23 **LABORATORY 23** *Specific Heat of Metals*

LABORATORY REPORT

Data Table

Mass Cup =			g	*Mass Stirrer* =			g
Metal	*Cup* + H_2O + *Stirrer*	*Cup* + *Stirrer* + H_2O + *Shot*	H_2O	*Shot*	T_o	T_m	T_e
	g	g	g	g	°C	°C	°C
	g	g	g	g	°C	°C	°C

Calculations Table

Metal	*Known Specific Heat*	*Measured Specific Heat*	*Percentage Error*
	Cal/g–C°	Cal/g–C°	
	Cal/g–C°	Cal/g–C°	

SAMPLE CALCULATIONS

1. $M_{\text{Water}} = M_{\text{Cup + Stirrer + Water}} - M_{\text{cup}} - M_{\text{stirrer}} =$
2. $M_{\text{shot}} = M_{\text{cup + stirrer + water + shot}} - M_{\text{cup + stirrer + water}} =$
3. $c_m = [M_w c_w (T_e - T_o) + M_c c_c (T_e - T_o) + M_s c_s (T_e - T_o)] / M_m (T_m - T_e)]$
4. $\%$ Error $= (|E - K|/\text{K})(100\%) =$

QUESTIONS

1. What is the accuracy of your results for the specific heats of the metals? Can you suggest any change in the original temperature of the metal that would improve the results?

2. Can you make a quantitative statement about the precision of your results? If you can make a statement about the precision, do so. If you cannot make any statement about the precision, suggest what measurements would allow you to do so.

3. Suppose the shot were wet and thus included some water at the same temperature as the shot when it was placed in the calorimeter. How would this affect the results?

4. What would the change in temperature of the water have been if 200 g of water had been used with each of the samples? Would this have tended to improve the results, tended to make the results worse, or tended to have no effect on the results?

Linear Thermal Expansion

OBJECTIVES

- ❑ Demonstrate that for the same temperature change the thermal expansion depends upon the type of metal.
- ❑ Determine the value of the linear coefficient of thermal expansion α for several metals and compare the results to known values.

EQUIPMENT LIST

- Linear expansion apparatus consisting of steam jacket containing a metal rod
- Steam generator, rubber tubing, and beaker to catch steam condensation
- Bunsen burner and stand, or electric heating plate
- Meter stick, thermometer (0–100°C)
- Ohmmeter if using micrometer screw-type indicator

THEORY

Most materials expand as their temperature increases. The expansion depends not upon the heat input, but rather upon the temperature change. Temperature expansion occurs in three dimensions, but we will investigate only one dimensional change in the length of a rod.

A metal rod of some initial length L_o at some initial temperature T_o is heated to some temperature T_1. The length of the rod will increase to a new length L_1. The change in length of the rod $\Delta L = L_1 - L_o$ is found to be proportional to the original length of the rod L_o and to the change in the temperature ΔT where $\Delta T = T_1 - T_o$. In equation form the result is

$$\Delta L = \alpha L_o \, \Delta T \qquad \text{(Eq. 1)}$$

where α is a constant called the **linear coefficient of thermal expansion.** Solving Equation 1 for the constant α gives

$$\alpha = \frac{\Delta L}{L_o \Delta T} = \frac{L_1 - L_o}{L_o \Delta T} \qquad \text{(Eq. 2)}$$

From Equation 2 it is clear that α is the fractional change in length per unit change in the temperature. Because the fractional change in length $\Delta L/L_o$ has no dimensions, the units of α are $(\text{C}^\circ)^{-1}$. Strictly

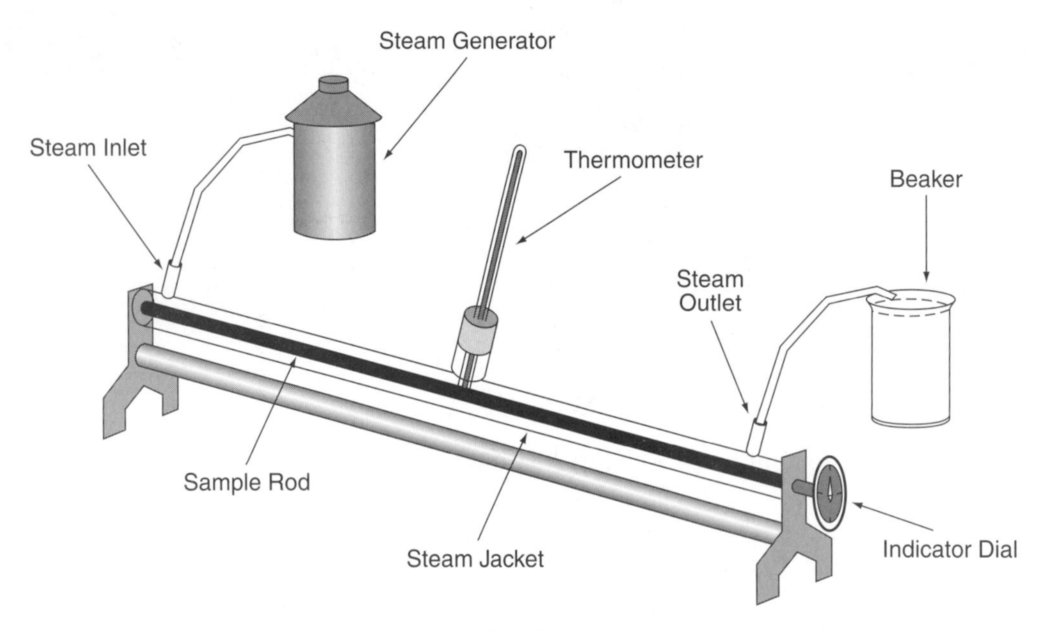

Figure 24-1 Experimental arrangement for thermal expansion apparatus.

speaking, α varies slightly with temperature. However, over the range of temperature used in this laboratory, we can assume α to be approximately constant.

The apparatus to be used in this laboratory is shown in Figure 24-1. It consists of a steam jacket containing a metal rod about 0.60 m long. The jacket is held by supports at either end. One of the end supports has a thumbscrew to keep that end of the rod fixed. The other end support contains an indicator to measure the change in length of the rod.

There are two types of indicators. One has a micrometer screw with a rotary dial of 100 divisions. Each division is 0.01 mm, and one complete turn of the dial is equivalent to a linear translation of 1 mm. Some micrometer screw indicators have a set of binding posts on each end support. We can use them to construct an electrical circuit to indicate when the micrometer screw makes contact with the rod. When using a micrometer screw-type indicator, remember to back the screw away from the rod before the rod is allowed to expand, to prevent damage to the micrometer screw.

The second type of indicator contains a plunger-activated dial that reads directly in 0.01 mm. One complete revolution of the dial corresponds to 1 mm linear displacement with a total of 3.5 mm displacement possible. When using this type of indicator, contact is made with the rod, and the rod is allowed to expand against the plunger, which always remains in contact with the rod.

EXPERIMENTAL PROCEDURE

Linear Thermal Expansion

1. Remove the rod from the steam jacket and measure the length of the rod with a meter stick. Measure to the nearest 0.001 m and record this length as L_o in the Data Table.
2. Replace the rod in the steam jacket and secure the jacket in the support ends. If using a device that has binding posts, connect the leads from an ohmmeter to each of the binding posts. Even if the apparatus does not have binding posts, you can use the ohmmeter to make contact with the end supports when taking a measurement.
3. Use a one-hole rubber stopper to place a thermometer in the opening provided for that purpose. The opening is in the center of the steam jacket. The thermometer bulb should just barely touch the rod. If the apparatus has been standing unused for several hours or more, record the temperature after the thermometer is in contact with the rod. If the steam jacket has been heated recently, run cool water

through the jacket until the entire system is at equilibrium at a temperature near room temperature. Record the temperature (to the nearest 0.1°) in the Data Table as T_o. Adjust the indicator dial until contact is made with the rod. If using the micrometer-type device, contact is indicated by the ohmmeter. Record the indicator dial setting as D_o in the Data Table.

4. If using the micrometer-type indicator, back the screw out several turns at this time. If using the plunger-type indicator, leave it in contact with the rod. *It is extremely critical that there be no disturbance of the rod between this reading and the final reading after the rod has been heated.*
5. Connect the steam supply to the steam jacket with a rubber hose. At the other end of the jacket, connect a hose from the steam outlet to a beaker to catch the steam condensation. Pass steam through the jacket for several minutes and monitor the temperature of the rod. When the temperature reaches its maximum value, record the temperature (to the nearest 0.1°) as T_1.
6. If using the plunger-type indicator, simply read the value on the indicator dial. If using the micrometer screw-type device, turn the screw in until it touches the rod as indicated by the ohmmeter. Record the reading as D_1 in the Data Table.
7. Repeat Steps 1 through 6 for other rods of a different metal. Be extremely careful not to burn yourself on the heated steam jacket. Before beginning the procedure, run cool water through the apparatus until the new rod and jacket are in equilibrium near room temperature.
8. From Appendix II, determine the known value of α for each of the rods measured and record them in the Data Table.

CALCULATIONS

Linear Thermal Expansion

1. Calculate the increase in length ΔL for each rod from $\Delta L = D_1 - D_o$ and record each of them in the Calculations Table.
2. Calculate the increase in temperature ΔT for each rod from $\Delta T = T_1 - T_o$ and record each of them in the Calculations Table.
3. Use Equation 2 to calculate the linear coefficient of thermal expansion α for each rod and record each of them in the Calculations Table.
4. Calculate the percentage error for each experimental value of α compared to the known value. Record them in the Calculations Table.

Name | Section | Date

24 LABORATORY 24 *Linear Thermal Expansion*

PRE-LABORATORY ASSIGNMENT

1. An object undergoes a change in length ΔL because of a change in its temperature. What are the three factors on which this change in length depends?

2. What are the units for the linear thermal coefficient of expansion α?

3. Most materials expand when the temperature is raised. What can you conclude about α for a material that contracts when its temperature is raised?

4. A copper rod has a length of 1.117 m when its temperature is 22.0°C. If the temperature of the rod is raised to 275.0°C, what is the new length of the rod? [The value of α for copper is $16.8 \times 10^{-6} (\text{C}^\circ)^{-1}$.] Show your work.

5. Iron rails are used to build a railroad track. If each rail is 10.000 m long when it is placed in the track at a temperature of 20.0°C, how much space must be left between the rails so that they just touch each other when the temperature is 40.0°C? [The value of a for iron is $11.4 \times 10\,(\text{C}°)^{-1}$.] Show your work.

6. A rod in a steam jacket is measured to have a length of 0.600 m at a temperature of 22.0°C. Steam is then passed through the jacket for several minutes until the rod is at a temperature of 98.0°C. The increase in the length of the rod is measured by a micrometer screw arrangement to be 1.19 mm. What is the linear thermal coefficient of expansion α for the rod? Show your work.

7. Considering the value of α found for the rod in Question 6, from what kind of metal is it probably made?

Name | Section | Date

Lab Partners

24

LABORATORY 24 *Linear Thermal Expansion*

LABORATORY REPORT

Data Table

Metal	*Known* α $(C^\circ)^{-1}$	L_o (m)	T_o (°C)	T_1 (°C)	D_o (m)	D_1 (m)

Calculations Table

Metal	ΔL (m)	ΔT (C°)	α_{exp} $(C^\circ)^{-1}$	% *Error*

SAMPLE CALCULATIONS

1. $\Delta L = D_1 - D_o =$
2. $\Delta T = T_1 - T_o =$
3. $\alpha = \dfrac{\Delta L}{L_o \Delta T} =$
4. $|E - K|/K\ (100\%) =$

QUESTIONS

1. What is the accuracy of your measurements of α? State clearly the basis for your answer.

2. We could measure the change in length ΔL with more accuracy if it were larger. We could do this by heating the rod directly with a Bunsen burner to a temperature considerably higher than 100°C. Would that be a reasonable alternative? In what way is the steam heat a more workable technique?

3. The original length L_o of the rod was measured only to the nearest 1 mm. Does this cause a significant error in the final result? If so, why does it? If not, why does it not?

4. Would the measured value of α have been the same or different if lengths were determined in inches instead of meters? State clearly the basis for your answer.

5. A brass washer has an inside diameter of 2.000 cm and an outside diameter of 3.000 cm at 20.0°C. A solid rod of aluminum has a diameter of 2.000 cm at 20.0°C and just fits inside the washer as shown in Figure 24-2. Both the washer and aluminum rod are raised to a temperature of 150.0°C. Will the rod still fit inside the washer? If so, how much smaller is the rod than the opening in the washer? If not, how much larger is the rod than the opening in the washer? Show your work.

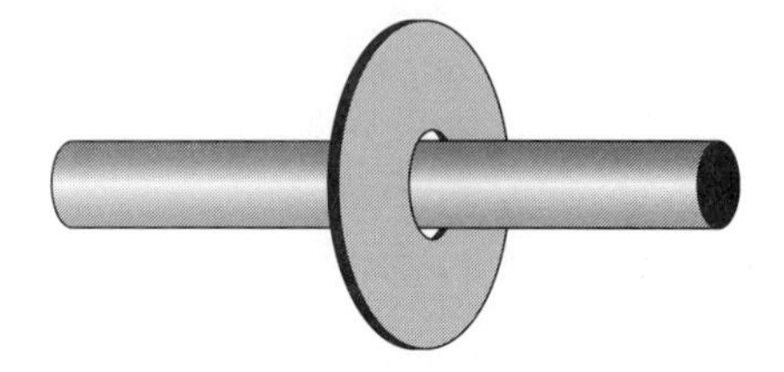

Figure 24-2 Brass washer on an aluminum rod.

The Ideal Gas Law

OBJECTIVES

- ❑ Demonstrate that the pressure P of a gas at a fixed temperature is proportional to the quantity $(1/V)$ where V is the gas volume.
- ❑ Demonstrate that the volume V of a gas at a fixed pressure is proportional to the temperature T of the gas.
- ❑ Determine an experimental value for the constant that relates Celsius temperature T_C to the absolute temperature T and compare the experimental value for this constant with its known value of 273.15.

EQUIPMENT LIST

- Gas law apparatus, Vaseline or stopcock grease, and string
- Thermometer (0–100°C) and vernier calipers
- 600 mL beaker and tongs to fit the beaker
- Masses to allow application of up to 6 kg in 1 kg increments
- Electric heating plate or Bunsen burner (If a Bunsen burner is used, a stand to hold beaker while it is heated is also needed.)
- One large container of ice for the class

THEORY

Consider a gas containing N molecules confined to a volume V, at a pressure P, and temperature T. In the most general case, all of these quantities vary over a wide range, and the equation of state that relates them is very complex. If only gases at low density are considered, the equation of state is greatly simplified. A gas that satisfies these conditions is referred to as an ideal gas. Although there are no true ideal gases, most real gases approximate an ideal gas near room temperature and atmospheric pressure. The **ideal gas law** is

$$PV = Nk_B T \qquad \text{(Eq. 1)}$$

The SI units of pressure P are N/m^2, and the SI units of volume V are m^3. The constant $k_B = 1.38 \times 10^{-23}$ J/K is called Boltzmann's constant. In Equation 1 the Kelvin temperature scale must be used. The relationship between temperature in **Celsius** T_C and the **Kelvin** temperature T is

$$T = T_C + 273.15 \qquad \text{(Eq. 2)}$$

In this laboratory, we will verify Equation 1 in two different ways. Consider first the case of a fixed amount of gas (thus N is constant) at some given constant temperature. Under these conditions the quantity Nk_BT is a constant designated as C_1. In terms of C_1 Equation 1 can be rewritten as

$$P = C_1(1/V) \qquad \text{(Eq. 3)}$$

Consider the apparatus shown in Figure 25-1. The apparatus can be constructed from a large plastic syringe (either 35 cc or 60 cc volume) marked in 1 cc increments. The needle is removed, and the end of the syringe is melted closed (airtight). The outer cylinder of the syringe is mounted on a small wooden support. A small flat wooden platform is attached to the top of the plunger. These devices can be constructed easily and cheaply. They are also available from Pasco Scientific without the wooden support.

The sliding rubber seal of the syringe allows a fixed amount of gas in the cylinder to vary in volume V as the pressure P is changed. Total pressure P is from the atmospheric pressure P_a and any additional pressure called the gauge pressure P_g. In Figure 25-1(b) the gauge pressure is the weight Mg placed on the platform of the plunger per unit area A of the cylinder. This is

$$P_g = F/A = Mg/A \qquad \text{(Eq. 4)}$$

Using $P_g + P_a$ for P in Equation 3 gives

$$P_g = C_1(1/V) - P_a \qquad \text{(Eq. 5)}$$

Equation 5 states that there is a linear relationship between the gauge pressure P_g and the quantity $1/V$ with C_1 as the slope and $-P_a$ as the intercept. This aspect of the ideal gas law is known as **Boyle's law.** In the first part of the laboratory the volume V of a gas sample will be measured as a function of the gauge pressure P_g. Experimental values for C_1 and P_a will then be obtained from a fit of the data to Equation 5.

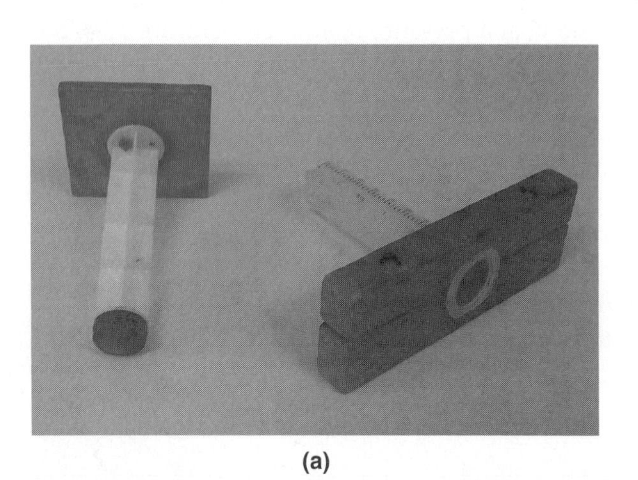

(a)

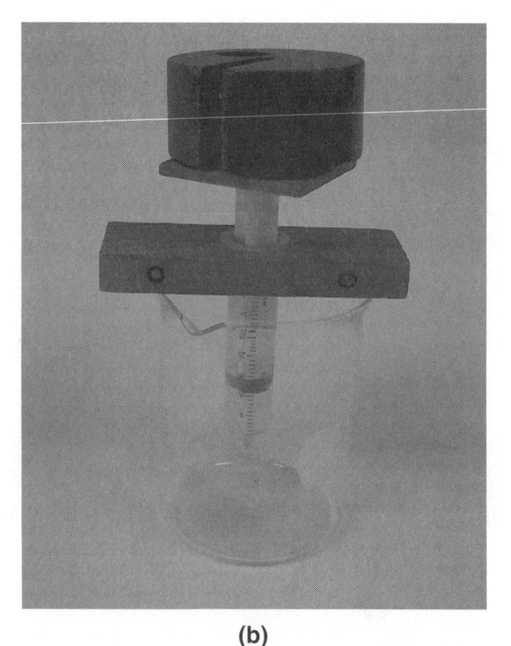

(b)

Figure 25-1 Syringe with needle removed and end closed. The sliding seal allows a fixed amount of air to change volume as pressure or temperature is changed.

Consider now the case of a fixed amount of gas at some constant pressure P. Under these conditions the quantity Nk_B/P is a constant designated as C_2. In terms of the constant C_2 and the Celsius temperature T_C, Equation 1 can be rewritten as

$$V = C_2 T_C + C_2 (273.15) \qquad \text{(Eq. 6)}$$

Equation 6 states that there is a linear relationship between the gas volume V and the temperature T_C with C_2 as the slope, and C_2 (273.15) as the intercept. This aspect of the ideal gas law is essentially the **law of Charles and Gay-Lussac.** In the second part of the laboratory we will measure the volume of a sample of gas as a function of the temperature T_C. We will obtain experimental values for C_2 and C_2 (273.15) by a fit of the data to Equation 6. Examination of Equation 6 shows that

$$K = \frac{\text{Intercept}}{\text{Slope}} = 273.15 \qquad \text{(Eq. 7)}$$

From Equation 7 we can obtain an experimental value for K from the intercept and slope of the fit to the volume versus temperature data. The negative of the value for the constant ($-K$) is an experimental value for absolute zero on the Celsius scale.

EXPERIMENTAL PROCEDURE

Constant Temperature

1. Pull the plunger from the syringe and place a short length of string into the cylinder with a loose end hanging out. Put a thin coat of Vaseline or stopcock grease on the rubber tip of the plunger. Replace the plunger in the syringe. The string allows a small air leak to adjust the volume of air trapped by the plunger. Adjust the volume of air to 25 cc if using the 35 cc syringe, and 45 cc if using the 60 cc syringe. Then withdraw the string. ($1\ \text{cc} = 1\ \text{cm}^3 = 1 \times 10^{-6}\ \text{m}^3$)
2. Place the support block of the syringe across the opening of the beaker as shown in Figure 25-1 and place mass M on the plunger platform. Use values for M from 1 kg to 6 kg in increments of 1 kg. Determine the volume of the gas trapped in the syringe for each value of M. To determine the volume accurately, push down slightly on the plunger and allow the plunger to spring back before taking each measurement of volume. Record the values of V for each M in Data Table 1.
3. Use the vernier calipers to determine the inside diameter D of the syringe cylinder and record it in Data Table 1.
4. Determine the temperature T_C of the room air and record it in Data Table 1.
5. Standard atmospheric pressure at sea level is $1.013 \times 10^5\ \text{N/m}^2$. Determine the present local barometric pressure in these units and record that value in Data Table 1.

Constant Pressure

1. Using a string as before, adjust the volume V of the syringe to 25 cc if using the 35 cc syringe, and 45 cc if using the 60 cc syringe.
2. Place tap water in the beaker and place the syringe support on the beaker. The syringe cylinder should be almost completely immersed in the water. Place a 2 kg mass on the plunger platform. Place the beaker in a position to be heated and bring the water to a boil. *The beaker must be securely supported while being heated.* Allow at least 5 minutes after the water begins to boil for the air in the syringe to come to equilibrium with the boiling water. Measure the volume V and the temperature of the water T_C. Record them in Data Table 2.

3. Remove the 2 kg mass from the plunger. Use the beaker tongs to carefully remove the beaker from the source of heat and set it on the laboratory table. Replace the 2 kg mass on the plunger. Add small pieces of ice to the water and stir the mixture after each addition of ice until all the ice melts. Monitor the temperature of the water at all times. Determine the volume V of the syringe for water temperatures of about 75°C, 50°C, and 25°C. For each temperature at which the volume is determined, there should be no ice in the water when the volume is measured. Allow a few minutes at each temperature to establish equilibrium between the water and the air in the syringe. Record the volume V and the temperature T_C (to the nearest 0.1°C) in Data Table 2.
4. For the final temperature add enough ice until ice and water are at equilibrium near 0°C. Some water may have to be discarded. Allow at least five minutes for the air in the syringe to come to equilibrium with the ice and water and record the volume V of the trapped air and the temperature T_C (to the nearest 0.1°C) in Data Table 2.

CALCULATIONS

Constant Temperature

1. Calculate the cross-sectional area A of the syringe from the measured diameter D of the cylinder where $A = \pi D^2/4$ and record in Calculations Table 1.
2. Calculate the force $F = Mg$ with $g = 9.80\,\text{m/s}^2$ for each mass M and record in Calculations Table 1.
3. Use Equation 4 to calculate gauge pressure P_g for each value of F and record in Calculations Table 1.
4. For each value of V calculate the quantity $1/V$ and record in Calculations Table 1.
5. Perform a linear least squares fit with P_g as the vertical axis and $1/V$ as the horizontal axis. Record the slope as C_1, the negative of the intercept as $(P_a)_{exp}$, and r the correlation coefficient in Calculations Table 1.
6. Calculate the percentage error for the value of $(P_a)_{exp}$ compared to the current atmospheric pressure and record the results in Calculations Table 1.

Constant Pressure

1. Perform a linear least squares fit with V as the vertical axis and T_C as the horizontal axis. Record the slope, intercept, and correlation coefficient of the fit in Calculations Table 2.
2. Use Equation 7 to calculate an experimental value of the constant K from the slope and intercept of the fit. Record this value as K_{exp} in Calculations Table 2.
3. Calculate the percentage error in the value of K_{exp} compared to the known value for K, which is 273.15. Record this percentage error in Calculations Table 2.

GRAPHS

1. Make a graph of the data in Calculations Table 1 with P_g as the vertical axis and $1/V$ as the horizontal axis. Also show on the graph the straight line obtained by the linear least squares fit to the data.
2. Make a graph of the data in Calculations Table 2 with V as the vertical axis and T as the horizontal axis. Choose a scale from approximately −300°C to 100°C that will allow the data to be extrapolated back to $V = 0$. Also show on the graph the straight line obtained from the fit to the data extrapolated to $V = 0$.

Name | Section | Date

25 LABORATORY 25 *The Ideal Gas Law*

PRE-LABORATORY ASSIGNMENT

1. What is the ideal gas law?

2. What are the conditions under which a real gas approximately obeys the ideal gas law?

3. What are the SI units for pressure and volume? What is the value of Boltzmann's constant k_B?

4. Which temperature scale must always be used in the ideal gas law?

5. The temperature in a room is measured to be $T_C = 24.5°C$. What is the Kelvin temperature of the room? Show your work.

6. If the volume of the room in Question 5 is 50.0 m^3, and the pressure is 1.013×10^5 N/m^2, what is the value of N, the number of molecules in the room? Show your work.

7. A gas at constant temperature has a volume of 25.0 m^3 and the pressure of the gas is 1.50×10^5 N/m^2. What is the volume of the gas if the pressure of the gas is increased to 2.50×10^5 N/m^2 while the temperature remains fixed? Is this an example of Boyle's law or Charles' law? Show your work.

8. A gas at constant pressure has a volume of 35.0 m^3, and its temperature is 20.0°C. What is its volume if its temperature is raised to 100.0°C at the same pressure? Is this an example of Boyle's law or Charles' law? Show your work.

Name | Section | Date

Lab Partners

25 LABORATORY 25 *The Ideal Gas Law*

LABORATORY REPORT

Data Table 1

M (kg)	V (m^3)
$D =$	m
$T_C =$	°C
$P_a =$	N/m^2

Calculations Table 1

F (N)	P_g (N/m^3)	$1/V$ (m^{-3})
$A =$ m^2	$r =$	
$C_1 =$ N−m	$(P_a)_{exp} =$ N/m^2	
Percentage error =		

Data Table 2

T_C (°C)	V (m^3)

Calculations Table 2

Slope = $m^3/C°$
Intercept = m^3
$r =$
$K_{exp} =$
Percentage error =

SAMPLE CALCULATIONS

1. $F = Mg =$
2. $A = \pi D^2/4 =$
3. $P_g = Mg/A =$
4. $1/V =$
5. $K_{exp} = -(\text{Intercept})/(\text{Slope}) =$
6. $\% \text{ Error} = |K - E|/K \times (100\%) =$

QUESTIONS

1. How well do your data confirm the ideal gas law? State your answer to this question as quantitatively as possible.

2. What is the accuracy of your value of $(P_a)_{exp}$ compared to the actual atmospheric pressure?

3. What is the accuracy of your value for K?

4. Consider the value of the constant C_1 in Calculations Table 1. It should be equal to Nk_BT. Equate the measured value of C_1 to Nk_BT and solve the resulting equation for N, the number of molecules in the syringe when the temperature was held constant.

5. Consider the value of the slope in Calculations Table 2. It should be equal to Nk_B/P. Equate the measured value of the slope to Nk_B/P and solve for N, the number of molecules in the syringe when the pressure was held constant. Remember that the value of P_g for this part of the laboratory was caused by the 2.00 kg mass. Use the known value of the atmospheric pressure P_a.

Equipotentials and Electric Fields

OBJECTIVES

- ❑ Investigate the location and shape of equipotential surfaces near oppositely charged electrodes and construct electric field lines perpendicular to the equipotentials.
- ❑ Compare experimentally determined electric field lines with known patterns for line charge, two line charges of opposite sign, and parallel plates.
- ❑ Determine the dependence of the magnitude of E on distance from a line of charge.

EQUIPMENT LIST

- Corkboard and push pins to attach power supply and voltmeter to electrodes
- Carbon-impregnated resistance paper with a grid
- Conducting paint or conducting pen (either silver-based or carbon-based)
- Direct current power supply (20 V, low current)
- High impedance voltmeter (preferably digital)

THEORY

Consider two electrodes of arbitrary shape some distance apart carrying equal and opposite charges. A potential difference of 20 V exists between the electrodes with the negative electrode at zero and the positive electrode at +20 V. In the space surrounding these electrodes, there are points at the same potential. There will be some points with potential +10 V. There will be other points with potential +15 V, and still other points with potential +5 V. In three dimensions all points with the same potential will form a surface. In fact, there will exist an infinite number of surfaces because the 20 V total potential difference can be divided into an infinite number of steps. Each surface with the same value of potential (voltage) is called an **equipotential surface.** In this laboratory we will determine the **equipotentials** associated with a few often used electrode configurations.

In addition to equipotential surfaces in the region around charged electrodes there is also present an electric field. By definition the **electric field** is a vector field that can be represented by lines drawn from the positively charged electrode to the negatively charged electrode. The direction of the electric field line at any point in space is the direction of the force that would be exerted on a positive test charge placed at that point in space. To ensure that the test charge does not disturb the other charges, the test

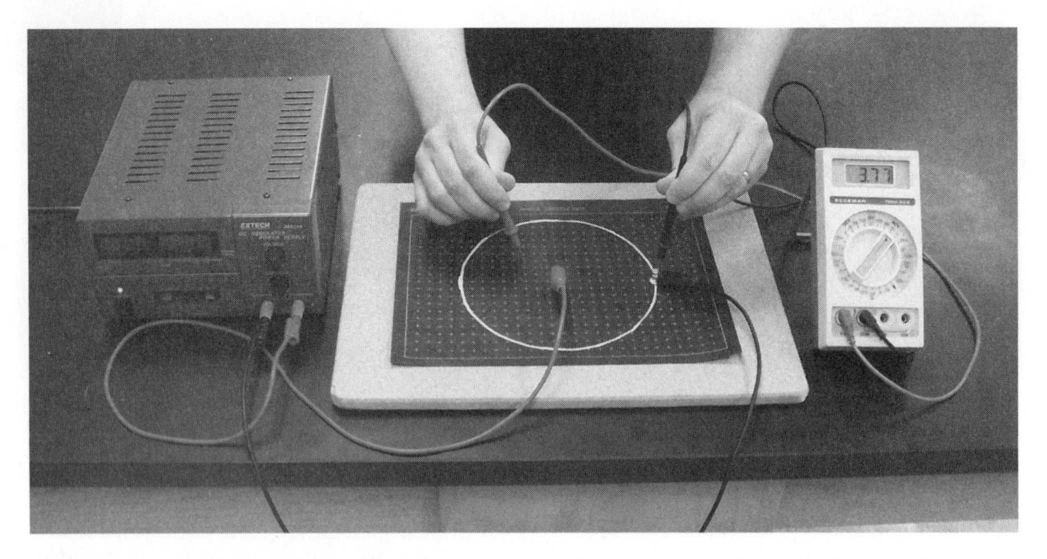

Figure 26-1 Equipotential mapping apparatus shown here using a battery rather than a power supply.

charge must be small. In fact, in the exact definition, the limit must be taken as the test charge approaches zero. The magnitude of the electric field is the force per unit charge on the positive test charge as the magnitude of the test charge approaches zero. The units of electric field are N/C. The number of field lines per unit area at a given point is a measure of the magnitude of the electric field. Thus a region where there are a large number of lines per unit area is a region of large electric field.

The geometrical relationship that electric field lines have with equipotential surfaces is that electric field lines are everywhere perpendicular to the equipotential surfaces. Electrodes themselves are equipotential surfaces, so electric field lines must intersect electrodes perpendicularly. This is a helpful guide to determine the shape of electric fields around an electrode arrangement.

For two points separated by a very small distance Δx with potential difference ΔV between the points, the electric field E is

$$E = -\frac{\Delta V}{\Delta x} \qquad \text{(Eq. 1)}$$

Equation 1 shows that another proper unit for the electric field is V/m. Because the electrodes on the carbon paper are limited to two dimensions, they represent a slice taken through a real three-dimensional electrode configuration. Therefore, the equipotentials mapped in this laboratory will be lines rather than surfaces. Any two-dimensional electrode arrangement can be produced by drawing the desired electrode shape on the carbon paper with a conducting pen and then attaching a direct current power supply between the electrodes. We will choose the electrode to which the negative terminal of the power supply is attached to be the zero of potential, and will take all measurements relative to that electrode. We will use a voltmeter to find the points on the paper in the region of the electrodes that are at some given value of potential. When enough points are located to establish the shape of the equipotential, the equipotential line can be constructed by joining the points with a smooth curve. The student must decide the number of data points needed to establish the shape.

EXPERIMENTAL PROCEDURE

1. Use a conductive ink pen to draw the three electrode configurations shown in Figure 26-2. Note the following cautions: (a) Place the conductive paper on a hard surface, not on the corkboard to draw the electrodes. (b) Make sure that the ink flows smoothly and evenly when drawing electrodes, and that a solid line is obtained.

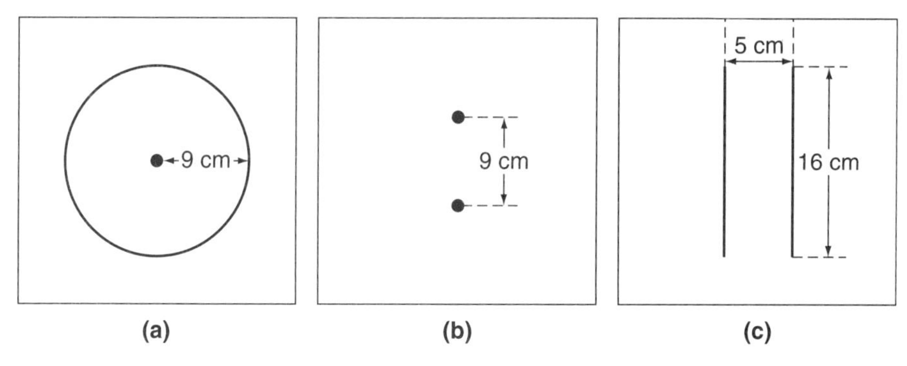

Figure 26-2 Electrode configurations to be mapped.

2. Draw the three electrode configurations pictured in Figure 26-2 and described below. For each configuration use one clean sheet of carbon paper and arrange the electrodes as nearly centered on the paper as possible to avoid edge effects. (a) Line of charge perpendicular to the paper and guard cylinder of radius 9.0 cm—draw a small dot at the center of the paper and then draw a circle of radius 9.0 cm centered on the dot. Be sure the dot is centered on one of the grid markings. (b) Two lines of opposite charge perpendicular to the paper—draw two small dots 9.0 cm apart symmetrically located on the paper. (c) Parallel plate capacitor—draw two straight lines 16.0 cm long and 5.0 cm apart symmetrically located on the paper.
3. For each electrode configuration in turn, place metal push pins in the electrodes and connect the two leads from the power supply to the pins. For the parallel plate and two-line charge configuration there is symmetry, and the assignment of which electrode is negative and which is positive is arbitrary. For the line charge and guard ring, choose the dot as positive and the circle as negative.
4. In each case, set the potential difference between the electrodes to be 20.0 V. Connect the voltmeter between the electrodes with the negative voltmeter lead connected to the negative power supply output, and the positive voltmeter lead connected to the positive power supply output. Once the 20.0 V is set it should remain fixed.
5. For each electrode configuration, make sure that all connections are secure and that the push pins are pressed firmly into the corkboard to ensure good contact with the electrodes. To check that the electrodes themselves have the proper conductivity, connect one lead of the voltmeter to one of the electrode push pins, and then using the other lead of the voltmeter as a probe, touch it to various parts of the same electrode. For all properly conductive electrodes, the maximum voltage between any two points on the same electrode should be less than 0.2 V.
6. Determine the equipotentials by connecting the negative voltmeter lead to the electrode push pin that is connected to the negative output of the power supply. As illustrated in Figure 26-3, the other voltmeter lead then serves as a probe, and it is used to measure the potential at any point on the paper by touching the probe to the paper at that point. Be sure that the probe used has a sharp point, and that the probe

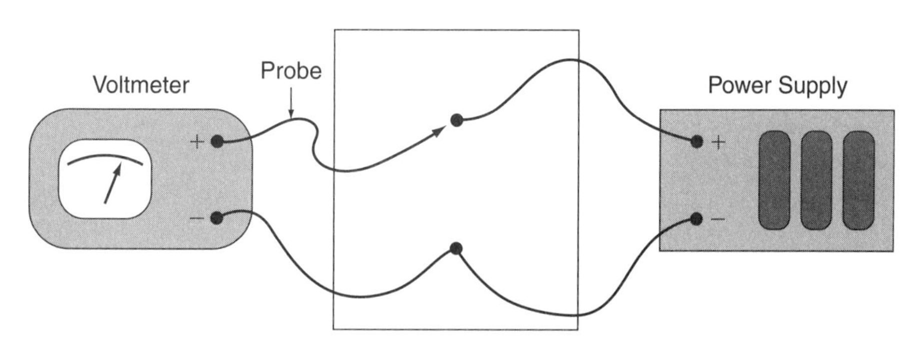

Figure 26-3 Power supply and voltmeter connections for mapping equipotentials.

is held perpendicular to the paper so that only the point of the probe touches the paper. A given equipotential line (for example, the 10.0 V equipotential) is mapped by moving the probe around to find the points at which the voltmeter reading is 10.0 V, and then connecting these points to produce a smooth curve or line. For each equipotential, obtain enough points to clearly define the shape of that equipotential line.

7. Use the procedure in Step 6 to map in pencil the following equipotential lines for each electrode configuration:
   (a) Line charge—15.0, 10.0, 6.50, 4.50, 3.50, 2.50, 1.50, and 0.75 V.
   (b) Two line charges—16.0, 13.0, 11.5, 10.0, 8.50, 7.00, and 4.00 V.
   (c) Parallel plates—4.00, 8.00, 12.0, and 16.0 V.
8. Electric field lines were stated to be everywhere perpendicular to equipotential surfaces. Because our electrodes are confined to the plane of the paper, the equipotentials are lines, but it is true that the electric field lines are everywhere perpendicular to these equipotential lines. For each set of electrodes, draw a set of lines that are perpendicular to the measured equipotential lines. These are the electric field lines. Place arrows on them to indicate direction from positive to negative charge. Distinguish them from the equipotential lines, either by dotting the lines or by drawing them in a different color.
9. Make the following measurements for the electrode configuration of the line of charge. Measure the change in potential at the distances r from the line of charge listed in the Data Table. Tape the two voltmeter probes together with a small piece of insulating material holding the points of the probes apart at a fixed distance, Δx, of about 0.0030 m. Place the probes symmetrically about the grid positions on the paper at the values listed in the Data Table so that the gap between the probes is centered on each value of r in turn. It is critical that the gap be as precisely centered on each r as possible. Record the values of ΔV and the value of Δx in the Data Table.

CALCULATIONS

1. For the measurements made in Step 9 above, calculate the approximate value of E at each point as $\Delta V/\Delta x$. Record these values of E in the Calculations Table.
2. Perform a linear least squares fit to this data with E as the vertical axis and $1/r$ as the horizontal axis. Record the slope, intercept, and correlation coefficient r in the Calculations Table.

GRAPHS

1. Construct accurate drawings on 1 cm by 1 cm graph paper showing the electrodes and your measured equipotentials and electric field lines for each electrode configuration.
2. Make a graph of the data for E versus $1/r$ for the line charge data. Also show on the graph the straight line obtained in the least squares fit.

Name Section Date

26 LABORATORY 26 *Equipotentials and Electric Fields*

PRE-LABORATORY ASSIGNMENT

1. Electric field lines are drawn (a) from positive charges to negative charges (b) from negative charges to positive charges (c) from the largest charge to the smallest charge (d) from the smallest charge to the largest charge.

2. The points where the potential is the same (in three-dimensional space) have the same voltage. (a) True (b) False

3. The points where the potential is the same (in three-dimensional space) lie on a surface. (a) True (b) False

4. The relationship between the direction of the electric field lines and the equipotential surfaces is (a) field lines are everywhere parallel to surfaces (b) field lines always intersect each other (c) field lines are everywhere perpendicular to surfaces (d) field lines always make angles between 0° and 90° with surfaces.

5. Why are the measured equipotentials lines instead of surfaces for this laboratory?

6. If two electrodes have a source of potential difference of 100 V connected to them, how many equipotential surfaces exist in the space between them?

7. Why is it important to center the electrodes on the resistance paper for this laboratory?

8. In the performance of this laboratory, what is the recommended maximum allowed potential difference from one end of an electrode to the other end?

9. On what basis are you to decide how many points to measure for each equipotential for a given electrode configuration?

Name　　　　Section　　　　Date

Lab Partners

26 LABORATORY 26 *Equipotentials and Electric Fields*

LABORATORY REPORT

Data Table

r (m)	ΔV (V)
0.0150	
0.0200	
0.0300	
0.0400	
0.0500	
0.0600	
0.0700	
0.0800	
$\Delta x =$	m

Calculations Table

$E = -\dfrac{\Delta V}{\Delta x}$ (V/m)		$1/r$ (m^{-1})
Slope =	Intercept =	$r =$

SAMPLE CALCULATIONS

1. $E = \Delta V/\Delta x =$

2. $1/r =$

QUESTIONS

1. Although the magnitudes of the equipotentials and electric fields for electrode configurations (a) and (b) are consistent with interpretation of those arrangements as line charges, the shapes of the equipotentials and electric field lines are the same as those for a point charge and a dipole charge. Compare your graphs with those in your textbook for a point charge and a dipole. Comment on their similarities and differences, if any.

2. According to theory, the electric field E for a line of charge should be proportional to $1/r$ where r is the distance from the line of charge. Considering the graph of the data and the value of the correlation coefficient, comment on whether or not your data for the line of charge confirms this dependence.

3. Examine your data for the dipole. Point A in Figure 26-4 is halfway between the positive electrode and the 16.00 V equipotential, point C is halfway between the negative electrode and 4.00 V, and B is halfway between the 8.50 V and 11.50 V equipotentials. Measure the values of Δx indicated below and calculate the $E = \Delta V/\Delta x$ at A, B, and C.

A $\Delta V = 4.00$ Δx = distance from center of + to 16.0 V = __________ E_A = __________

B $\Delta V = 3.00$ Δx = distance from 8.50 to 11.50 equipotentials = __________ E_B = __________

C $\Delta V = 4.00$ Δx = distance center of negative to 4.00 V = __________ E_C = __________

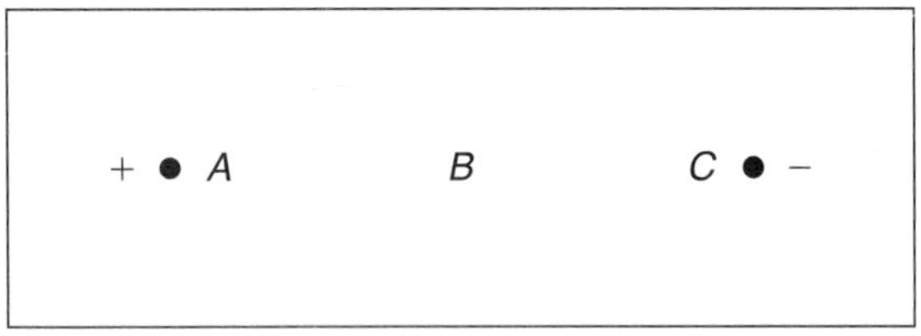

+ ● A   B   C ● −

Figure 26-4 Point A near positive electrode, C near negative electrode, and B at center.

4. Are the results for the E field at these points A, B, and C in the preceding question consistent with what you would expect for the relative values at these points? State your reasoning.

5. According to theory, the E field in the region between the parallel plates should be constant. Calculate the E field from Equation 1 at points A, B, and C defined by Figure 26-5 below using the same process as in Question 3.

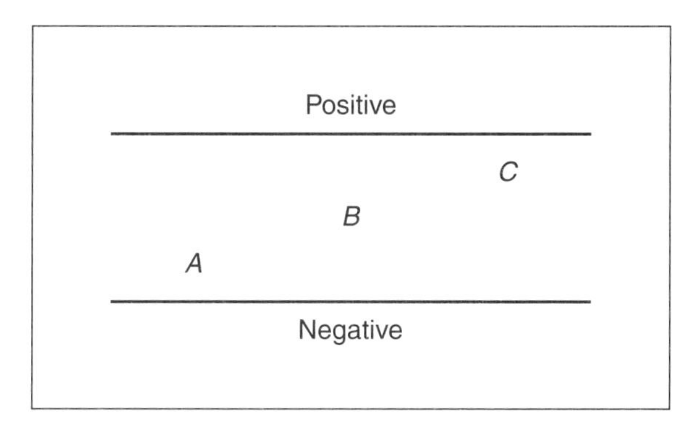

Figure 26-5 Point A near negative electrode displaced from the center, C near the positive electrode displaced in the other direction, and B at center in the middle.

$E_A =$ ________ V/m $E_B =$ ________ V/m $E_C =$ ________ V/m

Are the values for the E field at points A, B, and C consistent with the theory within the experimental uncertainty? State your reasoning.

Capacitance Measurement with a Ballistic Galvanometer

OBJECTIVES

- ❑ Determine the deflection of a ballistic galvanometer for several capacitors of known capacitance charged to a known voltage.
- ❑ Demonstrate that for a fixed voltage the galvanometer deflection is proportional to the capacitance of each capacitor.
- ❑ Determine the capacitance of unknown capacitors from galvanometer deflection.
- ❑ Experimentally determine the capacitance of series and parallel combinations of capacitors and compare the results with theoretical predictions.

EQUIPMENT LIST

- Ballistic galvanometer, support stand, telescope, and scale
- Direct current power supply (0–20 V), direct current voltmeter (0–20 V)
- Five or six known capacitors in range 0.5–3.0 μF (can be in form of decade box)
- Three capacitors to serve as unknowns (1–2 μF), switch (single pole, double throw)

THEORY

Figure 27-1 shows a capacitor of **capacitance** C with a **voltage** V across the plates. The figure illustrates the fact that a capacitor, with a **charge** of Q, has a charge of $+Q$ on one plate and $-Q$ on the other plate. Therefore, the net charge on a capacitor is always zero. The relationship between capacitance C, voltage V, and charge Q is

$$C = \frac{Q}{V} \qquad \text{(Eq. 1)}$$

The units of capacitance (Coulomb/Volt) have been given the name farad with symbol F.

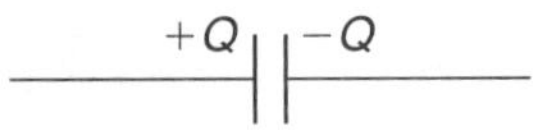

+Q −Q

Figure 27-1 Capacitor with charge Q has $+Q$ on one plate and $-Q$ on the other.

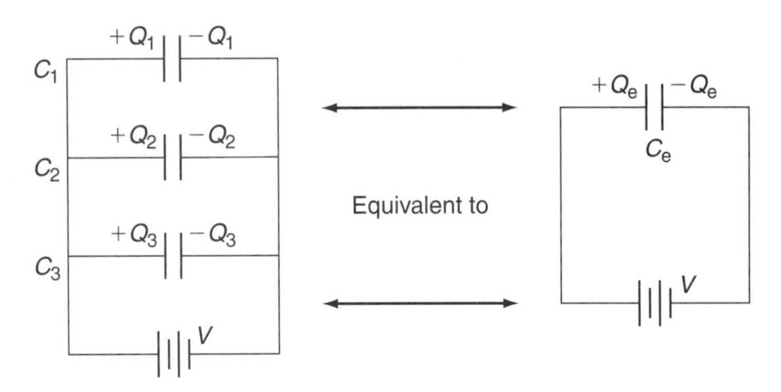

Figure 27-2 Circuit diagram for three capacitors in parallel.

In Figure 27-2 three capacitors C_1, C_2, and C_3 are shown connected in parallel to a battery of voltage V. Let Q_1, Q_2, and Q_3 stand for the charges on the capacitors, and let V_1, V_2, and V_3 stand for the voltage across the capacitors. There exists an equivalent capacitance C_e with voltage V_e and charge Q_e that would have the same effect in a circuit as the three parallel capacitors. For parallel combination of C_1, C_2, and C_3 we can demonstrate that the following relationships hold.

$$V = V_e = V_1 = V_2 = V_3 \qquad \text{(Eq. 2)}$$

$$Q_e = Q_1 + Q_2 + Q_3 \qquad \text{(Eq. 3)}$$

$$C_e = C_1 + C_2 + C_3 \qquad \text{(Eq. 4)}$$

Equation 4 states that for capacitors in parallel the equivalent capacitance is simply the sum of the individual capacitances. The extension of Equation 4 to the case of any number of capacitors in parallel should be clear.

Three capacitors C_1, C_2, and C_3 are shown connected in series in Figure 27-3. Again let Q_1, Q_2, and Q_3 stand for the charges on the three capacitors, and let V_1, V_2, and V_3 stand for the voltage across the three capacitors. For the case of series connection if C_e is the equivalent capacitance, Q_e is its charge, and V_e is its voltage, the following relationships hold.

$$Q_e = Q_1 = Q_2 = Q_3 \qquad \text{(Eq. 5)}$$

$$V = V_e = V_1 + V_2 + V_3 \qquad \text{(Eq. 6)}$$

$$\frac{1}{C_e} = \frac{1}{C_1} + \frac{1}{C_2} + \frac{1}{C_3} \qquad \text{(Eq. 7)}$$

The extension of Equation 7 to the case of any number of capacitors in series should be clear.

A current is a flow of charge Q in some time interval Δt. A **galvanometer** is a device for detecting a current. In most cases galvanometers are designed to respond to the current itself $Q/\Delta t$. A **ballistic galvanometer** is designed to detect the total charge Q that flows through the galvanometer during some very short time interval Δt. A galvanometer acts as a ballistic galvanometer when the time constant of its motion is large compared to the time Δt during which the charge Q flows.

We will perform measurements with a ballistic galvanometer with a total deflection, when a charge Q is passed through the galvanometer, that is proportional to the magnitude of the charge Q. In this laboratory,

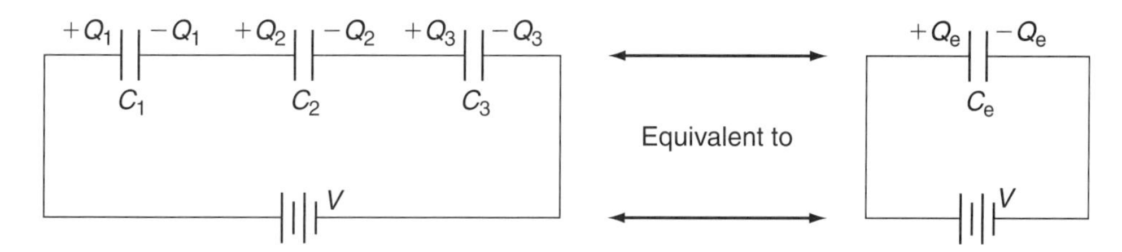

Figure 27-3 Circuit diagram for three capacitors in series.

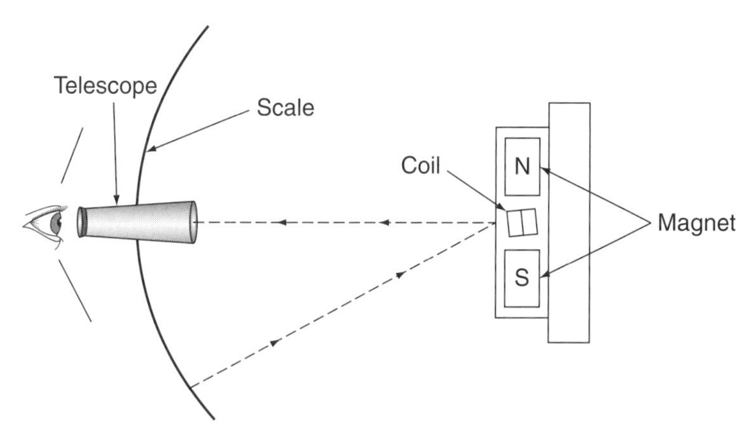

Figure 27-4 Experimental arrangement for the ballistic galvanometer.

several capacitors of known capacitance will be charged to the same voltage. When they are discharged through the ballistic galvanometer, the deflection produced provides a calibration of the galvanometer deflection in terms of the capacitance.

A sketch of the experimental arrangement for a ballistic pendulum is shown in Figure 27-4. The deflection of the coil in the galvanometer is read on a scale in front of the device by a telescope as reflected from a plane mirror on the coil.

EXPERIMENTAL PROCEDURE

Calibration

1. Do not touch the galvanometer until given explicit instructions to do so by the instructor. The gold suspensions in the galvanometers are delicate and expensive. The galvanometer may have already been set up and adjusted by the instructor. When given permission, proceed to the next step.
2. Construct the circuit shown in Figure 27-5 using the known capacitor with the largest value. Adjust the telescope to focus on the scale. You can adjust the angle of the telescope to align with the scale. Note that the scale is labeled red on one side of zero and black on the other side.

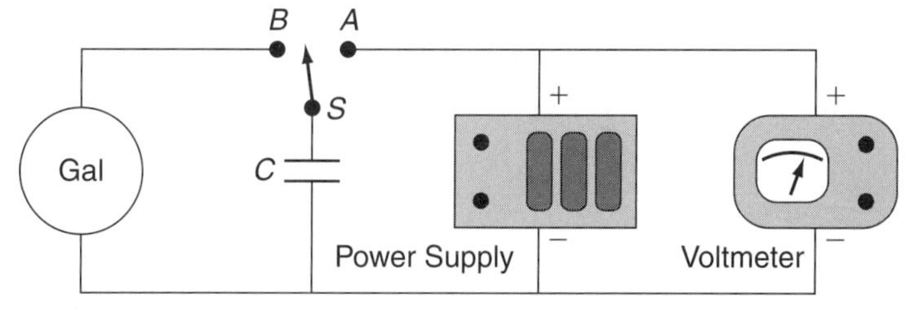

Figure 27-5 Circuit diagram to charge C and then discharge it through galvanometer.

3. Using the largest known capacitor, charge the capacitor to some known voltage by throwing switch S to position A. Discharge the capacitor through the galvanometer by throwing the switch to position B. Adjust the voltage applied to the capacitor until it produces approximately 90% deflection for this capacitor. Record that value of voltage as V in Data and Calculations Tables 1 and 2. Make all measurements with this voltage.
4. Measure the deflection for each of the known capacitors when they have been charged with voltage V and then discharged through the galvanometer. Perform three trials for each capacitor. Record the values of the deflection D and the known value of each capacitor in Data and Calculations Table 1.

Unknown Capacitance

1. Label the three unknown capacitors C_1, C_2, and C_3. For each of the unknown capacitors, determine the deflection of the galvanometer when they have been charged to voltage V and then discharged through the galvanometer. Perform three trials for each capacitor. Record the value of the deflection D for each trial.
2. Connect capacitors C_1 and C_2 in parallel and place them in the circuit in the position of C in Figure 27-5. Determine the deflection when the combination is charged with voltage V and then discharged through the galvanometer. Record the deflections in Data and Calculations Table 2. Repeat with capacitors C_2 and C_3 in parallel.
3. Connect capacitors C_1 and C_2 in series and place them in the circuit in the position of capacitor C in Figure 27-5. Determine the deflection when the combination is charged with voltage V and then discharged through the galvanometer. Record the deflections in Data and Calculations Table 2. Repeat with capacitors C_2 and C_3 in series.

CALCULATIONS

Calibration

1. Calculate the mean $\overline{D}$ and standard error α_D for the three trials for each capacitor and record them in Data and Calculations Table 1.
2. Perform a linear least squares fit to the data with the C as the vertical axis and $\overline{D}$ as the horizontal axis. Record the slope K, the intercept I, and the correlation coefficient r in Data and Calculations Table 1.

Unknown Capacitance

1. Calculate the mean $\overline{D}$ and standard error α_D for each of the three trials. Record them in Data and Calculations Table 2.
2. For the individual capacitors, the series combinations, and the parallel combinations, calculate the experimental values for the capacitance from $C_{exp} = K\overline{D} + I$ where K is the slope and I is the intercept of the least squares fit. The slope K has units of μF/deflection, and the intercept has units of μF. Record these values of the experimental capacitance in Data and Calculations Table 2.
3. Calculate a theoretical value for the series and parallel combinations of the capacitors using Equations 4 and 7. In the calculations, use the measured values for the individual capacitors. Record these theoretical values for the series and parallel combinations in Data and Calculations Table 2 in the appropriate places.
4. Calculate the percentage difference between the experimental and theoretical values for the series and parallel combinations and record them in Data and Calculations Table 2.

GRAPHS

1. Make a graph with the capacitance C as the vertical axis and the deflection $\overline{D}$ as the horizontal axis. Also show on the graph the straight line obtained by the linear least squares fit to the data.

Name | Section | Date

LABORATORY 27 *Capacitance Measurement with a Ballistic Galvanometer*

PRE-LABORATORY ASSIGNMENT

1. What is the definition of capacitance?

2. What are the units of capacitance?

3. A capacitor is said to have a charge of 10 μC. What is the charge on the positively charged plate? What is the charge on the negatively charged plate?

4. A 1.50 μF capacitor has a voltage across the plates of 6.00 V. What is the charge on the capacitor? Show your work.

5. Which of the following are true for three capacitors C_1, C_2, and C_3 in parallel with a battery of voltage V? More than one answer may be true.

   (a) $V = V_1 = V_2 = V_3$ **(b)** $Q_e = Q_1 = Q_2 = Q_3$ **(c)** $Q_e = Q_1 + Q_2 + Q_3$ **(d)** $V = V_1 + V_2 + V_3$

6. Which of the following are true for three capacitors C_1, C_2, and C_3 in series with a battery of voltage V? More than one answer may be true.

   (a) $V = V_1 = V_2 = V_3$ **(b)** $Q_e = Q_1 = Q_2 = Q_3$ **(c)** $Q_e = Q_1 + Q_2 + Q_3$ **(d)** $V = V_1 + V_2 + V_3$

7. Three capacitors of capacitance 5.00 μF, 8.00 μF, and 11.00 μF are connected in parallel. What is the equivalent capacitance of the combination? Show your work.

8. Three capacitors of capacitance 5.00 μF, 8.00 μF, and 11.00 μF are connected in series. What is the equivalent capacitance of the combination? Show your work.

Name | Section | Date

Lab Partners

27 LABORATORY 27 *Capacitance Measurement with a Ballistic Galvanometer*

LABORATORY REPORT

Data and Calculations Table 1

C (μF)	D_1	D_2	D_3	$\overline{D}$	α_D
$V =$ volts	$K =$ μF/defl	$I =$ μF	$r =$		

Data and Calculations Table 2

Capacitor	D_1	D_2	D_3	$\overline{D}$	α_D	C_{exp}	C_{theo}	% Diff
C1								
C2								
C3								
C_1 & C_2 Parallel								
C_2 & C_3 Parallel								
C_1 & C_2 Series								
C_2 & C_3 Series								
$V =$ volts								

SAMPLE CALCULATIONS

1. $C_{exp} = K\overline{D} + I =$
2. (Parallel) $C_{theo} = C_1 + C_2 =$
3. (Series) $(1/C_{theo}) = (1/C_1) + (1/C_2) =$
4. % Diff =
5. % Error (if actual values for unknown are given) =

QUESTIONS

1. What is the precision of your measurements of the unknown capacitors? State clearly the basis for your answer.

2. If possible, obtain values for the unknown capacitors. Determine the accuracy of your measurements of the capacitance if these values are available.

3. Based on the precision and accuracy of your results, is there any evidence for a systematic error in the measurements? State clearly the basis for your answer.

4. How well do your data confirm the theoretical equations for the parallel combination of capacitors? Answer the question as quantitatively as possible.

5. How well do your data confirm the theoretical equation for the series combination of capacitors? Answer the question as quantitatively as possible.

6. Suppose that you were given an unknown capacitor with a value of about twice as large as the largest known capacitor that you measured. What simple change in the procedure would allow you to determine an approximate value for its capacitance using the value for K that you have already determined?

Measurement of Electrical Resistance and Ohm's Law

OBJECTIVES

- ❑ Define the concept of electrical resistance using measurements of the voltage across and current in a wire coil.
- ❑ Investigate the dependence of the resistance on the length, cross-sectional area, and resistivity of the wire.
- ❑ Investigate the equivalent resistance of series and parallel resistors.

EQUIPMENT LIST

- Resistance coils (standard set available from Sargent-Welch or Central Scientific consisting of 10 m and 20 m length of copper and German silver wire)
- Direct current ammeter (0–2 A), direct current voltmeter (0–30 V, preferably digital readout)
- Direct current power supply (0–20 V at 1 A)

THEORY

If a **voltage** V is applied across an element in an electrical circuit, the **current** I in the element is determined by a quantity known as the **resistance** R. The relationship between these three quantities serves as a definition of resistance.

$$R = \frac{V}{I} \qquad \text{(Eq. 1)}$$

The units of resistance are volt/ampere, which are given the name ohm. The symbol for ohm is Ω. Some circuit elements obey a relationship known as **Ohm's Law.** For these elements the quantity R is a constant for different values of V. If a circuit element obeys Ohm's Law, when the voltage V is varied the current I will also vary, but the ratio V/I should remain constant. In this laboratory we will perform measurements on five coils of wire to investigate if they obey Ohm's Law. We also will determine the resistance of the coils.

The resistance of any object to electrical current is a function of the material from which it is constructed, the length, the cross-sectional area, and the temperature of the object. At constant temperature the resistance R is given by

R_1 R_2 R_3

Figure 28-1 Resistors in series.

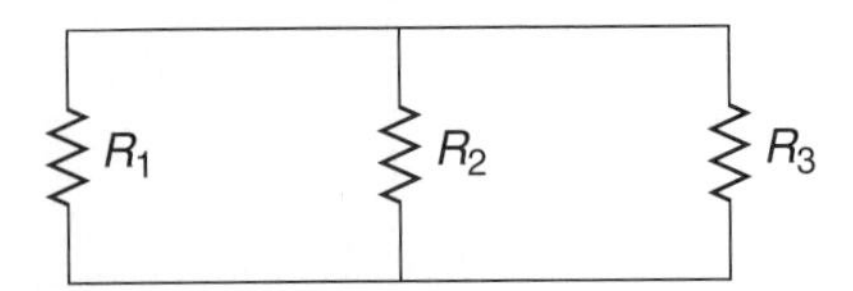

Figure 28-2 Resistors in parallel.

$$R = \rho \frac{L}{A} \qquad \text{(Eq. 2)}$$

where R is the resistance (Ω), L is the length (m), A is the cross-sectional area (m^2), and ρ is a constant dependent upon the material called the **resistivity** (Ω–m). Actually ρ is a function of temperature, and if the temperature of the coils of wire rises as a result of the current in them, this may be a source of error.

Circuit elements in an electrical circuit can be connected in series or parallel. Three resistors (R_1, R_2, and R_3) are connected in series as shown in Figure 28-1. For resistors in series the current is the same for all the resistors, but the voltage drop across each resistor is different. For resistors in series the equivalent resistance R_e of the three resistors is given by

$$R_e = R_1 + R_2 + R_3 \qquad \text{(Eq. 3)}$$

The same three resistors are shown connected in parallel in Figure 28-2. For resistors in parallel the current is different in each resistor, but the voltage across each resistor is the same. In this case the equivalent resistance R_e of the three resistors in terms of the individual resistors is given by

$$\frac{1}{R_e} = \frac{1}{R_1} + \frac{1}{R_2} + \frac{1}{R_3} \qquad \text{(Eq. 4)}$$

One of the objectives of this laboratory will be to observe the behavior of resistors in series and parallel.

EXPERIMENTAL PROCEDURE

1. Connect the ammeter A, the voltmeter V, and the power supply PS to the first resistor as shown in Figure 28-3. The basic circuit is the power supply in series with a resistor. To measure the current in the resistor, the ammeter is placed in series. To measure the voltage across the resistor, the voltmeter is placed in parallel.
2. Vary the current through resistor R_1 in steps of 0.250 A up to 1.000 A. For each specified value of the current, measure the voltage across the resistor and record the values in Data Table 1. The resistors will heat up and may be damaged by allowing current in them for long periods of time. Measurements should be made quickly at each value of the current. *APPLY VOLTAGE ONLY WHEN DATA ARE BEING TAKEN.*
3. Repeat Step 2 for each of the five resistors. For each resistor the ammeter must be in series with that resistor and the power supply, and the voltmeter must be in parallel with the resistor. Record all values in Data Table 1.
4. Connect the first four resistors in series to measure the equivalent resistance of the combination. Use two values of current, 0.500 A and 1.000 A, and measure the value of the voltage for each of these values of current. Record the voltage in Data Table 2.

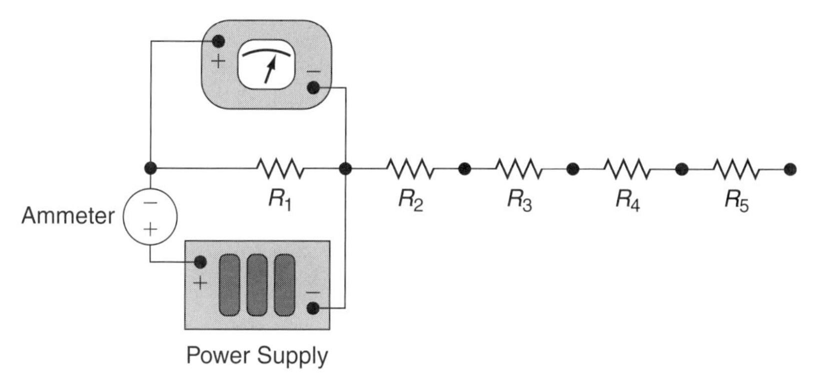

Figure 28-3 Measurement of current and voltage for resistor R_1.

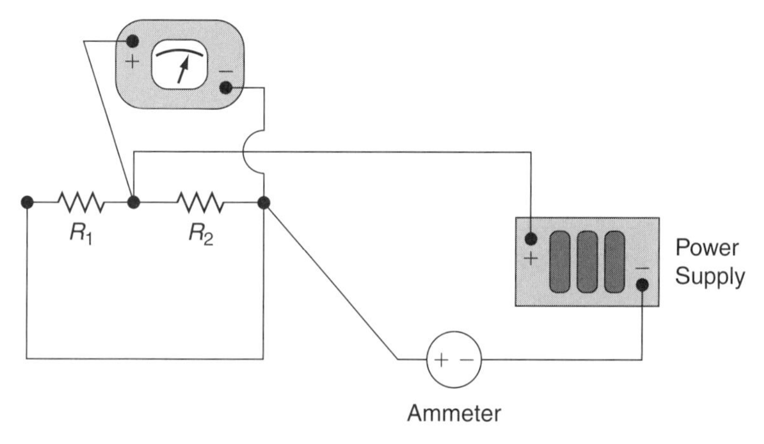

Figure 28-4 Resistors R_1 and R_2 in parallel.

5. Measure the voltage across the combination of R_2, R_3, and R_4 in series for currents of 0.500 A and 1.000 A and record the values in Data Table 2.
6. Connect R_1 and R_2 in parallel as shown in Figure 28-4 and measure the voltage across the combination for current values of 0.500 A and 1.000 A and record in Data Table 2.
7. Connect R_1 and R_3 in parallel as shown in Figure 28-5 and measure the voltage for current values of 0.500 A and 1.000 A and record in Data Table 2.
8. Connect R_2 and R_3 in parallel and perform the same measurements as described in Steps 6 and 7. Record the results in Data Table 2.

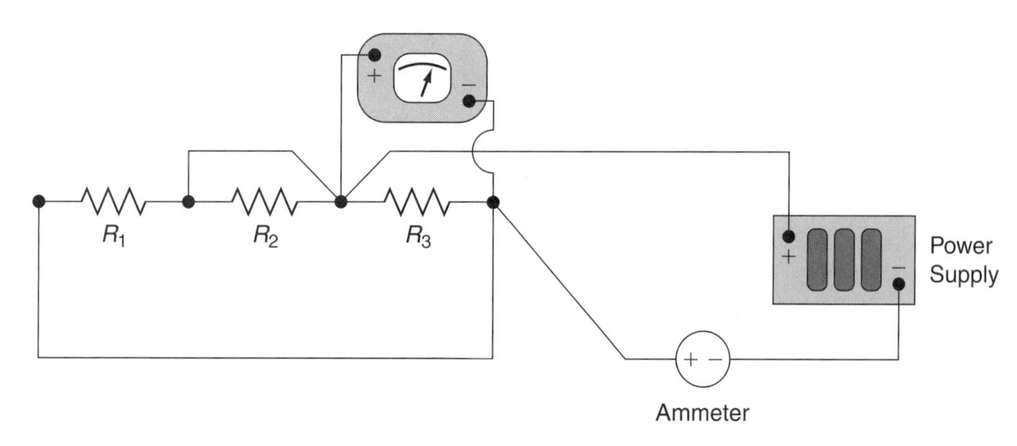

Figure 28-5 Resistors R_1 and R_3 in parallel.

CALCULATIONS

1. The first four coils are made of copper with resistivity of $\rho = 1.72 \times 10^{-8}$ Ω–m. The fifth coil is made of an alloy called German silver with resistivity of $\rho = 28.0 \times 10^{-8}$ Ω–m. The first, second, and fifth coils are 10.0 m long, and the third and fourth coils are 20.0 m long. The diameters of the first, third, and fifth coils are 0.0006439 m, and the diameters of the second and fourth coils are 0.0003211 m. Use these values in Equation 2 to calculate the value of the resistance for each of the five coils and record the results in Calculations Table 1 as the theoretical values for the resistance R_{theo}.
2. If Equation 1 is solved for V, the result is $V = IR$. There is a linear relationship between the voltage and the current, and the slope of V versus I will be the resistance R. Perform a linear least squares fit to the data in Data Table 1 with V as the vertical axis and I as the horizontal axis. Record in Calculations Table 1 the slope of the fit for each resistor as the experimental value for the resistance R_{exp}. Also record the value of the correlation coefficient r for each of the fits.
3. Calculate the percentage error in the values of R_{exp} compared to the values of R_{theo} for the five resistors and record the results in Calculations Table 1.
4. For the data of Data Table 2 calculate the values of the equivalent resistance for the various series and parallel combinations listed in the table as the value of the measured voltage divided by the appropriate current. Calculate and record the mean of the two trials as $(\overline{R_e})_{exp}$ in Calculations Table 2.
5. Equations 3 and 4 give the theoretical expressions for equivalent resistance for series and parallel combinations of resistance. Calculate a theoretical value for the equivalent resistance for each series and parallel combination measured in Data Table 2. For the values of the individual resistances R_1, R_2, and R_3 in Equation 3 and 4, use the experimental values determined from the fit to the data on the individual resistors. Record this theoretical value for the equivalent resistance in each case as $(R_e)_{theo}$ in Calculations Table 2.
6. Calculate the percentage difference between the values of $(\overline{R_e})_{exp}$ and $(R_e)_{theo}$ for each of the series and parallel combinations measured and record the results in Calculations Table 2.

GRAPHS

1. Construct graphs of the data in Data Table 1 with V as the vertical axis and I as the horizontal axis. Use only one piece of graph paper for all five resistors, making five small graphs on that one sheet. Choose different scales for each graph if needed, but make the five graphs as large as possible while still fitting on one page. Also show on each small graph the straight line for the linear least squares fit.

Name ______ Section ______ Date ______

28 LABORATORY 28 *Measurement of Electrical Resistance and Ohm's Law*

PRE-LABORATORY ASSIGNMENT

1. Three resistors R_1, R_2, and R_3 are connected in series with $R_1 < R_2 < R_3$. Choose all correct answers below. The total resistance of the combination is (a) less than R_1, (b) less than R_3, (c) greater than R_3, (d) greater than 3 R_3, (e) equal to $R_1 + R_2 + R_3$.

2. Two resistors R_1 and R_2 are connected in parallel with $R_1 < R_2$. Choose all correct answers below. The total resistance of the combination is (a) less than R_1, (b) less than R_2, (c) greater than R_2, (d) greater than 2 R_2, (e) equal to $(R_1R_2)/(R_1 + R_2)$.

3. A wire of length L_1 and diameter d_1 has resistance R_1. A second wire of the same material has length $L_2 = 2\ L_1$ and diameter $d_2 = 2\ d_1$. The resistance of wire two is R_2. Choose the correct value for R_2. (a) $R_2 = R_1$, (b) $R_2 = 2R_1$, (c) $R_2 = \frac{1}{2}R_1$, (d) $R_2 = 4R_1$.

4. If a circuit element carries a current of 3.71 A, and the voltage drop across the element is 8.69 V, what is the resistance of the circuit element? Show your work.

   $R =$ ______ Ω

5. A resistor is known to obey Ohm's Law. When there is a current of 1.72 A in the resistor, it has a voltage drop across its terminals of 7.35 V. If a voltage of 12.0 V is applied across the resistor, what is the current in the resistor? Show your work.

   $I =$ ______ A

6. The resistivity of copper is 1.72×10^{-8} Ω–m. A copper wire is 15.0 m long, and the wire diameter is 0.0500 cm. What is the resistance of the wire? Show your work.

$R =$ ____________________ Ω

7. A wire of cross-sectional area 5.00×10^{-6} m^2 has a resistance of 1.75 Ω. What is the resistance of a wire of the same material and length as the first wire, but with a cross-sectional area of 8.75×10^{-6} m^2? Show your work.

$R =$ ____________________ Ω

8. Three resistors of resistance 20.0 Ω, 30.0 Ω, and 40.0 Ω are connected in series. What is their equivalent resistance? Show your work.

$R =$ ____________________ Ω

9. Three resistors of resistance 15.0 Ω, 25.0 Ω, and 35.0 Ω are connected in parallel. What is their equivalent resistance? Show your work.

$R =$ ____________________ Ω

Name | Section | Date

Lab Partners

28 LABORATORY 28 Measurement of Electrical Resistance and Ohm's Law

LABORATORY REPORT

Data Table 1

I (A)	V_{R1} (V)	V_{R2} (V)	V_{R3} (V)	V_{R4} (V)	V_{R5} (V)
0.250					
0.500					
0.750					
1.000					

Calculations Table 1

	R_1	R_2	R_3	R_4	R_5
R_{theo}					
R_{exp}					
r					
% Error R_{exp}					

Data Table 2

Combination	I (A)	V (V)	I (A)	V (V)
R_1 R_2 R_3 R_4 Series	0.500		1.000	
R_2 R_3 R_4 Series	0.500		1.000	
R_1 R_2 Parallel	0.500		1.000	
R_1 R_3 Parallel	0.500		1.000	
R_2 R_3 Parallel	0.500		1.000	

Calculations Table 2

Combination	$(R_e)_{exp}{}^1$ (Ω)	$(R_e)_{exp}{}^2$ (Ω)	$(\overline{R_e})_{exp}$ (Ω)	$(R_e)_{theo}$ (Ω)	*% Diff*
R_1 R_2 R_3 R_4 Series					
R_2 R_3 R_4 Series					
R_1 R_2 Parallel					
R_1 R_3 Parallel					
R_2 R_3 Parallel					

SAMPLE CALCULATIONS

1. $R_{theo} = \rho L/A =$
2. $(R_e)_{exp} = V/I =$
3. % Error =
4. (Series) $(R_e)_{theo} = R_1 + R_2 + R_3 + R_4 =$
5. (Parallel) $(R_e)_{theo} = (1)/(1/R_1 + 1/R_2) =$
6. % Difference =

QUESTIONS

1. Do the individual resistors you have measured obey Ohm's Law? In answering this question, consider the least squares fits and the graphs you have made for each resistor. Remember that linear behavior of V versus I is the proof of ohmic behavior.

2. Evaluate the agreement between the theoretical values for the individual resistances and the experimental values.

3. Does your agreement between the experimental and theoretical values of the series combinations of resistors support Equation 3 as the model for series combination of resistors? The agreement is not expected to be perfect, but you are to determine if the agreement is reasonable within the expected experimental uncertainty.

4. Evaluate the agreement between the experimental and theoretical values of the parallel combinations of resistors. Do the results support Equation 4 as the model for the parallel combination of resistors within the expected experimental uncertainty?

5. The first and second coils have the same length, and the third and fourth coils have the same length. They differ only in the cross-sectional area. According to theory, what should be the ratio of the resistance of the second coil to the first and the fourth coil to the third? Calculate these ratios for your experimental results and compare the agreement with the expected ratio.

6. The first and third coils have the same cross-sectional area, and the second and fourth coils have the same cross-sectional area. They differ only in length. According to theory, what should be the ratio of the resistance of the third coil to the first and the fourth coil to the second? Calculate these ratios for your experimental results and compare the agreement with the expected ratio.

Wheatstone Bridge

OBJECTIVES

- ❑ Investigate the principles of operation of a slide-wire form Wheatstone bridge and determine the resistance of several unknown resistors.
- ❑ Demonstrate the standard color code used to specify the value of commercially available resistors.

EQUIPMENT LIST

- Slide-wire Wheatstone bridge, decade resistance box (1000 Ω), galvanometer
- Switch with 10 kΩ resistor in parallel, direct current power supply
- Five color-coded resistors (in range 100 to 1000 Ω) to serve as unknowns
- Assortment of circuit wires, multimeter with resistance scale

THEORY

Consider a circuit that contains three resistors with values that are both known and adjustable, an unknown resistance, a power supply, and a **galvanometer** connected as shown in Figure 29-1. Current I from the power supply divides at junction J. The current in R_1 is I_1, and the current in R_3 is I_2, where $I = I_1 + I_2$. By experimentally varying the values of the resistances, a condition (known as the balance condition) can be achieved where there is no current in the galvanometer G. This circuit is known as a **Wheatstone bridge.**

When there is no current in the galvanometer, the current in R_1 must go through R_2, and both resistors have current of I_1. Similarly, when the balance condition holds, the current through R_4 must be the same as the current through R_3, and thus it is equal to I_2.

When there is no current in the galvanometer, there is no potential difference between points A and B, so these points are at the same potential. The change in potential from point J to point B (V_{JB}) is equal to the potential change from point J to point A (V_{JA}), and

$$V_{JA} = I_1R_1 = V_{JB} = I_2R_3 \quad \text{and thus} \quad I_1R_1 = I_2R_3 \qquad \text{(Eq. 1)}$$

Similarly, the potential change across R_2 (V_{AK}) is the same as the potential change across R_4 (V_{BK}) so that

$$V_{AK} = I_1R_2 = V_{BK} = I_2R_4 \quad \text{and thus} \quad I_1R_2 = I_2R_4 \qquad \text{(Eq. 2)}$$

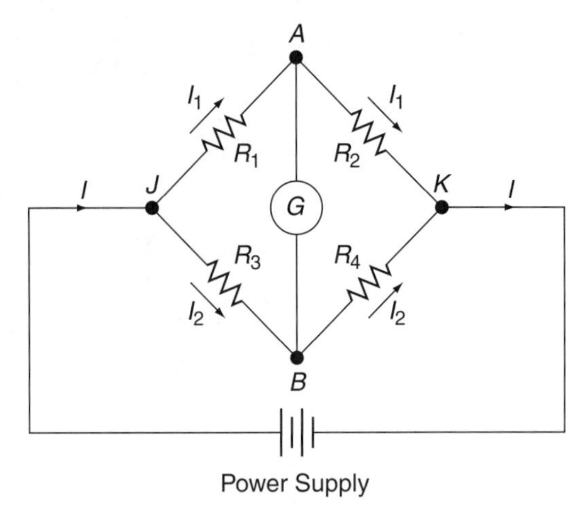

Figure 29-1 Wheatstone bridge circuit.

If Equation 1 is divided by Equation 2, the currents cancel, and it follows that

$$\frac{R_1}{R_2} = \frac{R_3}{R_4} \qquad \text{(Eq. 3)}$$

When a balance condition has been experimentally achieved, Equation 3 determines the value of an unknown resistance if three of the four values of resistance are known.

In this laboratory we will use a slide-wire form of the Wheatstone bridge shown in Figure 29-2. In that bridge, resistances R_3 and R_4 are replaced by a uniform wire between the points J and K, which has a sliding contact key at point B. Because the wire is of uniform cross-section, the resistance of the two portions of wire JB and BK are proportional to their lengths. The ratio of their lengths, JB/BK, is equal to the ratio of their resistances R_3/R_4. If R_1 is an unknown resistance R_U, and R_2 is a known variable resistor R_K, Equation 3 becomes

$$\frac{R_U}{R_K} = \frac{JB}{BK} \quad \text{or solving for } R_U, \quad R_U = R_K\frac{JB}{BK} \qquad \text{(Eq. 4)}$$

The 10 kΩ resistor and switch S in series with the galvanometer are designed to protect the galvanometer. Be sure that you are aware of their proper function as described in the procedure before completing the connection of the power supply to the circuit.

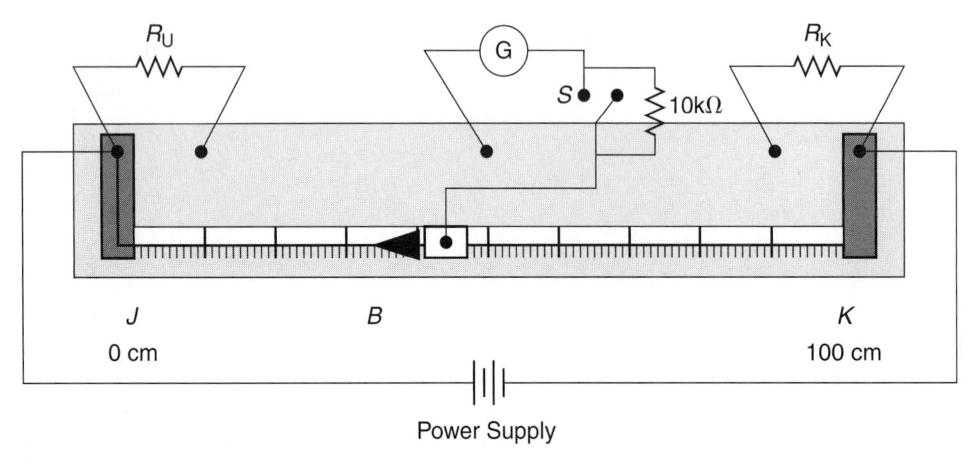

Figure 29-2 Slide-wire form of the Wheatstone bridge.

Table 29-1 *Resistor Color Code.*

First Two Bands		*Third Band*		*Fourth Band*	
Color	*Digits*	*Color*	*Exponent*	*Color*	*Precision*
Black	0	Black	0	Colorless	20%
Brown	1	Brown	1	Silver	10%
Red	2	Red	2	Gold	5%
Orange	3	Orange	3		
Yellow	4	Yellow	4		
Green	5	Green	5		
Blue	6	Blue	6		
Violet	7	Violet	7		
Gray	8	Silver	−2		
White	9	Gold	−1		

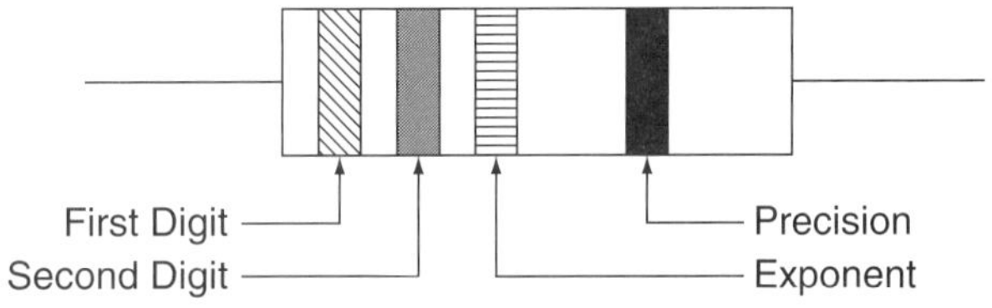

First Digit
Second Digit
Precision
Exponent

For resistors routinely used in electronic instrumentation, resistance is coded by a series of colored bands on the resistor. The key to the resistor color coding system is given in Table 29-1 above. The four bands are placed with three equally spaced bands close to one end of the resistor followed by a space, and then a fourth band. The first two bands are the first two digits in the value of the resistor, and the third band gives the exponent of the power of 10 to be multiplied by the first two digits. Thus a resistor with its first three bands labeled Yellow-Violet-Red has a value of $47 \times 10^2\,\Omega$, or $4700\,\Omega$.

EXPERIMENTAL PROCEDURE

1. Use the resistor code table to read the nominal values of the five unknown resistors and record them in Data Table 1. Record the smallest value as Unknown #1, and then the remaining ones in increasing order.
2. Adjust the power supply voltage to 1.50 V. Leave the power supply fixed at this value for all the measurements. All measurements should be made with this same voltage, which has been chosen so that the currents in all resistors of the circuit will be small. This ensures that there is no heating of the resistors. Any significant heating of the resistors could cause differential increases in resistance and would lead to errors.
3. Place the first unknown resistor in the Wheatstone bridge circuit in the position of R_U in the circuit of Figure 29-2. Place the resistance box in the position of R_K in Figure 29-2 and choose a value for R_K approximately equal to the nominal value that you read from the resistor code for this unknown resistor. Record the value of R_K in Data Table 1. (The value of R_K should be given to one place beyond the last digit of the resistance box because the uncertainty is not in the last digit set on the box, but in the digit beyond that. For example, if a resistance is set on the resistance box as $153\,\Omega$, it should be recorded as $153.0\,\Omega$ because there is clearly not one unit of uncertainty in the 3.)

4. The 10 kΩ resistor and switch S in series with the galvanometer are designed to protect the galvanometer from excessive current. Be sure that each attempt to find a balance condition starts with switch S open. This places the resistor in series with the galvanometer and limits the current.
5. With switch S open, move the sliding contact at B until a balance is achieved—i.e., zero current in the galvanometer. This is a rough balance.
6. With the system at rough balance, close switch S to achieve maximum sensitivity at the final balance condition. Balance is the point where there is no deflection of the meter, which may not be at zero of the meter if the galvanometer has a zero offset. Record in Data Table 1 the values of JB and BK, the length of the two sections of wire at balance. The Wheatstone bridge has 1 mm as the smallest marked division. Therefore, measurements of JB and BK should be made to the nearest 0.1 mm.
7. Using the same unknown resistor, repeat Steps 3 through 6 above with two other values of R_K, one value approximately 10% greater than the original R_K, and one value approximately 10% less than the original R_K.
8. Repeat Steps 3 through 7 for each of the other four unknown resistors.
9. Use the resistance scale on a multimeter to measure the value of each of the five unknown resistors and record those values in Data Table 2.

CALCULATIONS

1. Calculate and record the three measured values for each of the five unknown resistors in the Calculations Table.
2. Calculate and record the mean and standard error for the three measurements of each of the five resistors.

Name | Section | Date

LABORATORY 29 *Wheatstone Bridge*

PRE-LABORATORY ASSIGNMENT

1. When the Wheatstone bridge in Figure 29-1 is balanced, which of the following statements are true? (*More than one may be true.*) (a) There is no current in the unknown resistor. (b) There is no current in the galvanometer. (c) There is no voltage drop across the galvanometer. (d) The currents in R_1 and R_3 are the same. (e) The currents in R_1 and R_2 are the same. (f) The voltage across R_1 and R_3 is the same. (g) The voltage across R_1 and R_2 is the same.

2. The slide-wire form of the Wheatstone bridge makes use of the fact that (a) the resistance of a wire is equal to its length (b) the resistance of a wire is proportional to its cross-sectional area (c) the resistance of a wire is proportional to its length (d) the resistance of a wire is inversely proportional to its length.

3. A slide-wire Wheatstone bridge is used in the configuration shown in Figure 29-2 with the known resistor positioned as shown and equal to 10.00 Ω. The balance point is found to be at the position of 35.7 cm for the scale as shown in the figure. What is the value of the unknown resistance R_U? Show your work.

   $R_U =$ ____________ Ω

4. Does the measurement of resistance using the Wheatstone bridge depend upon the value of the power supply voltage used? If it does, explain why, and if it does not, then explain why not.

5. What is the purpose of the 10 kΩ resistor in parallel with the switch that is in series with the galvanometer in Figure 29-2?

6. A resistor is coded as shown below. What is the nominal value of the resistor, and what is the precision of this value?

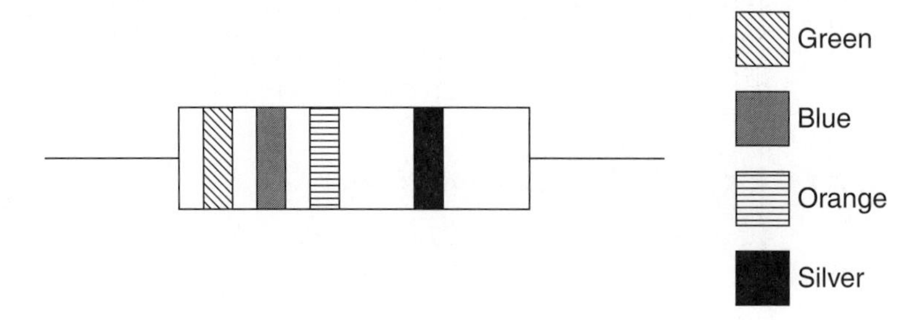

Name | Section | Date

Lab Partners

29 LABORATORY 29 Wheatstone Bridge

LABORATORY REPORT

Data Table 1

	R_K (Ω)	JB (cm)	BK (cm)
Unknown #1			
Coded Value			
(Ω)			
Unknown #2			
Coded Value			
(Ω)			
Unknown #3			
Coded Value			
(Ω)			
Unknown #4			
Coded Value			
(Ω)			
Unknown #5			
Coded Value			
(Ω)			

Calculations Table 1

R_U (Ω)	$\overline{R}_U$ (Ω)	α_{RU} (Ω)

Data Table 2

Unknown #	1	2	3	4	5
Resistance (Ω)					

SAMPLE CALCULATIONS

1. $BK = 100.00 - JB =$

2. $R_U = R_K \dfrac{JB}{BK} =$

QUESTIONS

1. Assume there is an uncertainty of ± 0.03 cm in locating the position B of the contact on the slide wire. What percentage error does this introduce in the determination of JB/BK and thus the measured value of resistance when B is at 50 cm? What percentage error results when B is at 10 cm?

2. Keeping in mind your answers to Question 1, why were the values of the known resistor chosen to be about equal to and ±10% on either side of the coded value for the unknown?

3. Compare each of your Wheatstone bridge values of the five unknown resistors with its coded value by calculating the percentage error assuming the coded values as correct. State the precision of the resistors based on the resistor color coding.

#1—Error = ____________% #2—Error = ____________% #3—Error = ____________%

#4—Error = ____________% #5—Error = ____________% Precision = ____________%

Are the percentage errors of the measurements within the precision of the resistors?

4. Compare your Wheatstone bridge values of the five unknown resistors with the values in Data Table 2 determined by using the resistance meter. Calculate the percentage errors for each resistor assuming Data Table 2 as correct. For each resistor, is the agreement better or worse than the agreement with the coded values?

#1—Error = ____________% #2—Error = ____________% #3—Error = ____________%

#4—Error = ____________% #5—Error = ____________%

5. Express the values of the standard error for each of the five unknown resistors as a percentage of the mean value.

#1 % std err = ____________% #2 % std err = ____________% #3 % std err = ____________%

#4 % std err = ____________% #5 % std err = ____________%

6. Considering the stated precision of the coded values and the various comparisons that have been done above, state whether or not your Wheatstone bridge measurements of these resistors represent more reliable values for the actual values of these resistors than the coded values. Be very specific about the facts upon which your opinion is based.

Bridge Measurement of Capacitance

OBJECTIVES

- ❑ Determine the capacitance of several unknown capacitors by establishing balance in a bridge circuit.
- ❑ Experimentally determine the capacitance of series and parallel combinations of capacitors and compare the results with theoretical predictions.

EQUIPMENT LIST

- Capacitor of accurately known value (in the range 0.5–1 μF)
- Three capacitors of unknown value (in the range 0.2–3 μF)
- Two decade resistance boxes (1000 Ω and 10,000 Ω)
- Sine wave generator (variable frequency, 5 V peak to peak amplitude)
- Alternating current voltmeter (digital readout, capable of measuring high frequency)

THEORY

A **capacitor** is a circuit element that consists of two conducting surfaces separated by an insulating material called a dielectric. When an alternating current of frequency f exists in a capacitor, the measure of opposition to that current is a quantity called the **capacitive reactance** X_C. The potential difference across the capacitor is given by IX_C where I is the current in the capacitor. Thus X_C plays a role for alternating current that is analogous to the role played by resistors for direct current. This quantity depends on the angular frequency $\omega = 2\pi f$ and on the capacitance C and is given by $X_C = 1/\omega C$. In this laboratory we will construct an **alternating current bridge** circuit with two capacitors and two resistors. When the bridge is balanced, the ratio of the value of the two capacitive reactances is equal to the ratio of the value of the two resistors. If the value of the ratio of the resistors and the value of one capacitor is known, the capacitance of the other capacitor can be determined.

Consider the circuit shown in Figure 30-1, which is a form of alternating current bridge. A capacitor of unknown value C_U and a capacitor of known value C_K are in the circuit along with resistors R_1 and R_2. A sine wave generator is applied between points A and D in the circuit. If an alternating current voltmeter between points B and C reads zero, the bridge is balanced and $V_{BC} = 0$. If V_{BC} is zero, then points B and C are at the same potential, and that means that the potential difference V_{AB} between points A and B is the

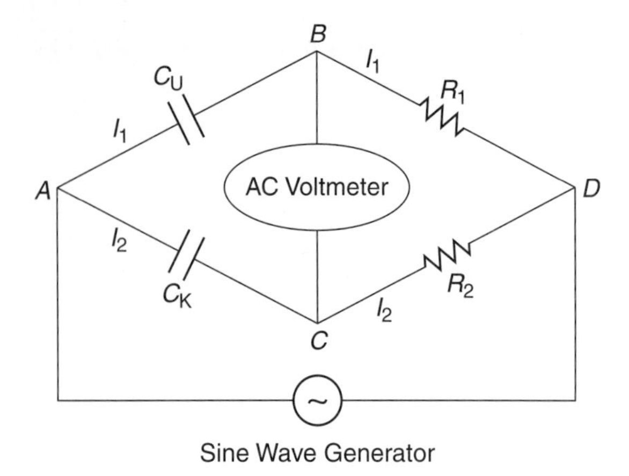

Figure 30-1 Alternating current bridge with two capacitors and two resistors.

same as V_{AC} between points A and C. Writing the potential differences in terms of the currents and the capacitive reactances gives

$$V_{AB} = I_1\left(\frac{1}{\omega C_U}\right) = V_{AC} = I_2\left(\frac{1}{\omega C_K}\right) \qquad \text{(Eq. 1)}$$

With the balance condition satisfied, the current in the resistor R_1 is I_1, the same as that in C_U. It is also true that the current in the resistor R_2 is I_2, the same as that in C_K, and that the potential differences V_{BD} and V_{CD} are equal. Writing these statements in terms of the currents and the resistances gives

$$V_{BD} = I_1R_1 = V_{CD} = I_2R_2 \qquad \text{(Eq. 2)}$$

If Equations 1 and 2 are each solved for the ratio of the currents I_1/I_2, the result is

$$\frac{I_1}{I_2} = \frac{\omega C_U}{\omega C_K} = \frac{R_2}{R_1} \qquad \text{(Eq. 3)}$$

Solving Equation 3 for the unknown capacitance C_U gives

$$C_U = C_K\frac{R_2}{R_1} \qquad \text{(Eq. 4)}$$

Equation 4 shows that if the value of C_K is known, a capacitor of unknown capacitance C_U can be determined by experimentally varying the ratio of the resistances until the ratio R_2/R_1 is found for which the bridge is in balance. Note that Equation 4 is independent of the frequency f, and that the currents I_1 and I_2 do not need to be determined.

Figure 30-2 shows two capacitors of capacitance C_1 and C_2 in parallel. They are equivalent to a single capacitance C_e. The expression for this equivalent capacitance C_e in terms of the capacitances C_1 and C_2 is

$$C_e = C_1 + C_2 \qquad \text{(Eq. 5)}$$

Figure 30-3 shows two capacitors of capacitance C_1 and C_2 in series. They are equivalent to a single capacitance C_e given by the expression

$$\frac{1}{C_e} = \frac{1}{C_1} + \frac{1}{C_2} \qquad \text{(Eq. 6)}$$

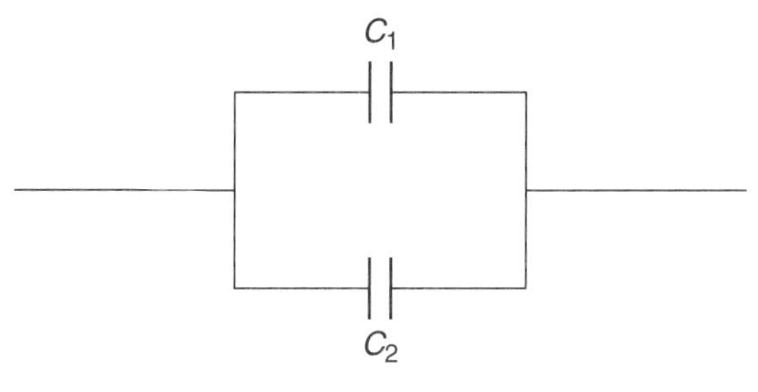

Figure 30-2 Two capacitors in parallel.

Figure 30-3 Two capacitors in series.

In this laboratory we will determine the capacitance of several capacitors with unknown capacitance by placing them in a bridge of the type shown in Figure 30-1. In addition, series and parallel combinations of two capacitors at a time will be determined and compared with the theoretical predictions of Equations 5 and 6.

EXPERIMENTAL PROCEDURE

1. Label the unknown capacitors as C_1, C_2, and C_3. Record the value of the known capacitor as C_K in the Data Table.
2. Construct the circuit shown in Figure 30-1 with capacitor C_1 in the position of C_U and the known capacitor in the position of C_K. Place the 1000 Ω resistance box set to value of 1000 Ω in the position of R_1. Place the 10,000 Ω resistance box in the position of R_2. Record the value of R_1 in the Data Table.
3. Plug in the sine wave generator and place the output of the generator between points *A* and *D*. Turn it up to maximum amplitude and set the frequency to 1000 Hz.
4. Place the voltmeter between points *B* and *C* in the circuit. Adjust the value of R_2 until the minimum voltage is read between points *B* and *C*. Record the value of R_2 that produces the minimum in the Data Table.
5. Repeat Steps 2 through 4 for the other two unknown capacitors C_2 and C_3.
6. Repeat Steps 2 through 4 for the following parallel combinations of capacitors placed in the position of C_U in the circuit in Figure 30-1: (a) C_1 and C_2 in parallel; (b) C_2 and C_3 in parallel.
7. Repeat Steps 2 through 4 for the following series combinations of capacitors placed in the position of C_U in the circuit in Figure 30-1: (a) C_1 and C_2 in series; (b) C_2 and C_3 in series.

CALCULATIONS

1. Use Equation 4 to calculate the experimental values for the unknown capacitors and for the series and parallel combinations. Record the results as C_{exp} in the Calculations Table.
2. Use Equations 5 and 6 to calculate the theoretical equivalent capacitance for the series and parallel combinations corresponding to the measured series and parallel combinations. Use the experimentally determined values for the capacitance of capacitors C_1, C_2, and C_3 in the calculations. Record those results as C_{theo} in the Calculations Table.
3. Calculate the percentage error of the experimental values of the series and parallel combinations compared to the theoretical values for those quantities. Record the results in the Calculations Table.

Name Section Date

30 LABORATORY 30 *Bridge Measurement of Capacitance*

PRE-LABORATORY ASSIGNMENT

1. Define capacitive reactance.

2. Describe the role that capacitive reactance of a capacitor plays when there is an alternating current in the capacitor.

Consider the circuit diagram of Figure 30-1 for Questions 3–7. Mark as true or false the statements concerning the conditions that hold when the bridge is balanced.

3. $V_{AB} = V_{BD}$ and $V_{AC} = V_{CD}$ (a) true (b) false

4. $V_{BC} = 0$ (a) true (b) false

5. $V_{AB} = V_{CD}$ and $V_{AC} = V_{BD}$ (a) true (b) false

6. $V_{AB} = V_{AC}$ and $V_{BD} = V_{CD}$ (a) true (b) false

7. The balance condition for the bridge is different for different frequencies. (a) true (b) false

8. A bridge like the one shown in Figure 30-1 is balanced. The known capacitor C_K has a capacitance of 1.17 μF. The value of R_1 is 1000 Ω, and the value of R_2 is 525 Ω. What is the value of the unknown capacitor C_U? Show your work.

9. A 3.75 μF capacitor is in series with a 6.85 μF capacitor. What is the equivalent capacitance C_e of the combination? Show your work.

10. A 5.75 μF capacitor is in parallel with a 3.82 μF capacitor. What is the equivalent capacitance C_e of the combination? Show your work.

Name | Section | Date

Lab Partners

30

LABORATORY 30 *Bridge Measurement of Capacitance*

LABORATORY REPORT

Data and Calculations Table

Capacitor	R_2 (Ω)	C_{exp} (μF)	C_{theo} (μF)	% *Error*
C_1				
C_2				
C_3				
C_1 & C_2 in parallel				
C_2 & C_3 in parallel				
C_1 & C_2 in series				
C_2 & C_3 in series				
$C_K =$ μF		$R_1 =$ Ω		

SAMPLE CALCULATIONS

1. $C_{exp} = C_K \dfrac{R_2}{R_1}$
2. (Parallel) $C_{theo} = C_1 + C_2 =$
3. (Series) $C_{theo} = (1)/(1/C_1 + 1/C_2)$
4. % Error =

QUESTIONS

1. To what extent do your data confirm the theoretical equation for the parallel combination of capacitors? State your answer as quantitatively as possible.

2. To what extent do your data confirm the theoretical equation for the series combination of capacitors? State your answer as quantitatively as possible.

3. The laboratory procedure instructs you to use the maximum amplitude of the sine wave generator. How would it affect the results if one-half the amplitude were used instead?

4. Suppose the bridge is rearranged into the form shown in Figure 30-4. What is the balance condition for this bridge?

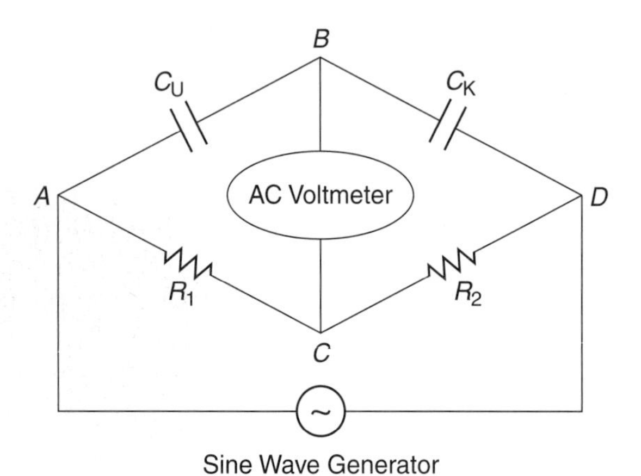

Figure 30-4 Different arrangement of the alternating current bridge.

5. Your lab instructor may have accurate values for your unknown capacitors or perhaps can make available to you a capacitance meter that will allow you to obtain a value for them. If you can obtain a reliable value for the capacitance of the unknowns, calculate the percentage error in your measurements.

Voltmeters and Ammeters

OBJECTIVES

- ❏ Determine the internal resistance R_g and current sensitivity K of a galvanometer.
- ❏ Construct a voltmeter and an ammeter by placing the appropriate values of resistance in series and parallel with the galvanometer.
- ❏ Compare the accuracy of the constructed voltmeter and ammeter with a standard voltmeter and a standard ammeter.

EQUIPMENT LIST

- Power supply (0–20 V direct current), galvanometer (D'Arsonal type, zero centered)
- Resistance box (variable in steps of 10 Ω between 2500 Ω and 3500 Ω)
- A resistor of about 330 Ω (either a 1% resistor or provide a resistance meter)
- Digital voltmeter (0–20 V direct current), digital ammeter (0–1.00 A direct current)
- Spool of #28 copper wire (one for the class), assorted leads

THEORY

Galvanometer Characteristics

The **D'Arsonal galvanometers** used in this laboratory are based upon the fact that a wire coil in the presence of a magnetic field experiences a torque when there is a current in the coil. This torque is exerted against a spring, and the deflection of a pointer attached to the coil is proportional to the current in the galvanometer.

Because the coil has a fixed resistance R_g, the deflection of the pointer will also be proportional to the voltage across the terminals of the galvanometer. Therefore, a galvanometer can be calibrated to serve as either a **voltmeter** or an **ammeter.**

A galvanometer is characterized by its resistance R_g and a constant K called the current sensitivity. K is the amount of current needed to deflect the galvanometer one scale division. It is expressed in units of A/div. Both R_g and K are taken to be unknown for the galvanometers used in the laboratory. We will determine them by a series of measurements described in a later section.

Conversion of the Galvanometer into a Voltmeter

The galvanometer deflects full scale for a value of current given by $I_g = KN$ where N is the number of scale divisions. The voltage V_g across the galvanometer terminals that produces a full-scale deflection is given by $V_g = I_g R_g = KN R_g$. If it is desired to measure a larger voltage than V_g, it is necessary to place a resistor R_V in series with the galvanometer so that most of the voltage is across R_V and the rest across the galvanometer. Figure 31-1 illustrates this idea.

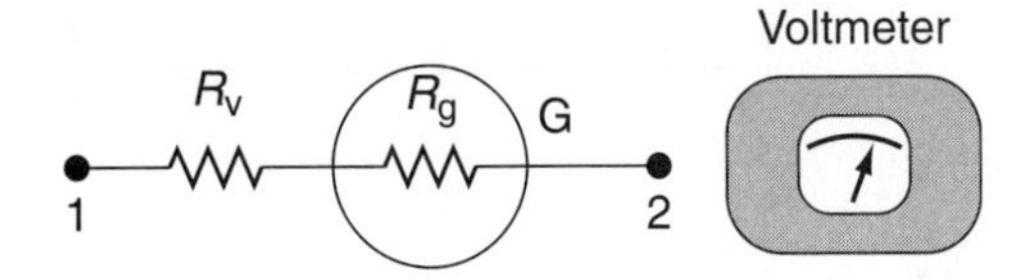

Figure 31-1 Combination of galvanometer and series resistor forms a voltmeter.

If a voltage of V_{FS} between terminals 1 and 2 in Figure 31-1 results in a current in the galvanometer equal to KN, then the series combination of R_V and the galvanometer acts as a voltmeter of full-scale voltage V_{FS}. In equation form this is

$$I = KN = \frac{V_{FS}}{R_V + R_g} \qquad \text{(Eq. 1)}$$

Solving Equation 1 for R_V leads to the following expression:

$$R_V = \frac{V_{FS}}{KN} - R_g \qquad \text{(Eq. 2)}$$

Equation 2 can be used to solve for the value of R_V needed to turn a galvanometer of given K, N, and R_g into a voltmeter of full-scale voltage V_{FS}.

Because a voltmeter must be connected into a circuit in parallel, it will alter the original circuit as little as possible when it has as high a resistance as possible. The ideal voltmeter, therefore, has infinite resistance.

Conversion of the Galvanometer into an Ammeter

The galvanometer deflects full scale when the current is $I_g = KN$. If it is desired to measure a larger current than I_g, it is necessary to place a small shunt resistance R_A in parallel with the galvanometer to divert part of the current away from the galvanometer as shown in Figure 31-2. The current I comes in at terminal 1 and divides at the junction. The current in the galvanometer is I_g, and I_A is the current in the shunt resistor where $I = I_g + I_A$. Because R_g and R_A are in parallel, they have the same voltage across them or in equation form, $I_g\, R_g = I_A\, R_A$. Combining the two previous equations and assuming $I = I_{FS}$ when $I_g = KN$ gives

$$R_A = \frac{KNR_g}{I_{FS} - KN} \qquad \text{(Eq. 3)}$$

Equation 3 can be used to calculate the value of the resistor R_A needed to cause the parallel combination shown in Figure 31-2 to be an ammeter with a full-scale current of I_{FS}.

Because an ammeter must be connected into a circuit in series, it will alter the original circuit as little as possible when it has as low a resistance as possible. The ideal ammeter therefore has zero resistance.

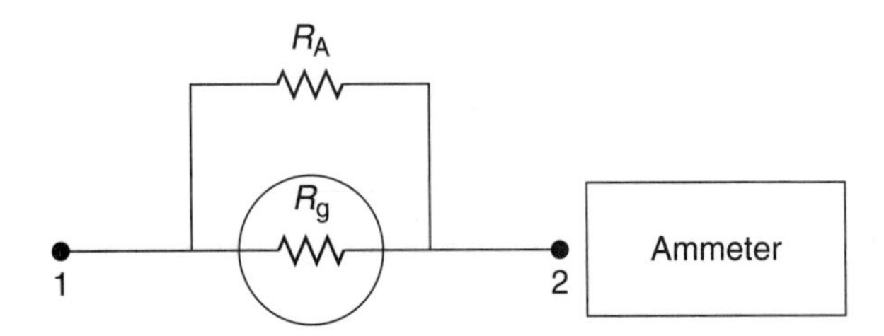

Figure 31-2 Galvanometer and shunt resistor in parallel form an ammeter.

EXPERIMENTAL PROCEDURE

Determine R_g and K

1. Connect the galvanometer, power supply, and decade resistance box in series, and then connect the voltmeter in parallel with the power supply as shown in Figure 31-3. Set the resistance box R_1 to a value of 2500 Ω and adjust the power supply voltage carefully until the galvanometer deflects full scale. Record the voltmeter reading as V and the number of large divisions into which the scale is divided as N in Data and Calculations Table 1.
2. A resistor in parallel with a device is called a shunt resistor because it diverts part of the current that was originally going through the device. Use a composition resistor with a value of approximately 330 Ω as a shunt resistor. Use the ohmmeter to measure an accurate value for the shunt resistor R_s and record it in Data and Calculations Table 1. With the power supply voltage set exactly as above, connect R_s in parallel with the galvanometer (Figure 31-4). The deflection of the galvanometer will now be less than full scale.
3. Leave the power supply voltage set and adjust the value of the resistance box to a somewhat lower value needed to cause the galvanometer to again deflect full scale. Make small adjustments and watch carefully so that the galvanometer does not deflect beyond full scale. A large abrupt decrease in the value of the resistance box could divert enough current through the galvanometer to damage it. Record the value of the resistance box setting that gives a full-scale deflection as R_2 in Data and Calculations Table 1.

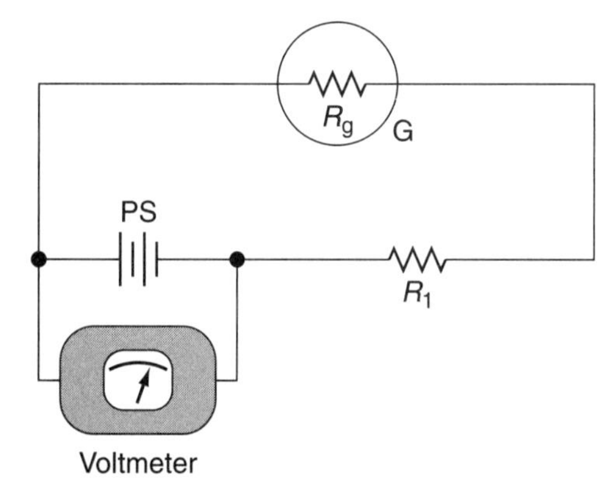

Figure 31-3 Original circuit.

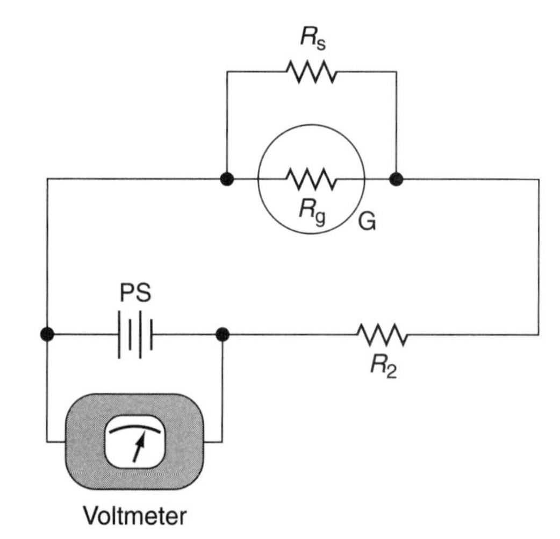

Figure 31-4 Original circuit plus shunt.

4. Turn the power supply to zero and remove the shunt resistor. The circuit is again like Figure 31-3; but now select a value of 3000 Ω for R_1, the resistance box, and repeat the procedure described in Step 1. Record the value of V needed to produce a full-scale deflection for this resistance in Data and Calculations Table 1.
5. Repeat Steps 2 and 3 above, inserting the same shunt resistor R_s. Determine the value of R_2 needed to produce a full-scale deflection with the shunt resistor in place, and record it in Data and Calculations Table 1.
6. Turn the power supply to zero and remove the shunt resistor. Set the resistance box to a value of $R_1 = 3500\,\Omega$ and repeat Steps 1, 2, and 3, recording the values of V and R_2 in Data and Calculations Table 1.

Galvanometer into a Voltmeter

1. Complete the Calculations section of this laboratory to determine the values of K and R_g.
2. Calculate the value of R_V needed to turn your galvanometer into a voltmeter that reads full-scale deflection for 5.00 V ($V_{FS} = 5.00$ V). In this calculation use the mean values of K and R_g from Data and Calculations Table 1. Record this value of R_V in Data and Calculations Table 2.
3. Connect one side of the galvanometer to one side of the resistance box set to the value of R_V. Between the other terminals of the galvanometer and the resistance box is what we will call the **experimental voltmeter** that reads 5.00 V full scale.
4. Compare the experimental voltmeter with the standard voltmeter by connecting them in parallel across the output of the power supply as shown in Figure 31-5. Turn the power supply up slowly until the experimental voltmeter reads exactly 1.00 V and record the value read by the standard voltmeter at this point. Make this same comparison at 2.00, 3.00, 4.00, and 5.00 V as read on the experimental voltmeter and record all results in Data and Calculations Table 2.
5. Following the steps in 1 through 3, calculate the value of R_V needed to make a voltmeter of 10.0 V full-scale deflection. Construct such a voltmeter and compare it to the standard voltmeter at 2.00, 4.00, 6.00, 8.00, and 10.00 V. Record all the results in Data and Calculations Table 2.
6. Following the same procedure, construct a voltmeter that reads 15.0 V full scale and compare it with the standard voltmeter at 3.00, 6.00, 9.00, 12.0, and 15.0 V. Record the results in Data and Calculations Table 2.
7. Calculate the percentage error of the experimental voltmeter readings compared to the standard voltmeter and record the results in Data and Calculations Table 2.

Galvanometer into an Ammeter

1. Calculate the value of R_A needed to turn your galvanometer into an ammeter that reads 1.00 A full scale. For values of R_g and K use the mean value in Data and Calculations Table 1. Record the value of R_A in Data and Calculations Table 3.
2. Number 28 copper wire has a resistance of 0.00213 Ω/cm. Calculate the length of #28 copper wire needed to have a resistance equal to R_A. Record that value in Data and Calculations Table 3.
3. Cut a piece of #28 copper wire a few centimeters longer than the length calculated in Step 2. If the wire being used has an insulating coating, cut away a few centimeters on each end of the wire. Attach the

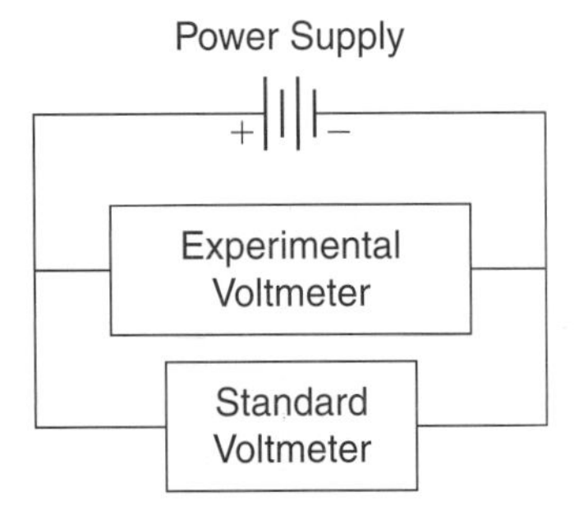

Figure 31-5 Experimental and standard voltmeter in parallel with power supply.

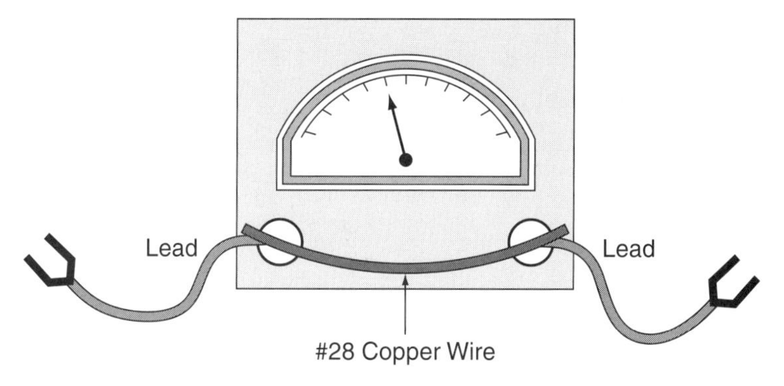

Figure 31-6 Galvanometer with #28 copper wire shunt resistor.

wire between the posts of the galvanometer so that the length of wire between where one end touches one post and the other end touches the other post is equal to the length calculated in Step 2. At the same time that the wire is attached between the posts, attach a short lead to each galvanometer post as shown in Figure 31-6. The two loose ends of the two leads are now an ammeter that reads 1.00 A full scale. Refer to it as the experimental ammeter.

4. After making sure that the power supply is turned completely to zero, place the experimental ammeter in series with a standard ammeter and with the power supply. Very slowly turn the supply up until the experimental ammeter reads 0.200 A. Record the reading of the standard ammeter in Data and Calculations Table 3. Continue this process, comparing the experimental ammeter to the standard ammeter at 0.400, 0.600, 0.800, and 1.000 A.
5. Calculate the percentage errors of the experimental ammeter readings compared to the standard ammeter readings and record the results in Data and Calculations Table 3.

CALCULATIONS

R_g and K

1. By applying Ohm's Law to the circuits in Figure 31-3 and Figure 31-4 when the applied voltage V is the same and the galvanometer is at full-scale deflection in both cases, it can be shown that the resistance of the galvanometer R_g is given by

$$R_g = \frac{R_s}{R_2}(R_1 - R_2) \qquad \text{(Eq. 4)}$$

Calculate the three values of R_g determined by the three trials in Data and Calculations Table 1. Also calculate the mean $\overline{R}_g$ and standard error α_{Rg} for these measurements. Record all calculated values in Data and Calculations Table 1.

2. The constant K is defined as the current needed to produce a deflection of one scale division, and the deflection in the above procedure was N scale divisions. Ohm's Law applied to the circuit of Figure 31-3 leads to the following:

$$K = \frac{V}{N(R_1 + R_g)} \qquad \text{(Eq. 5)}$$

Determine K the galvanometer current sensitivity from the values of V and R_1 in Data and Calculations Table 1 and the calculated values of R_g in Data and Calculations Table 1. For each calculation of K, use the value of R_g determined from the V and R_1 being used to calculate K. Also calculate the mean $\overline{K}$ and standard error α_K for the three values of K. Record all calculated quantities in Data and Calculations Table 1.

Name Section Date

31 LABORATORY 31 *Voltmeters and Ammeters*

PRE-LABORATORY ASSIGNMENT

1. Describe the principle on which the operation of a D'Arsonal type galvanometer is based.

2. A galvanometer has (a) a meter deflection proportional to the current in the galvanometer (b) a meter deflection proportional to the voltage across the galvanometer (c) a fixed resistance (d) all of the above are true.

3. The galvanometer constant K is (a) the current for full-scale deflection (units A) (b) the current for deflection of one scale division (units A/div) (c) the total current times the number of scale divisions (units A/div) (d) the reciprocal of the galvanometer resistance R_g (units 1/A).

4. In the procedure for determination of R_g and K when the shunt resistor R_S is placed in parallel with the galvanometer, what happens to galvanometer deflection, and why does it happen?

5. To construct a voltmeter of a given full-scale deflection from a galvanometer, the appropriate resistance must be placed in (a) series (b) parallel with the galvanometer.

6. To construct an ammeter of a given full-scale deflection from a galvanometer, the appropriate resistance must be placed in (a) series (b) parallel with the galvanometer.

7. A galvanometer has $R_g = 150\,\Omega$ and $K = 0.750 \times 10^{-4}$ A/div. The galvanometer has five divisions for a full-scale reading (i.e., $N = 5$). What value of resistance is needed, and how must it be connected to the galvanometer to form a full-scale voltmeter of 20.0 V? Show your work.

8. To form the galvanometer of Question 7 into an ammeter of 2.50 A full scale, what value of resistance is needed, and how must it be connected? Show your work.

9. To measure voltage, a voltmeter is placed in a circuit in (a) series (b) parallel. The resistance of an ideal voltmeter is ________________.

10. To measure the current, an ammeter is placed in a circuit in (a) series (b) parallel. The resistance of an ideal ammeter is ________________.

Name   *Section*   *Date*

Lab Partners

31 LABORATORY 31 *Voltmeters and Ammeters*

LABORATORY REPORT

Data and Calculations Table 1

R_1 (Ω)	V (V)	R_2 (Ω)	R_g (Ω)	K (A/div)	$\overline{R}_g$ (Ω)	α_{Rg} (Ω)	$\overline{K}$ (A/div)	α_K (A/div)
$N=$					$R_s=$ Ω			

Data and Calculations Table 2

$V_{FS}=5.00$ V $R_V=$ Ω	Experimental	1.00 V	2.00 V	3.00 V	4.00 V	5.00 V
	Standard	V	V	V	V	V
	% Error					
$V_{FS}=10.00$ V $R_V=$ Ω	Experimental	2.00 V	4.00 V	6.00 V	8.00 V	10.0 V
	Standard	V	V	V	V	V
	% Error					
$V_{FS}=15.00$ V $R_V=$ Ω	Experimental	3.00 V	6.00 V	9.00 V	12.0 V	15.0 V
	Standard	V	V	V	V	V
	% Error					

Data and Calculations Table 3

$I_{FS} = 1.00$ A		$R_A =$ Ω		*Length* $R_A =$ cm	
Experimental	0.200 A	0.400 A	0.600 A	0.800 A	1.000 A
Standard	A	A	A	A	A
% Error					

SAMPLE CALCULATIONS

1. $R_g = (R_s/R_2)(R_1 - R_2) =$
2. $K = (V)/(N(R_1 + R_g)) =$
3. $R_V = (V_{FS}/KN) - R_g =$
4. $R_A = (KN\ R_g)/(I_{FS} - KN) =$
5. Length of wire =
6. % Error =

QUESTIONS

1. Considering the standard error of your measurements, comment on the precision of your measurements of R_g and K. Express the standard error as a percentage of the mean.

2. Consider the 5 V experimental voltmeter. Does it tend to read too high or too low compared to the standard voltmeter?

3. Presumably a different value of R_V would give better agreement between the 5 V experimental voltmeter and the standard voltmeter. Would R_V need to be a larger or smaller resistance? Explain your answer.

4. Does the experimental ammeter tend to read too high or too low?

5. Presumably by a change in the value of R_A, the experimental voltmeter could be made to show better agreement with the standard ammeter. Does R_A need to be a larger or smaller resistance to accomplish this? Explain your answer.

EXTRA CREDIT QUESTION. It is stated in the laboratory instructions that the same voltage V is applied to both circuits in Figure 31-3 and Figure 31-4, and that the galvanometer deflects full scale in both cases. Thus the galvanometer current is $I_g = KN$ for both circuits. In Figure 31-4 let the current in R_s be called I_s. From the application of Ohm's Law to these circuits, derive the expression given as Equation 1 for R_g.

Potentiometer and Voltmeter Measurements of the emf of a Dry Cell

OBJECTIVES

- ❑ Investigate the principles of operation of a potentiometer.
- ❑ Compare the emf of dry cells determined by a potentiometer with the emf measured by a voltmeter.
- ❑ Determine the internal resistance of a dry cell.

EQUIPMENT LIST

- Slide-wire potentiometer, galvanometer, standard cell
- Single-pole (single-throw) switch with a 10 kΩ resistor in parallel
- Single-pole (double-throw) switch, dry cells (with emfs in range 1.5 to 1.3 V)
- Direct current power supply (voltage and current determined by form of potentiometer)
- Voltmeter (resistance at least 10 MΩ), student-type voltmeter (resistance ≈ 3 kΩ)
- Decade resistance box (100 kΩ maximum), assorted connecting leads, tap key

THEORY

Part 1. Dry Cell Characteristics

A new **dry cell** produces an **electromotive force (emf)** between 1.50 and 1.60 V with an internal resistance in the range of a fraction to a few ohms. The symbol for emf is ε, and the symbol for internal resistance is R_i. The voltage that appears across the dry cell terminals when the dry cell is providing current I to a circuit is called the **terminal voltage** V. It is given by

$$V = \varepsilon - IR_i \qquad \text{(Eq. 1)}$$

When current flows, the terminal voltage is always less than the emf of a cell by the amount of the voltage drop across R_i. If no current is provided by the cell, its terminal voltage equals its emf ($V = \varepsilon$ if $I = 0$).

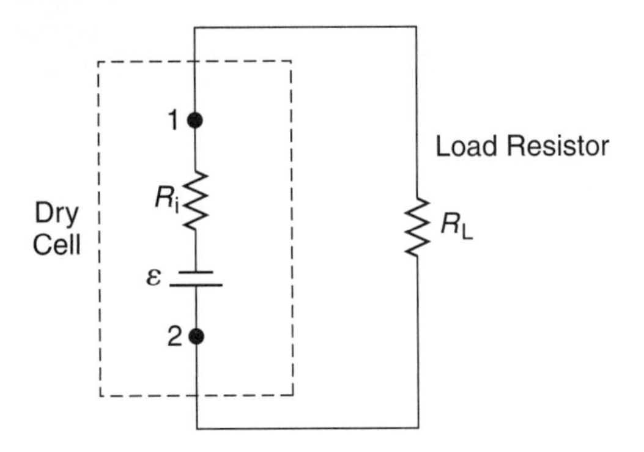

Figure 32-1 Dry cell with a load resistance.

A resistor (Figure 32-1) placed across the terminals of a dry cell is called a load resistor R_L. The current drawn from the dry cell in this case is determined by the values of R_L and R_i. The terminal voltage is the voltage between the dry cell terminals. It is given by

$$V = \varepsilon\left(\frac{R_L}{R_i + R_L}\right) \quad \text{(Eq. 2)}$$

As a dry cell ages, its emf decreases, and its internal resistance increases. A very old dry cell might have, for example, an emf of only 1.30 V and an internal resistance of several thousand ohms. When R_i becomes this large, it means that for any significant current drawn from the cell, most of the voltage drop occurs across the internal resistance. As a result, the terminal voltage of the cell becomes very small, and the cell is essentially nonfunctional.

Part 2. Voltmeter Measurements

When a **voltmeter** is placed across the terminals of a dry cell, the input resistance of the voltmeter is effectively a load resistor. The voltage measured by the voltmeter is essentially the same as that given by Equation 2 with R_L replaced in that equation by R_V, the voltmeter resistance. The equation that results is

$$V = \varepsilon\left(\frac{R_V}{R_i + R_V}\right) \quad \text{(Eq. 3)}$$

The value of R_V is strongly dependent upon the quality of the voltmeter. A typical student voltmeter used in this laboratory might have R_V as low as 3000 Ω. If such a meter is used to measure the terminal voltage of a "good" dry cell with R_i of the order of 1 Ω, the terminal voltage would be (3000/3001) = 0.9997 of the emf. For many purposes this would be an acceptable estimate of the emf. On the other hand, if this same voltmeter were used to measure the terminal voltage of a cell with R_i of 5000 Ω, the terminal voltage would be (3000)/(8000) = 0.375 of the emf. Clearly this is not a useful estimate of the emf.

A typical value of R_V for a modern voltmeter of good quality might be in the range of 10 to 20 MΩ. For such a voltmeter, the terminal voltage measured for both of the dry cells described above would be an extremely good approximation to the true emf.

Part 3. Measurements with a Potentiometer

The **potentiometer** is a device that compares a source of known emf to an unknown emf. The unknown emf is placed in the circuit in the opposite direction to a known voltage difference. The unknown emf is balanced against the known potential difference when there is no current in the unknown emf. The true emf is determined because there is no voltage drop across the internal resistance.

Figure 32-2 schematically shows the wiring of a simple slide-wire potentiometer. It consists of a wire (from A to B) of uniform cross-sectional area through which a constant current is maintained by the power supply. Because the wire is uniform in cross-sectional area, the potential change along the length from

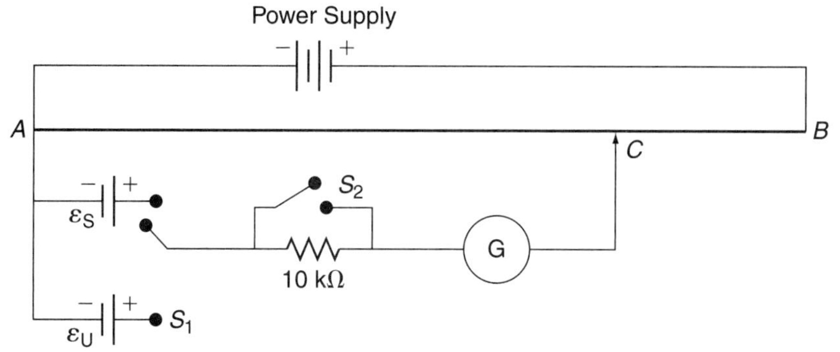

Figure 32-2 Simple slide-wire potentiometer.

A to *B* is proportional to the length of the wire. Switch S_1 allows either cell ε_S or ε_U to be connected in parallel with a portion of the wire *AB*. The cell designated ε_S is a standard cell for which the emf is accurately known, and ε_U stands for the cell of unknown emf that is to be determined.

The current in the wire *AB* is chosen to be large enough so that the potential change from *A* to *B* is greater than either ε_U or ε_S. With switch S_1 set so that ε_S is in the circuit, the slider can be moved along the wire until the point C is located where the voltage drop from *A* to *C* is exactly equal to ε_S. We accomplish this experimentally by finding the point *C* for which there is no current in the galvanometer that is in series with ε_S. After throwing switch S_1 to the other position, which places ε_U in the circuit, a new point *D* is found where the voltage drop from *A* to *D* is equal to ε_U. Again, this point is found experimentally by finding the point where there is no current in the galvanometer. Because the potential drop along the wire is proportional to the length of the wire, a ratio exists between the lengths of the wire, *AD* and *AC*, and the two emfs. This can be expressed as

$$\varepsilon_U = \varepsilon_S \left(\frac{AD}{AC} \right) \qquad \text{(Eq. 4)}$$

The potentiometer can be made more accurate if its total length is extended by including additional coils of wire each of the same length (Figure 32-3). We will give specific instructions below for making measurements with a potentiometer of this type, which has a slide wire 1 m in length and 15 other 1 m coils of wire in series with the slide wire. There is a banana-plug receptacle at the junction of each coil of wire, so that any number of the 15 coils of wire can be incorporated in the circuit along with the chosen length of the slide wire. The potentiometer is designed to be standardized to ensure that each meter of wire has a voltage drop of 0.100 V. Thus the maximum emf that can be measured with this standardization is 1.600 V.

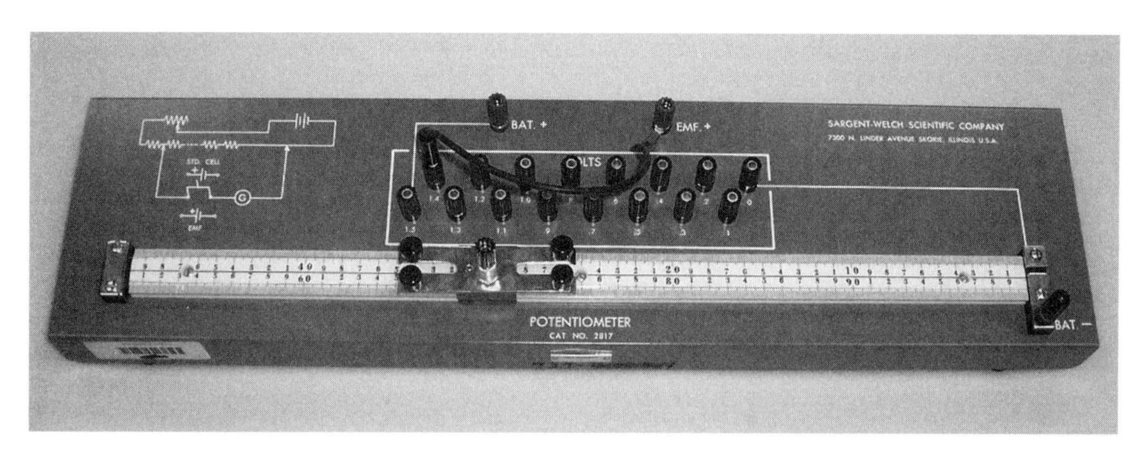

Figure 32-3 Slide-wire potentiometer with a total length of 16 m of wire.

EXPERIMENTAL PROCEDURE

Dry Cell emf by Potentiometer

1. Before making any measurements with the potentiometer, note that special care is required in the use of the standard cell. The cell should remain upright at all times and never be allowed to turn on its side or upside down. Be certain that the cell is placed in the circuit with the correct polarity, and that during initial adjustment, the 10 kΩ resistor is placed in series with the standard cell by having switch S_2 open. This serves to limit the current drawn from the standard cell.
2. The circuit diagram for the potentiometer described here is shown in Figure 32-4. Connect the circuit as shown, but have the circuit approved by your instructor before the power supply is turned on. *(Note that the circuit diagram below and the detailed procedure given below apply to a 16 m long potentiometer with provisions to calibrate it directly in volts. If another type of potentiometer is provided, consult your instructor for specific instructions in its use.)*
3. Following the steps detailed below for standardizing and using the potentiometer, measure the emf of three unknown dry cells. The three unknown dry cells should include one that is new and one that is old. For each unknown cell, make three independent measurements by standardizing the potentiometer and then measuring the emf of the cell three separate times.
4. Procedure to standardize the potentiometer:
   - **(a)** Throw switch S_1 to the standard cell position.
   - **(b)** Adjust the position of the banana plug and the slider until the total corresponds to the emf of the standard cell used. For example, if the standard cell has an emf of 1.0566 V, place the plug in position 1.0 and the slider at a length of 56.6 cm.
   - **(c)** With switch S_2 open, press and release the slider key, noting the galvanometer deflection. Adjust the power supply for minimum galvanometer deflection when the slider key is pressed.
   - **(d)** Close switch S_2 and press and release the slider key, noting the galvanometer deflection. Adjust the power supply for zero galvanometer deflection when the slider key is pressed.
   - **(e)** The potentiometer has now been standardized to 0.1000 V per meter of wire. The power supply should remain fixed at its setting for any measurements taken with this standardization.

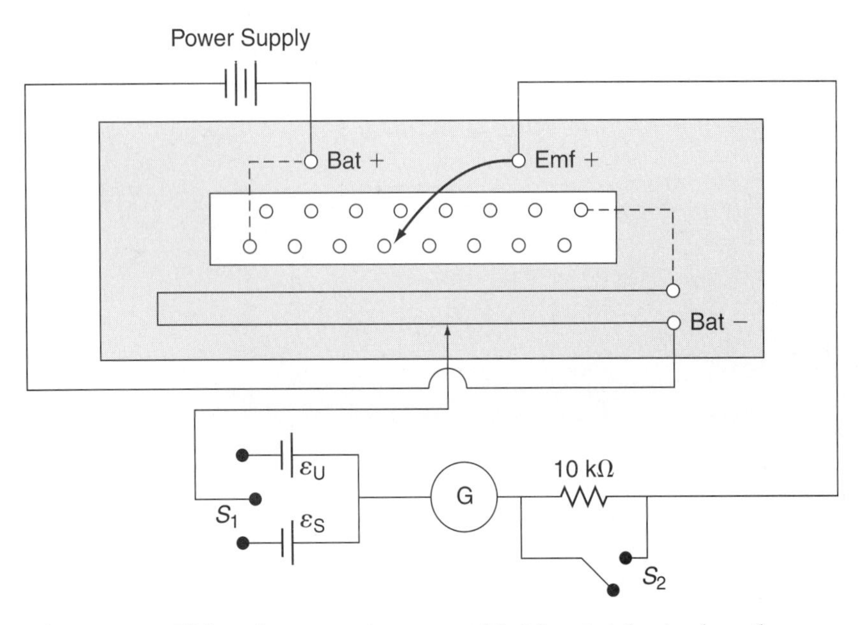

Figure 32-4 Slide-wire potentiometer with 16 m total wire length.

5. Procedure to measure an unknown emf:
   (a) Throw switch S_1 to unknown emf position.
   (b) With switch S_2 open, touch the slider key, noting the galvanometer deflection.
   (c) Move the traveling plug and the slider position until the point is found that gives the minimum galvanometer deflection when the slider key is pressed.
   (d) Close switch S_2, and then move the slider until the point is found that gives zero galvanometer deflection when the slider key is pressed.
   (e) Read the value of the unknown emf as the position of the traveling plug plus the slider position. For example, if the plug is at 1.4 and the slider at 35.5 cm, the emf of the unknown cell is 1.4355 V.

Dry Cell Voltage by Voltmeter

1. Use a voltmeter with an internal resistance of the order of 10 MΩ to measure the terminal voltage of each of your unknown dry cells in turn. Record the values in Data Table 2 under the column labeled High Resistance Voltmeter.
2. Use a student-type low resistance voltmeter to measure the terminal voltage of each of the unknown dry cells. Record the values in the Data Table under the column labeled Low Resistance Voltmeter.

Internal Resistance of a Dry Cell

1. Place the terminals of the high resistance voltmeter across the terminals of the dry cell with the lowest emf as measured by the potentiometer. Note the value of the voltmeter reading, which should be essentially the same as determined earlier by the potentiometer.
2. Set a decade resistance box with a 10 kΩ decade to the maximum value of 100 kΩ. Place the decade resistance box in parallel with the voltmeter and the dry cell with a tap key as shown in Figure 32-5. The resistance box is serving as a load resistor, and the voltmeter now reads the terminal voltage of the cell when the tap key is momentarily pressed. *Just touch the key momentarily and read the voltmeter at that instant. If the key is left pressed for any length of time, it will drain the cell as it provides current for the load resistor, thus making the results meaningless.*
3. Touch the tap key and note the value of the terminal voltage for 100 kΩ load resistor. If there is at least a 5% decrease in the voltmeter reading below the original emf of the cell, record this value of the terminal voltage and the value of 100 kΩ for the load resistance in Data Table 3. If the voltmeter reading does not fall by at least 5% when the 100 kΩ resistor is placed across the cell, lower the decade resistance box value and touch the tap key again to read the terminal voltage. When the terminal voltage is approximately 5% less than the cell emf, record the value of the load resistance and the terminal voltage in Data Table 3.
4. Continue this process of lowering the resistance box while noting the terminal voltage for values of the terminal voltage that are about 10% and then about 15% less than the emf of the cell. Record the exact value of the terminal voltage and the load resistance that gives that terminal voltage in Data Table 3.

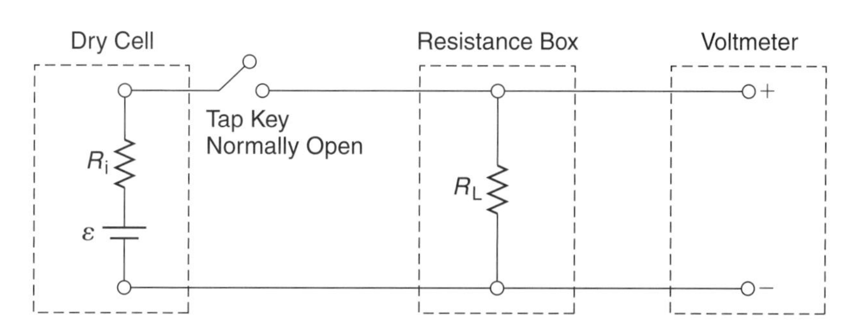

Figure 32-5 Circuit to measure internal resistance of the dry cell.

CALCULATIONS

Dry Cell emf by Potentiometer

1. Calculate the mean and standard error of the three determinations of the emf for each unknown dry cell and record them in Calculations Table 1.

Dry Cell Voltage by Voltmeter

1. Calculate the percentage error in the values of the terminal voltage measured with the high resistance voltmeter compared to the emf determined with the potentiometer, assuming the potentiometer values as correct. Use the mean values for potentiometer values of emf. Record the percentage errors in Calculations Table 2.
2. Repeat the calculations in Step 1 for the terminal voltage values measured with the low resistance voltmeter. Record the results in Calculations Table 2.

Internal Resistance of a Dry Cell

1. Equation 2 relates the terminal voltage of a dry cell to its emf, internal resistance, and load resistance. Using the data in Data Table 3 for the terminal voltage versus the load resistance, solve Equation 3 for the internal resistance and determine a value for the internal resistance from each of the three values of load resistance and terminal voltage. Record the values of R_i in Calculations Table 3.

Name

Section

Date

32

LABORATORY 32 *Potentiometer and Voltmeter Measurements of the emf of a Dry Cell*

PRE-LABORATORY ASSIGNMENT

1. What is the emf and the internal resistance of a typical new dry cell? What might be typical values for the emf and internal resistance of a very old dry cell?

2. A dry cell has an emf of 1.48 V and an internal resistance of 1.11 Ω. What is its terminal voltage when it is connected to a load resistance of 25.00 Ω? Show your work.

3. A voltmeter can be used to measure the terminal voltage of a dry cell. What determines whether or not this terminal voltage is a good approximation to the emf of the cell?

4. Upon what experimental condition is the operation of a potentiometer based, and what advantage does this experimental condition produce for the measurement of the emf?

5. In a measurement using a potentiometer, a known emf and an unknown emf are compared to the voltage drop (a) across a voltmeter (b) across a galvanometer (c) along a wire of uniform cross-section (d) across a power supply of fixed voltage.

6. A dry cell has an emf of 1.35 V and an internal resistance of 2000 Ω. What will be its terminal voltage when measured with a voltmeter whose input resistance is 10,000 Ω? Show your work.

7. What will be the terminal voltage of the dry cell of Question 6 when measured with a voltmeter of input resistance 10 MΩ? Show your work.

8. When the potentiometer described in this laboratory has been properly standardized, to what emf does 1.0000 m of wire correspond?

9. An unknown emf is measured with a properly standardized potentiometer of the type described in this laboratory. It balances when the traveling plug is at position 1.3, and the slider is 95.7 cm. What is the emf of the dry cell? Show your work.

Name ______ Section ______ Date ______

Lab Partners ______

32 LABORATORY 32 *Potentiometer and Voltmeter Measurements of the emf of a Dry Cell*

LABORATORY REPORT

Data Table 1

Unknown emf	*Trial 1* (V)	*Trial 2* (V)	*Trial 3* (V)
#			
#			
#			

Calculations Table 1

$\overline{\text{emf}}$ (V)	α_{emf} (V)

Data Table 2

Unknown emf	*High Resistance Voltmeter* (V)	*Low Resistance Voltmeter* (V)
#		
#		
#		

Calculations Table 2

% Error High R Meter	*% Error Low R Meter*

Data Table 3

ε	*Load Resistor* R_L (Ω)	*Terminal Voltage* (V)

Calculations Table 3

R_i (Ω)

SAMPLE CALCULATIONS

1. % Error =
2. $R_i = (\varepsilon R_L/V) - R_L =$

QUESTIONS

1. Consider the mean values of the emf of each cell as determined by the potentiometer. Comment on the precision of the measurements.

2. What can you say about the accuracy of these same measurements discussed in Question 1? What is the major factor controlling the accuracy of these measurements?

3. Consider the percentage differences in Calculations Table 2. Are the high resistance voltmeter readings a reasonable estimate of the emf?

4. Are there significant differences between the high resistance voltmeter measurements and the low resistance voltmeter measurements? Can you conclude anything about the internal resistance of any of the unknown dry cells from these data? State what you can about the internal resistance of each unknown dry cell.

5. The idea that a dry cell can be represented as a pure emf in series with a pure resistance R_i is a model. If that model is accurate for the dry cell for which you determined R_i in Calculations Table 3, the value of R_i should be approximately the same for the three determinations. Comment on the extent to which such a model seems appropriate to your measurements of the dry cell.

The RC Time Constant

OBJECTIVES

- ❑ Investigate the time needed to discharge a capacitor in an *RC* circuit.
- ❑ Measure the voltage across a resistor as a function of time in an *RC* circuit as a means to determine the *RC* time constant.
- ❑ Determine the value of an unknown capacitor and resistor from the measurements.

EQUIPMENT LIST

- Voltmeter (at least 10 MΩ resistance-digital readout), laboratory timer
- Direct current power supply (20 V), high quality unknown capacitor (5–10 μF)
- Unknown resistor (approximately 10 MΩ), single-pole (double-throw) switch
- Assorted connecting leads

THEORY

Consider the circuit shown in Figure 33-1 consisting of a capacitor C, a resistor R, a source of emf ε, and a switch S. If the switch S is thrown to point A at time $t = 0$ when the capacitor is initially uncharged, charge begins to flow in the series circuit consisting of ε, R, and C, and it flows until the capacitor is fully charged. The current I has an initial value of ε/R and decreases exponentially with time. The charge Q on the capacitor begins at zero and increases exponentially with time until it becomes equal to $C\varepsilon$. The equations that describe those events are

$$Q = C\varepsilon\,(1 - e^{-t/RC}) \quad \text{and} \quad I = \varepsilon/R\, e^{-t/RC} \qquad \text{(Eq. 1)}$$

The quantity ***RC*** is called the **time constant** of the circuit, and it has units of seconds if R is in ohms and C is in farads. After a period of time that is long compared to the time constant RC, the charge Q is equal to $C\varepsilon$, and the current in the circuit is zero.

If switch S is now thrown to position B, the capacitor discharges through the resistor. The charge on the capacitor and the current in the circuit both decay exponentially while the capacitor is discharging. The equations that describe the discharging process are

$$Q = C\varepsilon\, e^{-t/RC} \quad \text{and} \quad I = \varepsilon/R\, e^{-t/RC} \qquad \text{(Eq. 2)}$$

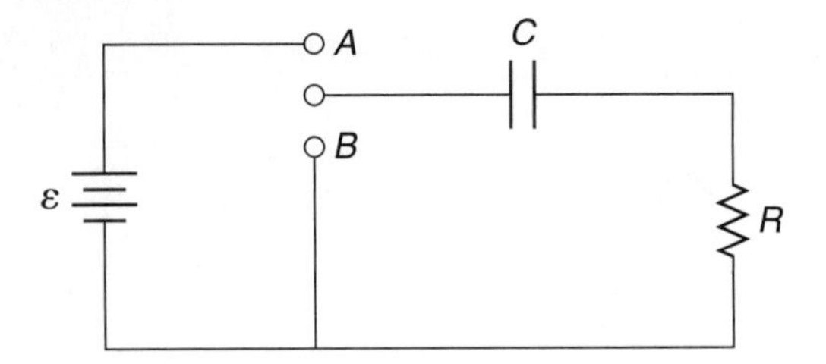

Figure 33-1 Simple series RC circuit.

The current in the discharging case will be in the opposite direction from the current in the charging case, but the magnitude of the current is the same in both cases.

Consider the circuit shown in Figure 33-2 consisting of a power supply of emf ε, a capacitor C, a switch S, and a voltmeter with an input resistance of R. If initially the switch S is closed, the capacitor is charged almost immediately to ε, the voltage of the power supply. When the switch is opened, the capacitor discharges through the resistance of the meter R with a time constant given by RC. With the switch open, the only elements in the circuit are the capacitor C and the voltmeter resistance R, and thus the voltage across the capacitor is equal to the voltage across the voltmeter. It is given by

$$V = \varepsilon\, e^{-t/RC} \qquad \text{(Eq. 3)}$$

Rearranging and taking the natural log of both sides of the equation gives

$$\ln(\varepsilon/V) = (1/RC)t \qquad \text{(Eq. 4)}$$

If the voltage across the capacitor is determined as a function of time, a graph of $\ln(\varepsilon/V)$ versus t will give a straight line with a slope of $(1/RC)$. Thus RC can be determined, and if R the voltmeter resistance is known, then C can be determined.

If an unknown resistor is placed in parallel with the voltmeter, it produces a circuit like that shown in Figure 33-3. The capacitor can again be charged and then discharged, but now the time constant will be equal to R_tC where R_t is the parallel combination of R and R_U. If the relationship between R, R_U, and R_t is solved for R_U, the result is

$$R_U = \frac{RR_t}{R - R_t} \qquad \text{(Eq. 5)}$$

Therefore, a measurement of the capacitor voltage as a function of time will produce a dependence like that given by Equation 4, except that the slope of the straight line will be $(1/R_tC)$. Thus if C is known and R_tC is found from the slope, then R_t can be determined. Using Equation 5, R_U can be found from R and R_t.

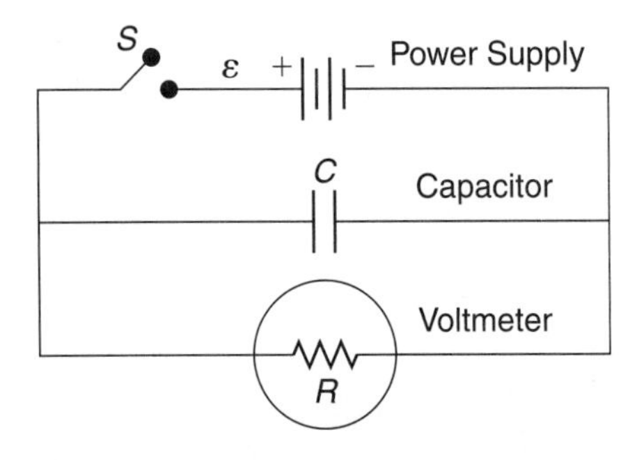

Figure 33-2 An RC circuit using a voltmeter as the resistance.

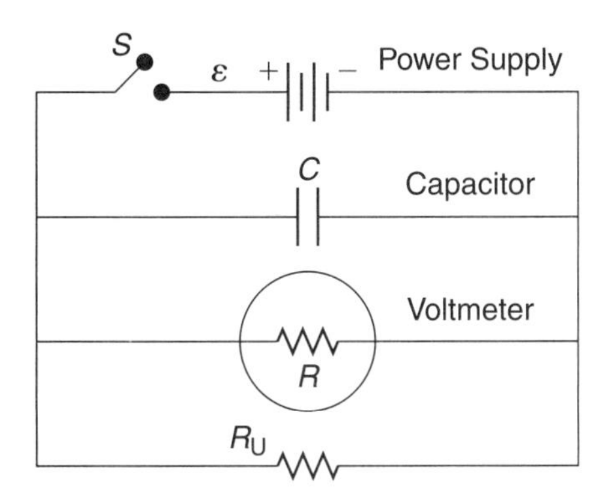

Figure 33-3 *RC* circuit using voltmeter and R_U in parallel.

EXPERIMENTAL PROCEDURE

Unknown Capacitance

1. Construct a circuit such as the one in Figure 33-2 using the capacitor supplied, the voltmeter, and the power supply. Have the circuit approved by your instructor before turning on any power. Obtain from your instructor the value of the input resistance of the voltmeter and record it in Data and Calculations Table 1 as *R*.
2. Close the switch, and adjust the power supply emf ε as read on the voltmeter to the value chosen by your instructor. Record the value of ε in Data and Calculations Table 1.
3. Open the switch and simultaneously start the timer.
4. The voltmeter reading will fall as the capacitor discharges. Let the timer run continuously, and for eight predetermined values of the voltage, record the time t at which the voltmeter reads these voltages. A convenient choice for voltages at which to measure t would be increments of 10%. For example, if $\varepsilon = 20.0$ V, then use voltage of 18.0, 16.0, 14.0, etc. Record the voltage V and times t in Data and Calculations Table 1.
5. Repeat Steps 2 through 4 two more times for Trials 2 and 3.

Unknown Resistance

1. Construct a circuit such as the one in Figure 33-3 using the same capacitor used in the last circuit and the unknown resistor supplied. Close the switch and adjust the power supply voltage to the same value used in the last procedure.
2. Repeat Steps 2 through 5 of the procedure above, and record all values in the appropriate places in Data and Calculations Table 2.

CALCULATIONS

Unknown Capacitance

1. Calculate the values of $\ln(\varepsilon/V)$ and record them in Data and Calculations Table 1.
2. Calculate the mean $\bar{t}$ and the standard error α_t for the three trials of t at each voltage and record them in Data and Calculations Table 1.
3. Perform a linear least squares fit of the data with $\ln(\varepsilon/V)$ as the vertical axis and $\bar{t}$ as the horizontal axis.
4. Record the value of the slope in Data and Calculations Table 1.

5. Calculate RC as the reciprocal of the slope. Record the value of RC in Data and Calculations Table 1.
6. Use the value of RC and the value of R to calculate the value of the unknown capacitor C and record it in Data and Calculations Table 1.

Unknown Resistance

1. Calculate the values of $\ln(\varepsilon/V)$ and record the values in Data and Calculations Table 2.
2. Calculate the mean $\bar{t}$ and the standard error α_t for the three trials of t at each voltage and record them in Data and Calculations Table 2.
3. Perform a linear least squares fit of the data with $\ln(\varepsilon/V)$ as the vertical axis and $\bar{t}$ as the horizontal axis.
4. Record the value of the slope in Data and Calculations Table 2.
5. Calculate the value of R_tC as the reciprocal of the slope. Record the value of R_tC in Data and Calculations Table 2.
6. Using the value of the capacitance C determined in the first procedure and the value of R_tC, calculate the value of R_t and record it in Data and Calculations Table 2.
7. Calculate the value of the unknown resistance R_U from the values of R_t and R. Record the value of R_U in Data and Calculations Table 2.

GRAPHS

1. For the data from Data and Calculations Table 1 graph the quantity $\ln(\varepsilon/V)$ as the vertical axis and $\bar{t}$ as the horizontal axis. Also show on the graph the straight line obtained from the linear least squares fit to the data.
2. For the data from Data and Calculations Table 2, graph the quantity $\ln(\varepsilon/V)$ as the vertical axis and $\bar{t}$ as the horizontal axis. Also show on the graph the straight line obtained from the linear least squares fit to the data.

Name ______ Section ______ Date ______

LABORATORY 33 The RC Time Constant

PRE-LABORATORY ASSIGNMENT

1. In a circuit such as the one in Figure 33-1 with the capacitor initially uncharged, the switch S is thrown to position A at $t = 0$. The charge on the capacitor (a) is initially zero and finally $C\varepsilon$ (b) is constant at a value of $C\varepsilon$ (c) is initially $C\varepsilon$ and finally zero (d) is always less than ε/R.

2. In a circuit such as the one in Figure 33-1 with the capacitor initially uncharged, the switch S is thrown to position A at $t = 0$. The current in the circuit is (a) initially zero and finally ε/R (b) constant at a value of ε/R (c) equal to $C\varepsilon$ (d) initially ε/R and finally zero.

3. In a circuit such as the one in Figure 33-2 the switch S is first closed to charge the capacitor, and then it is opened at $t = 0$. The expression $V = \varepsilon e^{-t/RC}$ gives the value of (a) the voltage on the capacitor but not the voltmeter (b) the voltage on the voltmeter but not the capacitor (c) both the voltage on the capacitor and the voltage on the voltmeter, which are the same (d) the charge on the capacitor.

4. For a circuit such as the one in Figure 33-1, what are the equations for the charge Q and the current I as functions of time when the capacitor is charging?

   $Q =$ ______ $I =$ ______

5. For a circuit such as the one in Figure 33-1, what are the equations for the charge Q and the current I as functions of time when the capacitor is discharging?

   $Q =$ ______ $I =$ ______

6. If a 5.00 μF capacitor and a 3.50 MΩ resistor form a series RC circuit, what is the RC time constant? Give proper units for RC and show your work.

   $RC =$ ______

7. Assume that a 10.0 μF capacitor, a battery of emf $\varepsilon = 12.0$ V, and a voltmeter of 10.0 MΩ input impedance are used in a circuit such as that in Figure 33-2. The switch S is first closed, and then the switch is opened. What is the reading on the voltmeter 35.0 s after the switch is opened? Show your work.

$V =$ ______________ V

8. Assume that a circuit is constructed such as the one shown in Figure 33-3 with a capacitor of 5.00 μF, a battery of 24.0 V, a voltmeter of input impedance 12.0 MΩ, and a resistor $R_U = 10.0$ MΩ. If the switch is first closed and then opened, what is the voltmeter reading 25.0 s after the switch is opened? Show your work.

$V =$ ______________ V

9. In the measurement of the voltage as a function of time performed in this laboratory, the voltage is measured at fixed time intervals. (a) true (b) false

Name   *Section*   *Date*

Lab Partners

33 LABORATORY 33 *The RC Time Constant*

LABORATORY REPORT

Data and Calculations Table 1

V (V)	t_1 (s)	t_2 (s)	t_3 (s)	$ln(\varepsilon/V)$	$\bar{t}$ (s)	α_t (s)

$\varepsilon =$ V	$R =$ Ω	$r =$	Intercept =
Slope = s^{-1}	$RC =$ s	$C =$ F	

Data and Calculations Table 2

V (V)	t_1 (s)	t_2 (s)	t_3 (s)	$ln(\varepsilon/V)$	$\bar{t}$ (s)	α_t (s)

$\varepsilon =$ V	$R =$ Ω	$r =$	Intercept $=$
Slope $=$ s^{-1}	$R_tC =$ s	$R_t =$ Ω	$R_U =$ Ω

SAMPLE CALCULATIONS

1. $\ln(\varepsilon/V) =$
2. $RC = 1/\text{slope} =$
3. $C = RC/R =$
4. $R_t = R_tC/C =$
5. $R_U = (R_tR)/(R - R_t) =$

QUESTIONS

1. Evaluate the linearity of each of the graphs. Do they confirm the linear dependence between the two variables that is predicted by the theory?

2. Ask your instructor for the values of the unknown capacitor and resistor. Calculate the percentage error of your measurement compared to the values provided. On this basis, evaluate the accuracy of your measurement of the capacitance and resistance.

3. Show that RC has units of seconds if R is in Ω and C is in F.

4. A capacitor of 5.60 μF and a 4.57 MΩ resistor form a series RC circuit. If the capacitor is initially charged to 25.0 V, how long does it take for the voltage on the capacitor to reach 10.0 V? Show your work.

Kirchhoff's Rules

OBJECTIVES

- ❑ Investigate what type of circuit to which Kirchhoff's rules must be applied.
- ❑ Apply Kirchhoff's rules to several circuits, solve for the currents in the circuit, and compare the theoretical values predicted by Kirchhoff's rules to measured values.

EQUIPMENT LIST

- Direct current voltmeter (digital, 0–20 V), direct current ammeter (digital, 0–1000 mA)
- Two sources of emf (direct current power supplies, up to 12 V)
- Three or four resistors or resistor boxes (range 500–1000 Ω)
- Digital ohmmeter (one for the class), connecting wires

THEORY

Consider the circuit in Figure 34-1. The circuit is labeled with all of the currents. The 2 Ω resistor, 8 Ω resistor, and 12 V power supply have current I_1, the 6 Ω resistor has current I_2, and the 3 Ω resistor has current I_3. This circuit is called a **single-loop circuit** because it can be reduced to a single resistor in series with the power supply. The 6 Ω resistor and the 3 Ω resistor are in parallel with an equivalent resistance of 2 Ω. That equivalent 2 Ω resistance is in series with the 12 V power supply and the other two resistors, reducing the circuit to a single-loop circuit. The total resistance across the 12 V power supply is 12 Ω and its current is therefore $I_1 = 1$ A. Applying Ohm's law to the remaining part of the circuit gives $I_2 = 1/3$ A and $I_3 = 2/3$ A.

Consider now the circuit of Figure 34-2. This laboratory is concerned with the fundamental difference between circuits of the type depicted in Figure 34-1 and circuits of the type depicted in Figure 34-2. The circuit in Figure 34-2 cannot be reduced to a single-loop circuit, but instead is called a **multi-loop circuit.** Before analyzing this circuit, first we will define some terms. A point at which at least three possible current paths intersect is defined as a **junction.** For example, points *A* and *B* in Figure 34-2 are junctions. A **closed loop** is any path that starts at some point in a circuit and passes through elements of the circuit (in this case resistors and power supplies), and then arrives back at the same point without passing through any circuit element more than once. By this definition there are three loops in the circuit of Figure 34-2: (1) starting at *B*, going through the 10 V power supply to *A*, and then down through the 20 V power supply back to *B*, (2) starting at *B*, up through the 20 V power supply, and then around the outside through the 10 Ω resistor and back to *B*, (3) completely around the outside part of the circuit. One can traverse a loop in either of two directions, but regardless of which direction is chosen, the resulting equations are equivalent.

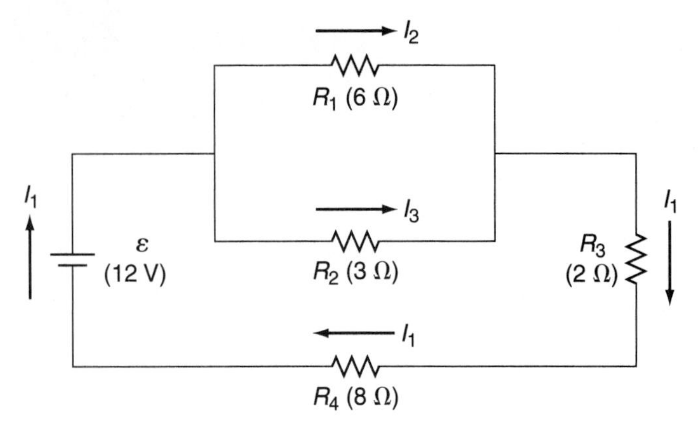

Figure 34-1 Single-loop circuit.

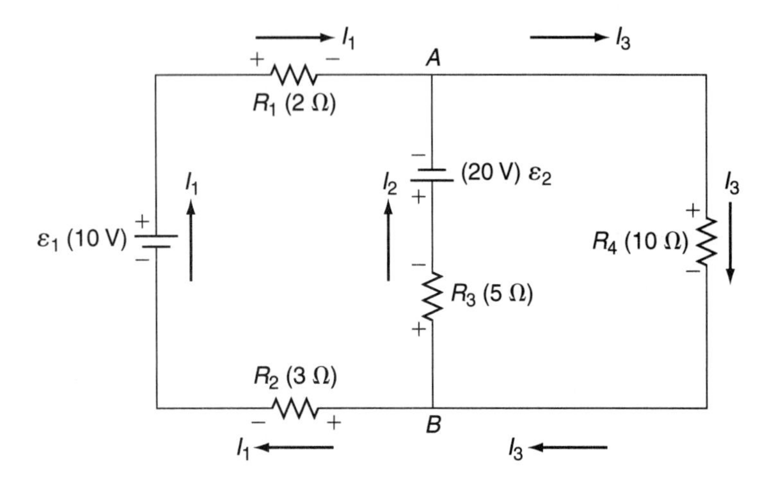

Figure 34-2 Multi-loop circuit.

The solution for the currents in a multi-loop circuit uses two rules developed by Gustav Robert Kirchhoff. The first of these rules is called **Kirchhoff's current rule** (KCR). It can be stated in the following way:

KCR—The sum of currents into a junction = the sum of currents out of the junction.

This rule actually amounts to a statement of conservation of charge. In effect it states that charge does not accumulate at any point in the circuit. The second rule is called **Kirchhoff's voltage rule** (KVR). It can be stated as:

KVR—The algebraic sum of the voltage changes around any closed loop is zero.

This rule is essentially a statement of the conservation of energy, which recognizes that the energy provided by the power supplies is absorbed by the resistors.

In a multi-loop circuit the values of the resistors and the power supplies are known. It is necessary to determine how many independent currents are in the circuit, to label them, and then to assign a direction to each current. Application of Kirchhoff's rules to the circuits, treating the assigned currents as unknowns, will produce as many independent equations as there are unknown currents. Solving those equations will determine the values of the currents.

In the application of KVR to a circuit, take care to assign the proper sign to a voltage change across a particular element. The value of the voltage change across an emf ε can be either $+\varepsilon$ or $-\varepsilon$ depending upon which direction it is traversed in the loop. If the emf is traversed from the (−) terminal to the (+) terminal, the change in voltage is $+\varepsilon$. However, when going from the (+) terminal to the (−) terminal, the change in voltage is $-\varepsilon$. In the laboratory we will measure the terminal voltage of the sources of emf. We will assume that those values approximate the emf.

When a resistor R with an assumed current I is traversed in the loop in the same direction as the current, the voltage change is $-IR$. If the resistor is traversed in the direction opposite that of the current, the voltage change is $+IR$. The sign of the voltage change across an emf is not affected by the direction of the current in the emf. The sign of the voltage change across a resistor is completely determined by the current direction.

Consider the application of Kirchhoff's rules to the multi-loop circuit of Figure 34-2. At the junction A currents I_1 and I_2 go into the junction, current I_3 goes out of the junction, and KCR states

$$I_1 + I_2 = I_3 \qquad \text{(Eq. 1)}$$

It might appear that applying KCR to the junction B would produce an additional useful equation, but in fact it would result in an equation that is identical to Equation 1.

Applying KVR to the loop that starts at B, goes through the 10 V power supply to A, and then down through the 20 V power supply back to B, gives the following equation with values of the resistances included.

$$-R_2 I_1 + \varepsilon_1 - R_1 I_1 + \varepsilon_2 + R_3 I_2 = 0 \quad \text{or} \quad -3 I_1 + 10 - 2 I_1 + 20 + 5\, I_2 = 0 \qquad \text{(Eq. 2)}$$

The signs used in Equation 2 and the circuit diagrams are consistent with the description given above for determining the signs of voltage changes. Applying KVR to the loop that starts at B and goes clockwise around the right side of the circuit gives

$$-R_3 I_2 - \varepsilon_2 - R_4 I_3 = 0 \quad \text{or} \quad -5 I_2 - 20 - 10 I_3 = 0 \qquad \text{(Eq. 3)}$$

Equations 1, 2, and 3 are the three needed equations for the three unknowns I_1, I_2, and I_3. The solution of these equations gives values for the currents of $I_1 = 2.800$ A, $I_2 = -3.200$ A, and $I_3 = -0.400$ A. The currents I_2 and I_3 are negative. This indicates that the original assumption of direction for these two currents was incorrect. The interpretation of the solution is that there is a current of 2.800 A in the direction indicated in the figure for I_1, a current of 3.200 A in a direction opposite to that indicated in Figure 34-2 for I_2, and a current of 0.400 A in a direction opposite to that indicated for I_3. This is a general feature of solutions using Kirchhoff's rules. Even if the original assumption of the direction of a current is wrong, the solution of the equations leads to the correct understanding of the proper direction by virtue of the sign of the current.

EXPERIMENTAL PROCEDURE

1. If using resistance boxes, choose $R_1 = 500\,\Omega$, $R_2 = 750\,\Omega$, and $R_3 = 1000\,\Omega$, but if using standard resistors, choose values as close as possible to the values listed and use the ohmmeter to measure the value of the resistors. Record those values in Data Table 1.
2. Using two power supplies and the resistors R_1, R_2, and R_3, construct a circuit like that shown in Figure 34-3 with $\varepsilon_1 = 10.0$ V and $\varepsilon_2 = 5.00$ V.
3. Measure the currents I_1, I_2, and I_3. Assuming that only one ammeter is available, the currents will have to be measured one at a time by placing the ammeter in the positions shown in the circuit diagram as a circle. Note that placing the ammeter in the circuit with the polarity shown in the circuit diagram will give positive readings when the current is in the direction assumed. If a modern digital ammeter is used for the measurements, the ammeter will give a positive reading if the current is in the direction assumed, and will give a negative reading if the current is in the opposite direction. If an ammeter is used that properly deflects in only one direction, the meter could be damaged if the current is in the opposite direction from that assumed. In this case, bring the voltage of the power supplies up slowly to see that the meter deflects in the proper direction.

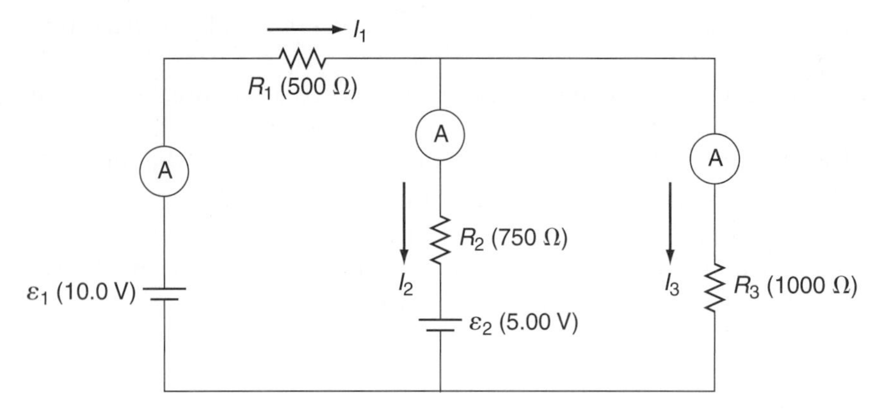

Figure 34-3 Experimental multi-loop circuit with three unknown currents.

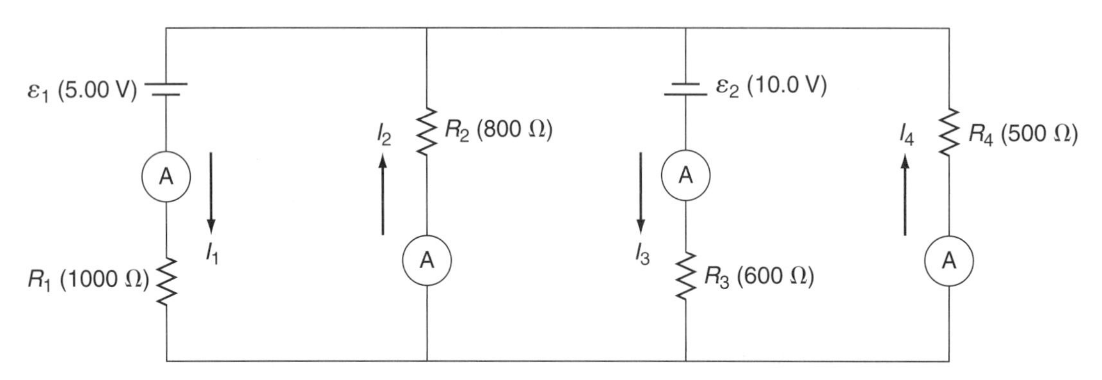

Figure 34-4 Experimental multi-loop circuit with four unknown currents.

4. Measure the emfs ε_1 and ε_2 with a voltmeter and record those values in Data Table 1. The terminal voltages of the power supplies are assumed to approximate the emfs.
5. Construct the circuit of Figure 34-4, which also has two power supplies but has four resistors. Choose values of $\varepsilon_1 = 5\,\text{V}$, $\varepsilon_2 = 10\,\text{V}$, $R_1 = 1000\,\Omega$, $R_2 = 800\,\Omega$, $R_3 = 600\,\Omega$, and $R_4 = 500\,\Omega$, or values as close to those as possible. Determine and record in Data Table 2 the values of the resistors, the four currents, and the emf of the power supplies.

CALCULATIONS

1. Apply Kirchhoff's rules to the circuit of Figure 34-3 for the actual values used in the circuit. Three equations in the three currents I_1, I_2, and I_3 will result. One equation will be a KCR equation, and two will be KVR equations. Record those three equations in the appropriate place in Calculations Table 1.
2. Solve the three equations for the values of I_1, I_2, and I_3 and record the values in Calculations Table 1.
3. Calculate the percentage error of the experimental values of the current compared to the theoretical values for each of the currents and record in Calculations Table 1.
4. Apply Kirchhoff's rules to the circuit of Figure 34-4 with the actual values that were used in your circuit. Four equations in the four currents I_1, I_2, I_3, and I_4 will result. One equation will be a KCR equation, and three will be KVR equations. Record the four equations in the appropriate place in Calculations Table 2.
5. Solve the four equations that you have written for the values of I_1, I_2, I_3, and I_4. Record those values in Calculations Table 2.
6. Calculate the percentage error of the experimental values of the current compared to the theoretical values for each of the currents and record in Calculations Table 2.

Name Section Date

34 LABORATORY 34 *Kirchhoff's Rules*

PRE-LABORATORY ASSIGNMENT

1. Consider the circuit in Figure 34-5. Choose any of the following statements about the circuit that are true. More than one may be correct. (a) It is a single-loop circuit. (b) It is a multi-loop circuit. (c) Assuming R_1 and R_2 were known, the currents could be determined, but only if Kirchhoff's rules were used. (d) Assuming R_1 and R_2 were known, the currents could be determined without the use of Kirchhoff's rules.

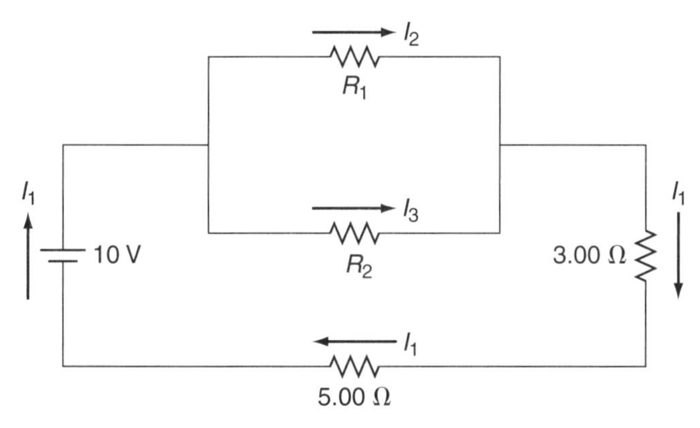

Figure 34-5 Circuit for Questions 1 to 4.

2. In the circuit of Figure 34-5, if $I_1 = 2.00$ A and $I_2 = 0.75$ A, what is the value of I_3?

For questions 3 and 4, assume that the value of R_1 and R_2 in Figure 34-5 are both 4.00 Ω.

3. What is the equivalent resistance of the circuit?

4. What is the current in the 5.00 Ω resistor?

5. Consider the circuit of Figure 34-6. Apply Kirchhoff's rules to the circuit and write three equations in terms of known circuit elements and the unknown currents shown in the figure.

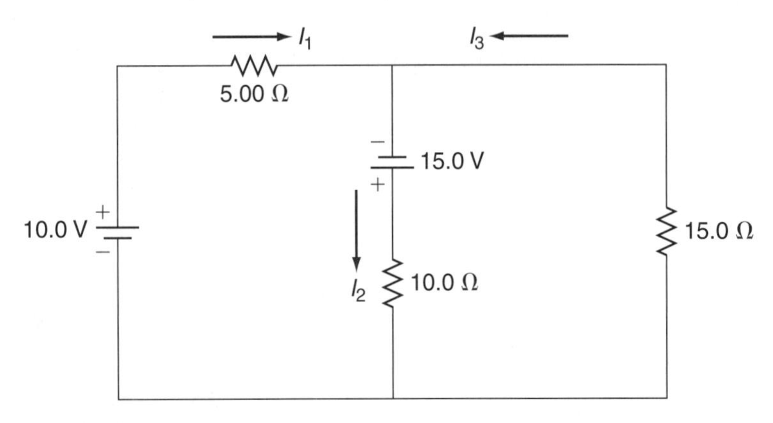

Figure 34-6 Multi-loop circuit.

6. Solve the three equations that you wrote in Question 5 for the values of the currents.

7. If any of the current values obtained in the solution to Question 6 are negative, explain the significance of a negative value for a current.

Name ______ Section ______ Date ______

Lab Partners ______

34 LABORATORY 34 *Kirchhoff's Rules*

LABORATORY REPORT

Data Table 1

Power Supply Voltages $\varepsilon_1 =$ ______ V $\varepsilon_2 =$ ______ V	
Resistor Values (Ω)	*Experimental Current* (mA)
$R_1 =$	$I_1 =$
$R_2 =$	$I_2 =$
$R_3 =$	$I_3 =$

Calculations Table 1

Kirchhoff's rules for the circuit **(1)** KCR— **(2)** KVR1— **(3)** KVR2—	
Theoretical Current (mA)	*% Error Experimental to Theoretical Current*
$I_1 =$	
$I_2 =$	
$I_3 =$	

Data Table 2

Power Supply Voltages $\varepsilon_1 =$ ______ V $\varepsilon_2 =$ ______ V	
Resistor Values (Ω)	*Experimental Current* (mA)
$R_1 =$	$I_1 =$
$R_2 =$	$I_2 =$
$R_3 =$	$I_3 =$
$R_4 =$	$I_4 =$

Calculations Table 2

Kirchhoff's rules for the circuit **(1)** KCR— **(2)** KVR1— **(3)** KVR2— **(4)** KVR3—	
Theoretical Current (mA)	*% Error Experimental to Theoretical Current*
$I_1 =$	
$I_2 =$	
$I_3 =$	
$I_4 =$	

SAMPLE CALCULATIONS

1. KCR, KVR1, and KVR2 provide three equations to be solved for I_1, I_2, and I_3 for first circuit.

2. KCR, KVR1, KVR2, and KVR3 provide four equations to be solved for I_1, I_2, I_3, and I_4 for second circuit.

QUESTIONS

1. In Figure 34-3, state the equation that relates the currents I_1, I_2, and I_3. Calculate the percentage difference between the experimental values of the two sides of the equation.

2. In Figure 34-4, state the equation that relates the currents I_1, I_2, I_3, and I_4. Calculate the percentage difference between the experimental values of the two sides of the equation.

3. Are the experimental values of the currents for the entire laboratory generally larger or smaller than the theoretical values expected for the currents?

4. An ideal ammeter has zero resistance. Real ammeters have small but finite resistance. Would ammeter resistance cause an error in the proper direction to account for the direction of your error indicated in Question 3? State your reasoning.

5. The connecting wires in the experiment are assumed to have no resistance, but in fact have a finite resistance. Would this error be in the proper direction to account for the direction of the error stated in your answer to Question 3? State your reasoning.

Magnetic Induction of a Current Carrying Long Straight Wire

OBJECTIVES

- ❑ Use a compass to determine the direction of the **B** field surrounding a current carrying long straight wire to confirm that it is consistent with the right-hand rule.
- ❑ Determine the induced voltage in a small inductor coil placed near the long straight wire as a relative measurement of the **B** field.
- ❑ Demonstrate that the magnitude of the **B** field surrounding a long straight wire decreases as $1/r$ where r is the perpendicular distance from the wire.

EQUIPMENT LIST

- Direct current power supply (low voltage, 2 A), direct current ammeter (2A)
- Sine wave generator (variable frequency up to 100 kHz, 5 V peak to peak)
- Alternating current digital voltmeter (frequencies up to 100 kHz)
- 100-mH inductor coil (length $\approx$ 1 cm and inside diameter $\approx$ 1 cm)
- Small compass, long straight wire apparatus (Consists of a frame on which a continuous strand of wire is wrapped for 10 loops. The 10 strands are taped together over a length of approximately 40 cm to approximate a wire with a current having 10 times the current as in a single strand of the wire. The apparatus can be placed with the long straight section parallel to the laboratory table or perpendicular to the table.)

THEORY

When a current I exists in an infinitely long straight wire, the lines of **magnetic induction B** are concentric circles surrounding the wire. At a perpendicular distance r from the wire, the **B** field is tangent to the circle as shown in Figure 35-1. The direction of the current I is perpendicular to the plane of the page and directed out of the page. The direction of the current is by definition the direction that positive charge would flow. The magnitude of the **B** field as a function of I and r is given by

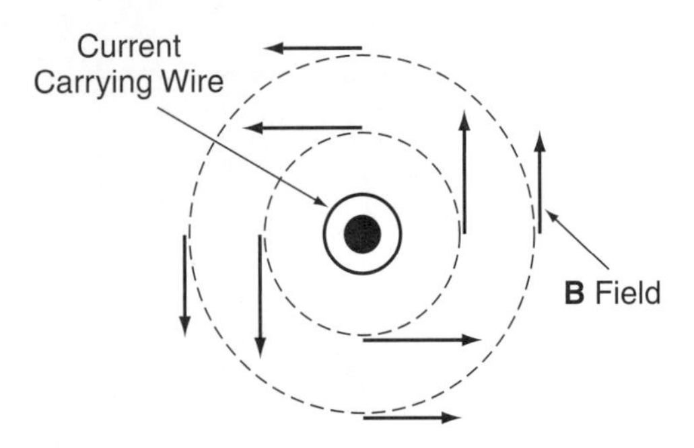

Figure 35-1 *B* field near a wire carrying current out of the page.

$$B = \frac{\mu_o I}{2\pi r} \qquad \text{(Eq. 1)}$$

where $\mu_o = 4\pi \times 10^{-7}$ weber/amp-m, *I* is in amperes, and *r* is in meters. The units of **B** are weber/m^2, which has been given the name Tesla.

The direction of the **B** field relative to the current direction is given by the following right-hand rule. If the thumb of the right hand points in the direction of the current, the four fingers of the right hand curl in the direction of the **B** field. This rule assumes that the **B** field forms circles, and the rule determines only in which direction to take the tangent to the circles as shown in Figure 35-1. In Figure 35-1 the lengths of the **B** vectors are shorter for the larger circles, which shows that the **B** field decreases with distance from the wire as predicted by Equation 1.

In a strict sense, the above statements apply only to an infinitely long straight wire. In this laboratory, the straight portion of the wire is of some finite length *L*. For measurements made at the center of the wire length within a perpendicular distance of *L*/4 from the wire, the finite wire will approximate an infinite wire. If the current in the long straight wire is constant in time, the **B** field created by that current will be constant in time. The direction of the **B** field will be determined by observing the effect of the **B** field on a small compass placed in the vicinity of the long straight wire.

If the current in the long straight wire is an alternating current produced by a sine wave generator, the **B** field surrounding the wire will also vary with time. If a small coil of self-inductance 100 mH is placed next to the wire, an alternating voltage will be induced in the coil, according to Faraday's law of induction. The **induced voltage** in the coil is proportional to the rate of change of the magnetic flux through the coil, and hence to the magnitude of the time-varying **B** field. The quantity actually measured is an alternating electric voltage, but its magnitude is proportional to the **B** field and will be taken to be a *relative* measurement of the **B** field at different distances from the wire.

EXPERIMENTAL PROCEDURE

Direction of the B Field

1. Connect the circuit shown in Figure 35-2 using the direct current power supply and the direct current ammeter. Arrange the long wire apparatus so that the outside long wire is in a horizontal plane along a north-south axis. Ask your instructor the direction of north in the laboratory room. Arrange the wire so that the direction of the current is from north to south. Determine the direction of the current by tracing the wires from the (+) terminal of the power supply. Have the circuit approved by your instructor to ensure that the current is in the proper direction.
2. Turn on the power supply and turn up the voltage until a current of 2.00 A is read on the ammeter. Do not exceed a current of 2.00 A.
3. Place the compass in the middle of the long wire section directly above the wire as close to the wire as possible. State the direction (north, south, east, northeast, etc.) that the compass needle points. Record your answer in Data Table 1.

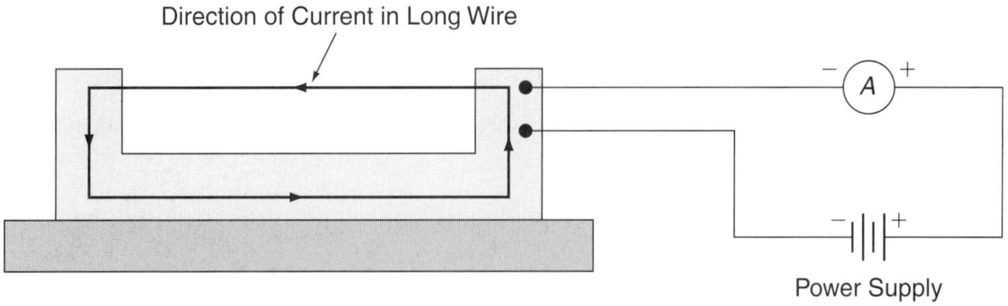

Figure 35-2 Long wire apparatus connected to direct current supply.

4. Repeat Step 3 with the compass immediately below the long wire section.
5. Stand the long wire apparatus on its end so that the current in the outside long wire is vertically downward. Place the compass next to the wire at the four positions indicated by the open circles in Figure 35-6 in the Laboratory Report section. The ⊗ represents the downward current viewed from above. In the open circles that represent the four compass positions, draw an arrow showing the direction that the compass needle points.

B Field as a Function of Distance

1. Connect the circuit shown in Figure 35-3 using the long wire apparatus and the sine wave generator. Turn the generator to maximum amplitude. Stand the long wire apparatus on its end so that the outside long wire is vertical. Place in the apparatus the platform that serves to hold the inductor coil.
2. Connect the inductor coil to the digital voltmeter. Twist the leads about 10 to 15 times before connecting them between the inductor coil and the voltmeter. This is extremely important because it will minimize the voltage that is induced in the leads themselves, and will ensure that the voltage induced is in the inductor coil. Place the inductor coil on the platform as shown in Figure 35-4. The axis of the inductor coil should be perpendicular to an imaginary line that is perpendicular to the

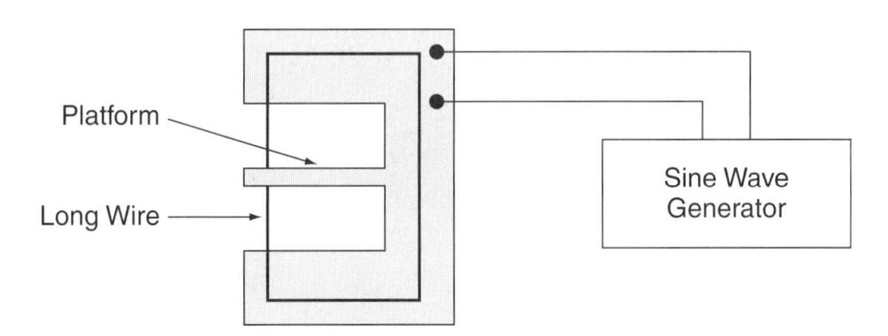

Figure 35-3 Long wire apparatus connected to the sine wave generator.

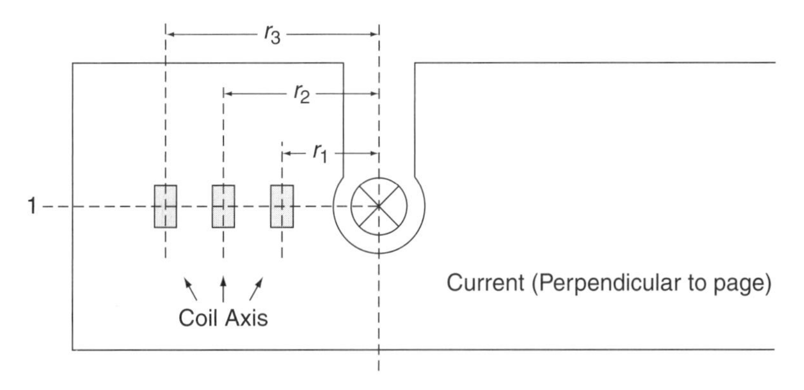

Figure 35-4 View looking down from above. Current alternates in and out of the page.

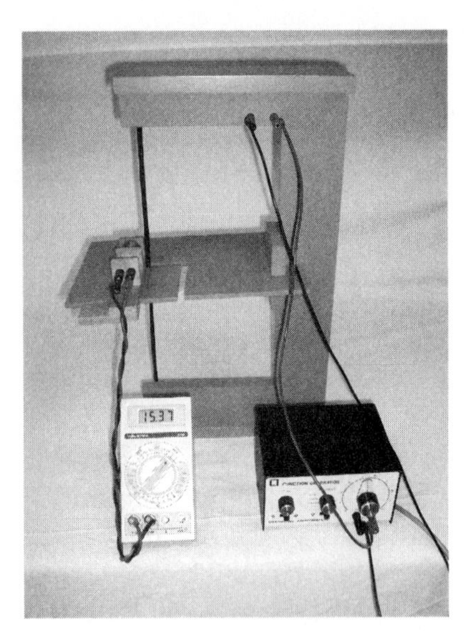

Figure 35-5 Homemade long wire apparatus with inductor coil in position.

current-carrying wire (shown as the dotted line labeled 1 in the figure). The inductor coil is shown at three different distances r_1, r_2, and r_3 from the wire. At each position of the inductor coil shown, the **B** field will alternate in opposite directions along the axis of the coil. The coil is chosen to be short (≈ 1 cm) and of small cross-section (diameter ≈ 1 cm) because for that choice, the **B** field direction is approximately along the coil axis and is approximately uniform over the cross-section of the coil.

3. The amplitude of the induced voltage on the digital voltmeter will depend upon the frequency of the generator. With the inductor about 3 cm from the wire, with its axis positioned as shown in Figure 35-5, vary the generator frequency until the maximum voltage is read on the digital voltmeter. Make all measurements at this frequency.
4. Measure the voltage induced in the inductor coil as a function of r, the distance from the center of the coil to the center of the wire. Take data from $r = 3.0$ cm to $r = 9.0$ cm in increments of 1 cm. Data are not taken for r less than 3 cm because at distances close to the wire, the **B** field is extremely nonuniform over the coil cross-section. Record the values of the voltage as trial one in Data and Calculations Table 2 under the column labeled B_1. If this was actually the **B** field, the units would be Tesla. The measured quantity is a voltage that is proportional to **B**, so no units are stated.
5. Repeat Step 4 two more times measuring the induced voltage at each r. Record the values of trials two and three in Data and Calculations Table 2 under B_2 and B_3.

CALCULATIONS

1. Calculate the mean and standard error for the three trials of B and record them as $\overline{B}$ and α_B in Data and Calculations Table 2.
2. Calculate the percent standard error at each point by calculating $\alpha_B/\overline{B}$ and expressing it as a percentage. Record the values in Data and Calculations Table 2.
3. Calculate the value of $1/r$ for each of the values of r and record them in Data and Calculations Table 2.
4. Perform a linear least squares fit to the data of $\overline{B}$ versus $1/r$ with $\overline{B}$ as the vertical axis and $1/r$ as the horizontal axis. Record the value of the slope, the intercept, and the correlation coefficient.

GRAPHS

1. Make a graph of the data with $\overline{B}$ as the vertical axis and $1/r$ as the horizontal axis. Also show on the graph the straight line obtained from the least squares fit.

Name ____________ Section ____________ Date ____________

35 LABORATORY 35 *Magnetic Induction of a Current Carrying Long Straight Wire*

PRE-LABORATORY ASSIGNMENT

1. State the right-hand rule that relates the direction of the **B** field near a long straight wire to the direction of the current in the wire.

2. The direction of current is defined to be the direction in which ____________ charges would flow.

3. State the equation that relates the magnitude of the **B** field near a long straight wire to the current I in the wire and the distance r from the wire.

   $B =$ ____________

4. There is a current of 10.0 A in a long straight wire. What is the magnitude of the **B** field 5.00 cm from the wire? Show your work.

5. When a current that is constant in time passes through a wire, the **B** field that is produced around the wire is (a) time varying (b) constant in time (c) negative (d) zero.

6. Why are measurements using the inductor coil not taken close to the wire? In other words, why do the measurements start 3.00 cm away from the wire?

7. If an inductor coil is placed near a long wire carrying a current that is constant in time, the voltage induced in the coil is (a) positive (b) negative (c) zero (d) nonzero.

Name | Section | Date

Lab Partners

35 LABORATORY 35 *Magnetic Induction of a Current Carrying Long Straight Wire*

LABORATORY REPORT

Data Table 1

With compass above the wire, compass direction =
With compass below the wire, compass direction =

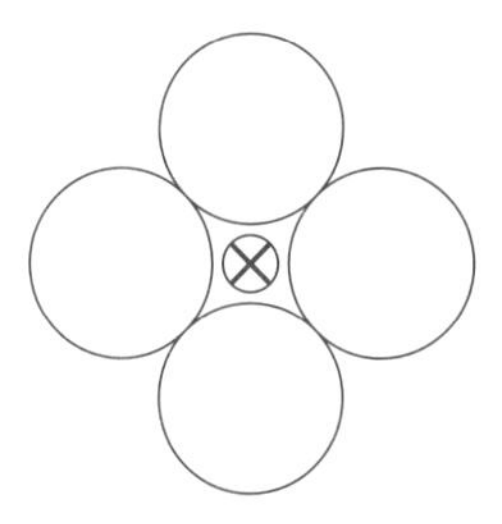

Figure 35-6 Indicate the compass direction at the positions shown.

Data and Calculations Table 2

r (cm)	B_1	B_2	B_3	$1/r$ (cm^{-1})	$\overline{B}$	α_B	% α_B
3.00							
4.00							
5.00							
6.00							
7.00							
8.00							
9.00							
Slope =		Intercept =			Corr. Coeff. =		

SAMPLE CALCULATIONS

1. $1/r =$

QUESTIONS

1. Describe the reasoning used to apply the right-hand rule to the situations in Data Table 1, what is predicted, and how your results do or do not agree with those predictions.

2. Evaluate the precision of the measurements of the induced voltage as a function of distance from the wire. Consider the percentage standard error of the measurements in your evaluation.

3. State the extent to which your measurements confirm the expectation that **B** is proportional to $1/r$. Give the evidence for your evaluation of this question.

4. When the direct current is 2.00 A in a single wire of the bundle of 10 wires, the total current in the bundle of wire that approximates the long straight wire is 20.0 A. What is the magnitude of the **B** field 3.00 cm from this long straight wire carrying a current of 20.0 A? What is the magnitude of the **B** field 9.00 cm from the wire carrying 20.0 A?

5. A constant current is in a long straight wire in the plane of the paper in the direction shown below by the arrow. Point X is in the plane of the paper above the wire, and point Y is in the plane of the paper but below the wire. What is the direction of the **B** field at point X? What is the direction of the **B** field at point Y?

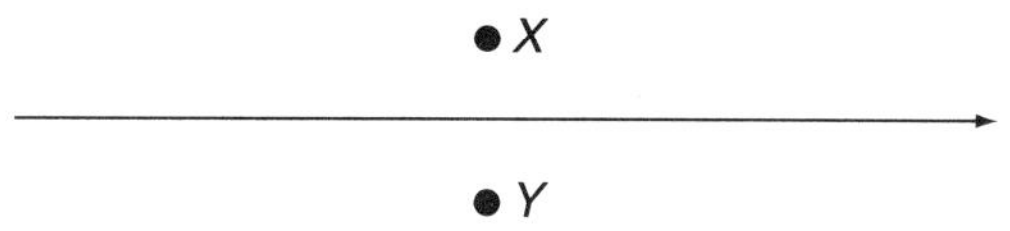

• X

————————————→

• Y

Direction at X = ______________________________

Direction at Y = ______________________________

Alternating Current LR *Circuits*

OBJECTIVES

- ❑ Investigate the phase angle of a generator current relative to the generator voltage.
- ❑ Demonstrate that real inductors consist of both inductance and resistance, and that they can be represented by a pure inductor L in series with a pure resistance r.
- ❑ Determine the value of L and r for an unknown inductor.

EQUIPMENT LIST

- Sine wave generator (variable frequency, 5 V peak to peak amplitude), resistance box
- A 100-mH inductor (resistance $\approx 350\ \Omega$ to serve as an unknown)
- Alternating current voltmeter (digital readout, high frequency capability), compass, protractor

THEORY

Consider the two circuits shown in Figure 36-1 in which a sine wave generator of frequency f is connected separately to resistor R and then to a pure inductance L. The generator is assumed to have a maximum voltage of V and will thus produce a maximum voltage of V across the **resistor** in circuit (a). It will also produce a maximum voltage of V across the inductor in circuit (b). The voltage across the resistor is related to the current by a relationship like that for direct current circuits, which is

$$V_R = IR \qquad \text{(Eq. 1)}$$

If L is the **inductance** (units H) and $\omega = 2\pi f$ is the **angular frequency** of the generator in rad/s, then the following relationship exists between the voltage V_L and the current I

$$V_L = I\omega L \qquad \text{(Eq. 2)}$$

The quantity ωL is called the inductive reactance, and it has units of Ω.

When an alternating current or voltage is measured in the laboratory on a meter, the number read for the current or voltage must be a time-averaged value. Meters are normally calibrated so that they respond to the root-mean-square value of the current or voltage. A **root-mean-square value** of voltage is designated

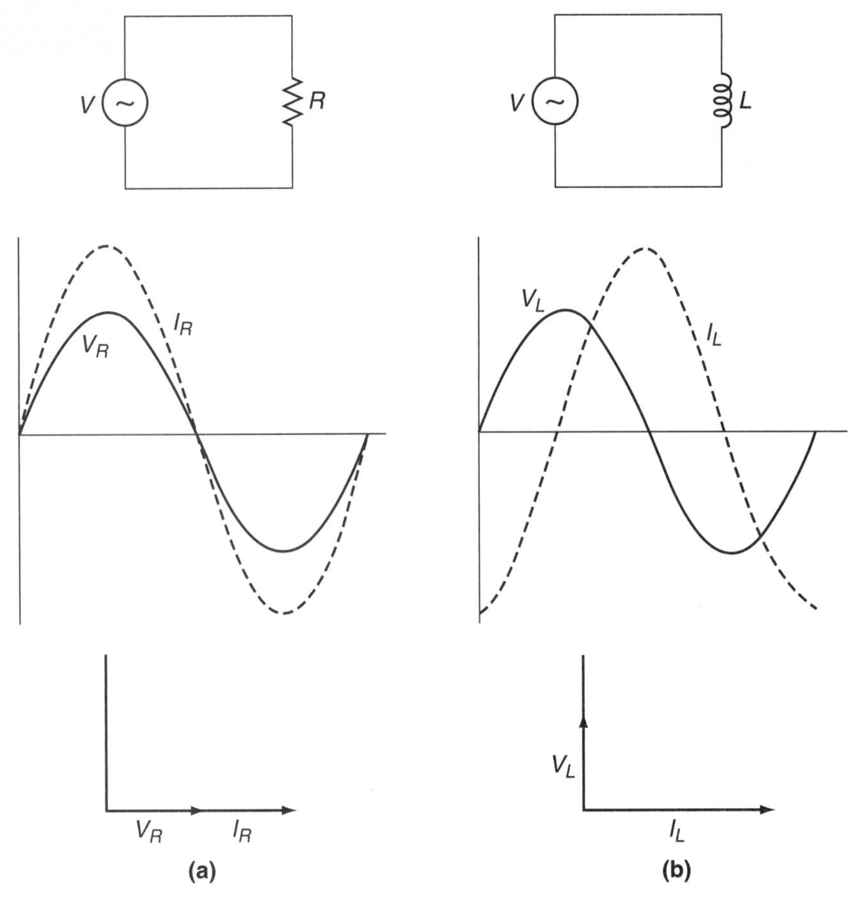

Figure 36-1 Generator and resistor and generator and inductor. Phase relationships between the voltage and current and phasor diagrams of the phase relationships.

as V_{rms}. The relationship between V_{rms} and V the maximum voltage is $V_{rms} = 0.707\,V$. *In this laboratory only voltage will be measured, and all the measurements will be rms values.*

Also shown in Figure 36-1 below each circuit is a graph of the current and voltage across the element for one full period. The graph for the case of the resistor indicates that the resistor current I_R and the resistor voltage V_R are in phase. For the inductor, the graph shows that the inductor current I_L and the inductor voltage V_L are 90° out of phase, with the voltage leading the current by 90°.

Shown at the bottom of Figure 36-1 is a diagram called a **phasor diagram.** Its purpose is also to show the phase relationship. The phasors are vectors drawn with length proportional to the value of the represented quantity, and they are assumed to be rotating counterclockwise with the frequency of the generator. At any time, a projection of one of the rotating vectors on the y axis is the instantaneous value of that quantity. Because the resistor current and voltage are in phase, the phasors are in the same direction. For the inductor, the vector representing the inductor voltage is 90° ahead of the vector representing the current.

Consider now the circuit obtained by placing a pure inductance L having no resistance and a resistor R in series with a sine wave generator of voltage V shown in Figure 36-2.

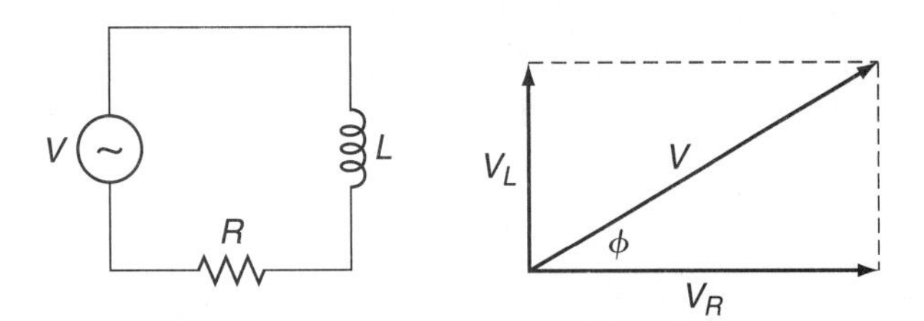

Figure 36-2 Series circuit of resistor and inductor and associated phasor diagram.

For this circuit, the current I is the same at every instant of time in all three circuit elements. Also given in Figure 36-2 is a phasor diagram in which only the voltages are shown. The phasor representing the current (which is not shown) would be in the direction of the phasor labeled V_R because the current and the resistor voltage are in phase. Note that the inductor voltage V_L is 90° ahead of the resistor voltage V_R, and the generator voltage is angle ϕ ahead of V_R. This phasor diagram shows that the generator voltage V is the vector sum of V_R and V_L. In equation form the phasor diagram states

$$V = \sqrt{V_L^2 + V_R^2} \quad \text{(Eq. 3)}$$

The phasor diagram shows that the phase angle ϕ is related to the voltages V_L and V_R, and thus to the resistance R and ωL through Equations 1 and 2. The relationship is given by

$$\tan\phi = \frac{V_L}{V_R} = \frac{\omega L}{R} \quad \text{(Eq. 4)}$$

Note that Equation 4 is strictly valid only for a pure inductor that has no resistance. Real inductors have both an inductance L and an internal resistance r, and can be represented by a pure inductance L in series with a pure resistance r. In Figure 36-3 a real inductor is shown in series with a resistor R and a generator of voltage V. The voltage between points A and B is the generator voltage V, and the voltage between A and C is the resistor voltage V_R. Between the points B and C is the combined voltage across the inductance L and the internal resistance r. This voltage will be referred to as V_{ind}. There is some voltage V_L across L, and some voltage V_r across r. However, there can be no direct measurement of V_L or V_r. The only quantity that can be measured is V_{ind}, which is the vector sum of V_L and V_r. A phasor diagram for the circuit is also shown in Figure 36-3.

Applying the law of cosines to the triangle formed by V, V_R, and V_{ind} leads to

$$\cos\phi = \frac{V^2 + V_R^2 - V_{ind}^2}{2VV_R} \quad \text{(Eq. 5)}$$

The phasor diagram in Figure 36-3 shows that voltages V_L and V_r can be determined from V, V_R, and ϕ by

$$V_L = V\sin\phi \quad \text{and} \quad V_r = V\cos\phi - V_R \quad \text{(Eq. 6)}$$

The current I is the same in all the elements of the circuit, and it can be related to the voltage across each element by the following equations:

$$V_L = I\omega L \quad V_R = IR \quad V_r = Ir \quad \text{(Eq. 7)}$$

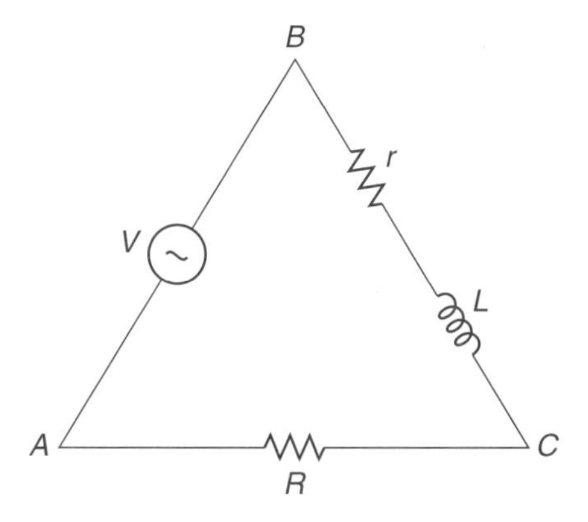

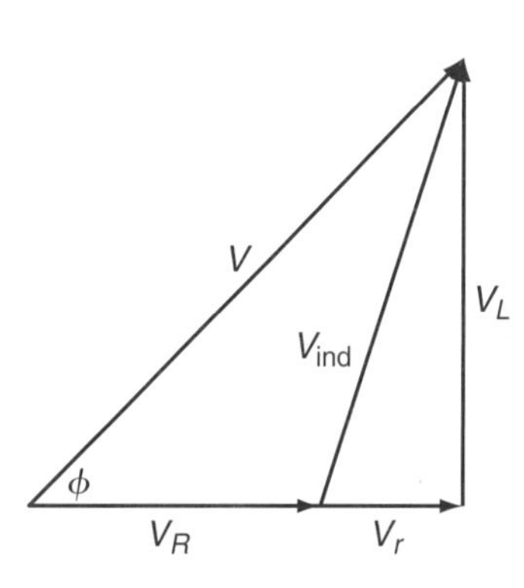

Figure 36-3 Series circuit of inductor with inductance L and internal resistance r, a resistor R, and a generator of voltage V. Also shown is the phasor diagram of the voltages.

With ϕ, V_L, and V_r determined from Equations 5 and 6, Equations 7 can be used to solve for ωL and r by eliminating I to get

$$\omega L = R\frac{V_L}{V_R} \qquad r = R\frac{V_r}{V_R} \qquad \text{(Eq. 8)}$$

EXPERIMENTAL PROCEDURE

1. Connect the inductor in series with the sine wave generator and a resistance box to form a circuit like that of Figure 36-3. Set the generator to maximum voltage and a frequency of 800 Hz. Set the resistance box to a value of 400 Ω and record that value as R and the frequency f in the Data Table.
2. Using the alternating current voltage scale on the voltmeter, measure the generator voltage V, the inductor voltage V_{ind}, and the resistor voltage V_R. Record these values in the Data Table.
3. Repeat Steps 1 and 2 for R of 600, 800, and 1000. Even though the voltage setting is left at the maximum setting, the generator output might change slightly in response to the changes in R. Therefore, be sure to measure all three voltages for each value of R.
4. Make careful note of the particular inductor used and the values of L and r determined. You may need to identify it and use it again in other laboratory exercises.

CALCULATIONS

1. From the known value of the frequency f, calculate and record in the Calculations Table the value of the angular frequency ω ($\omega = 2\pi f$).
2. Use the appropriate equations to calculate $\cos\phi$, ϕ, V_L, V_r, ωL, r, and L for each of the four cases. Record all values in the Calculations Table.
3. Calculate the mean and standard errors for the four values of r and the four values of L and record them in the Calculations Table as $\bar{r}$, $\bar{L}$, α_r, and α_L.

GRAPHS

1. Construct to scale a phasor diagram like the one shown in Figure 36-4 for each of the four cases. Use one sheet of graph paper and make four separate diagrams on the one sheet of paper. Choose a scale (for example, 1.00 V/cm) so that the diagrams are as large as possible, but that each one fits on one-fourth of the sheet of paper. First construct a vector along the x axis with a length scaled to the magnitude of V_R, as shown in Figure 36-4. Use a compass to construct an arc from the end of V_R with a

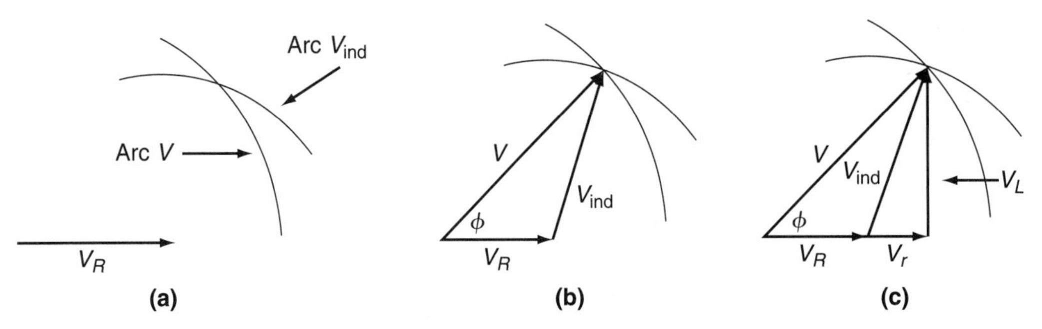

Figure 36-4 Phasor diagram construction.

radius the length of the scaled value of V_{ind}. Finally, construct an arc from the beginning of V_R with a radius the length of the scaled value of V. The intersection of the two arcs is the intersection of V_{ind} and V, and those two vectors can then be drawn in their proper direction as shown in part (b) of the figure. Finally, V_L and V_r can be constructed as shown in part (c) of the figure by dropping a perpendicular from the intersection of the arcs to the x axis and extending a vector from the end of V_R.

Name Section Date

36 LABORATORY 36 *Alternating Current* LR *Circuits*

PRE-LABORATORY ASSIGNMENT

1. For a resistor in a series alternating current circuit, the phase relationship between the current in the resistor and the voltage across the resistor is (a) the current leads the voltage by 90° (b) the voltage leads the current by 90° (c) the current is in phase with the voltage (d) the current is at some phase angle ϕ relative to the voltage (ϕ is dependent on the circuit parameters).

2. For an inductor in a series alternating current circuit, the phase relationship between the current in the inductor and the voltage across the inductor is (a) the current leads the voltage by 90° (b) the voltage leads the current by 90° (c) the current is in phase with the voltage (d) the current is at some phase angle ϕ relative to the voltage (ϕ is dependent on the circuit parameters).

3. For a generator in a series alternating current circuit, the phase relationship between the generator voltage and the current in the generator is (a) the current leads the voltage by 90° (b) the voltage leads the current by 90° (c) the current is in phase with the voltage (d) the current is at some phase angle ϕ relative to the voltage (ϕ is dependent on the circuit parameters).

4. If a generator has a maximum voltage of 5.00 V, what is the root-mean-square voltage of the generator? Show your work.

   V_{rms} = __________ V

5. A 2.50 mH inductor has an rms voltage of 15.0 V across it at a frequency $f = 200$ Hz. What is the rms current in the inductor? Show your work.

   I_{rms} = __________ A

6. A pure inductor L and a pure resistor R are in series with a generator of voltage V. The voltage across the inductor is $V_L = 10.0\,\text{V}$. The voltage across the resistor is 15.0 V. What is the voltage V of the generator? Show your work.

$V =$ ______________ V

7. A 500 Ω resistor and a real inductor with a pure inductance of L and an internal resistance of r are in series with a generator with a voltage of $V = 10.0\,\text{V}$ and an angular frequency of $\omega = 1000\,\text{rad/s}$. The voltage across the real inductor is measured to be 4.73 V, and the voltage across the 500 Ω resistor is measured to be 6.57 V. What is the value of L and r? (Hint—This is the measurement to be performed in this laboratory exercise. Use the appropriate equation to find ϕ, then the appropriate equations to find V_L and V_r, and then the appropriate equations to find ωL and r. Finally, find L from the known value of ω.) Show your work.

Name　　　　Section　　　　Date

Lab Partners

36 LABORATORY 36 *Alternating Current LR Circuits*

LABORATORY REPORT

Data Table

$f=$ Hz				
R (Ω)				
V (V)				
V_{ind} (V)				
V_R (V)				

Calculations Table

$\omega = 2\pi f =$ rad/s				
$\cos\phi$				
ϕ (degrees)				
V_L (V)				
V_r (V)				
ωL (Ω)				
r (Ω)				
L (H)				

$\bar{r} =$ Ω	$\alpha_r =$ Ω	$\bar{L} =$ H	$\alpha_L =$ H

SAMPLE CALCULATIONS

1. $\omega = 2\pi f =$
2. $\cos\phi = \dfrac{V^2 + V_R^2 - V_{\text{ind}}^2}{2VV_R} =$
3. $\phi = \cos^{-1}(\cos\phi) =$
4. $V_L = V\sin\phi =$
5. $V_r = V\cos\phi - V_R =$
6. $\omega L = (R)(V_L)/V_R =$
7. $r = (R)(V_r)/V_R =$
8. $L = (\omega L)/(\omega) =$

QUESTIONS

1. Comment on the precision of your measurement of L and r. State the evidence for your comments.

2. Examine the phasor diagrams that you have constructed. Using a protractor, measure the angle ϕ of the constructed triangle of V, V_R, and V_{ind}. Compare it with the calculated value of ϕ for each of the phasor diagrams. Calculate the percentage error in the value of ϕ from the diagram compared to the calculated value.

3. If your inductor was used in a series circuit with a resistance of $R = 10{,}000\ \Omega$ and a generator of $\omega = 100{,}000$ rad/s, what would be the phase angle ϕ? (Hint—The resistance of the inductor would be negligible.)

4. Consider the circuit that you measured with $R = 600\ \Omega$. Calculate the value of the current from each of the three Equations 7 and compare their agreement.

Alternating Current RC *and* LCR *Circuits*

OBJECTIVES

- ❑ Investigate the phase relationship between the voltage across the resistor V_R and the voltage across the capacitor V_C in an *RC* circuit.
- ❑ Determine the value of the capacitance of a capacitor in an *RC* circuit.
- ❑ Investigate the phase relationships among the voltages across the resistor, the capacitor, and the inductor in an *LCR* circuit.

EQUIPMENT LIST

- Sine wave generator (variable frequency, 5 V peak to peak amplitude), resistance box
- A 100-mH inductor of known L and r, 1.00-μF capacitor, compass, protractor
- Alternating current voltmeter (digital readout, capable of measuring high frequency)
- *(This laboratory assumes that either Laboratory 36 has previously been performed, and that the values of the inductance (L) and resistance (r) for that inductor coil have been recorded and retained, or else an inductor with accurate known values of L and r is provided.)*

THEORY

RC Circuit

A series circuit consisting of a capacitor C, a resistor R, and a sine wave generator of frequency f is shown in Figure 37-1. Also shown in the figure is a **phasor diagram** for the generator voltage V, the voltage across the resistor V_R, and the capacitor voltage V_C. It is assumed that all voltages discussed in this laboratory are root-mean-square values. The voltages V_R and V_C are 90° out of phase, and the voltages V, V_R, and V_C form a right triangle as shown in Figure 37-1. Therefore, the equation relating the magnitudes of the measured voltages in an **RC circuit** is

$$V = \sqrt{V_C^2 + V_R^2} \qquad \text{(Eq. 1)}$$

Note that Equation 1 is valid for only a pure capacitor with no resistive component. If measurements on a real capacitor show agreement with Equation 1, it would indicate that the capacitor has no significant resistive component.

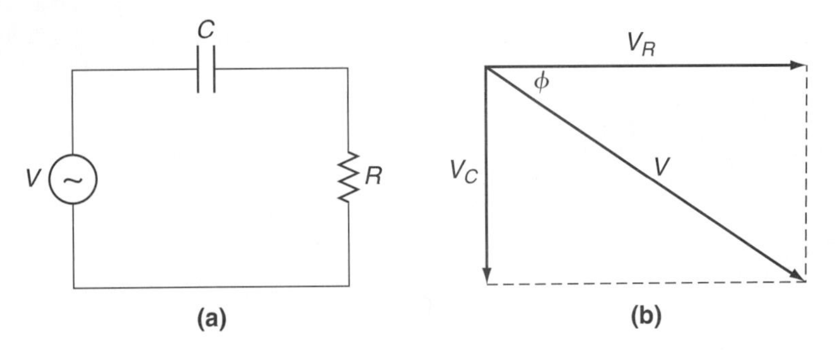

Figure 37-1 *RC* circuit and the phasor diagram for the voltage across each element.

In an *RC* circuit the current *I* is the same in each element of the circuit, and the relationships between the voltage and the current for the resistor and capacitor are

$$V_R = IR \quad \text{and} \quad V_C = I(1/\omega C) \tag{Eq. 2}$$

The quantity $1/\omega C$ is called the **capacitive reactance,** and it has units of ohms. If the current is eliminated between the two equations in 2, an equation for *C* is given by

$$C = \left(\frac{1}{\omega R}\right)\left(\frac{V_R}{V_C}\right) \tag{Eq. 3}$$

Thus a value for the capacitance of an unknown capacitor can be determined from Equation 3 if ω and *R* are known and V_R and V_C are measured.

LCR Circuit

Consider a series **LCR** circuit shown in Figure 37-2 with a generator of voltage *V*, a resistor *R*, a capacitor *C*, and an inductor having inductance *L* and resistance *r*. Note that the capacitor is assumed to have no resistance. Also shown in Figure 37-2 is the phasor diagram for the voltages V, V_R, V_C, V_L, and V_r. The figure shows that V_L and V_C are 180° out of phase, and V_R and V_r are in phase. The quantities $V_L - V_C$, V, and $V_R + V_r$ form a right triangle and

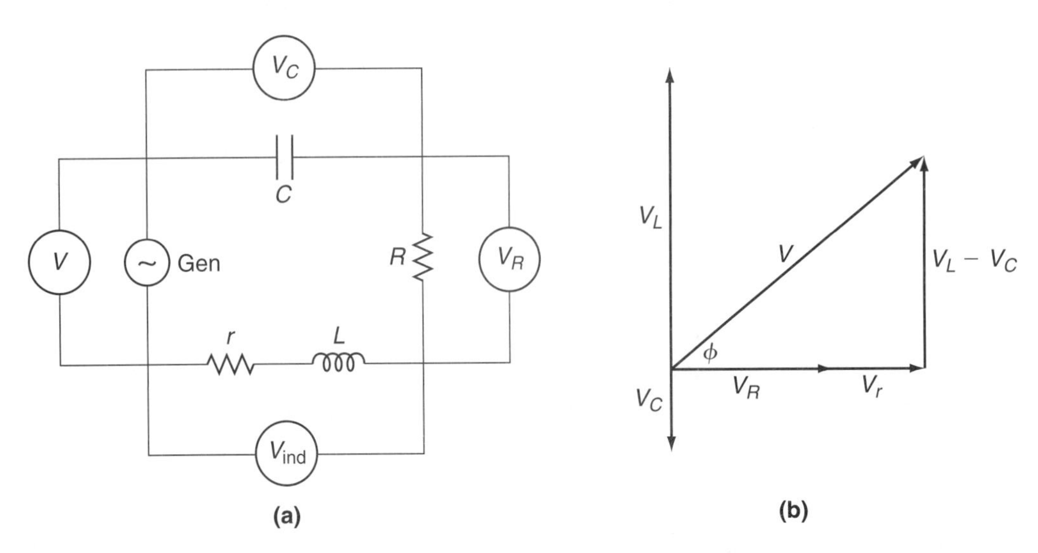

Figure 37-2 *LCR* circuit with voltmeter in the four positions to measure voltage across each element of the circuit. Also shown is a phasor diagram of all the relevant voltages.

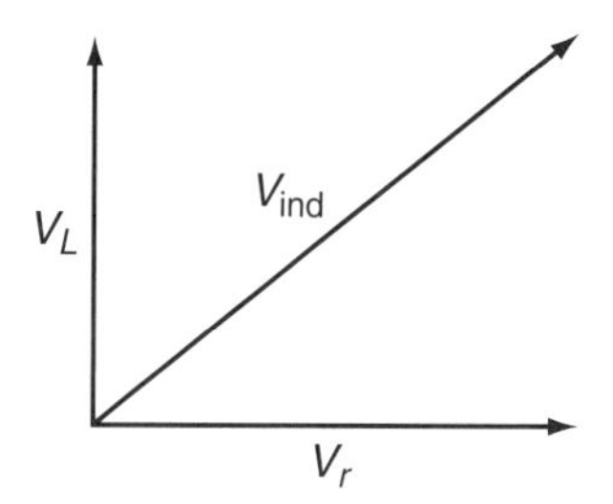

Figure 37-3 Phasor diagram for the voltages across the components of the inductor.

$$V = \sqrt{(V_L - V_C)^2 + (V_R + V_r)^2} \quad \text{(Eq. 4)}$$

It is not possible to measure either V_L or V_r directly. The only voltage associated with the inductor that can be measured experimentally is shown in Figure 37-2 as V_{ind}, and it is the vector sum of the voltages V_L and V_r. The relationship between V_{ind}, V_L, and V_r is shown in Figure 37-3. The figure shows that the voltages V_{ind}, V_L, and V_r obey the relationship

$$V_{\text{ind}}^2 = V_L^2 + V_r^2 \quad \text{(Eq. 5)}$$

The current I is the same in each element of the circuit, and V_L and V_r can be expressed as

$$V_L = I\,(\omega L) \qquad \text{and} \qquad V_r = I\,r \quad \text{(Eq. 6)}$$

The quantity ωL is called the inductive reactance. It has units of ohms. If Equations 5 and 6 are combined, and the current I is eliminated, it can be shown that

$$V_L = V_{\text{ind}} \frac{\omega L}{\sqrt{(\omega L)^2 + r^2}} \qquad \text{and} \qquad V_r = V_{\text{ind}} \frac{r}{\sqrt{(\omega L)^2 + r^2}} \quad \text{(Eq. 7)}$$

Assuming that ω, L, and r are known, Equation 7 can be used to determine V_L and V_r if V_{ind} is measured. These values of V_L and V_r combined with measured values of V_R and V_C can be used in Equation 4 to verify the relationship between these quantities and the measured generator voltage V.

EXPERIMENTAL PROCEDURE

RC Circuit

1. Connect the capacitor provided in series with the sine wave generator and a resistance box to form a circuit like that shown in Figure 37-1. Set the generator output to maximum voltage and set the frequency to 250 Hz. Record the value of f in Data Table 1. Set the resistance box to a value of 300 Ω and record that value as R in Data Table 1.
2. Use the alternating voltage scale on the voltmeter to measure and record in the Data Table the generator voltage V, the capacitor voltage V_C, and the voltage across the resistor V_R.
3. Repeat the above procedure for three other cases using $R = 500$, 700, and 900 Ω. The generator output might change slightly in response to changes in R. Therefore, measure all three voltages for each value of R.
4. Obtain a value for the capacitance of your capacitor from your instructor. Record this known value as C_k in Data Table 1.

LCR Circuit

1. Construct a series *LCR* circuit like the one shown in Figure 37-2 using the same capacitor used previously, an inductor for which the values of L and r are known, a resistance box, and the sine wave generator. Record the values of L and r in Data Table 2.
2. Set the generator output to maximum voltage and set the frequency to 800 Hz. Set the resistance box to a value of 200 Ω. Record the values of f and R in Data Table 2.
3. Using the alternating voltage scale on the voltmeter, measure the generator voltage V, the capacitor voltage V_C, the resistor voltage V_R, and the inductor voltage V_{ind}. Record these values in Data Table 2.
4. Repeat the procedure of Steps 1 through 3 with the generator frequency set to $f = 600$ Hz and the resistance box set to $R = 200\ \Omega$.
5. Repeat the procedure of Steps 1 through 3 two more times, once with $R = 300\ \Omega$ and $f = 600$ Hz, and again with $R = 300\ \Omega$ and $f = 800$ Hz.

CALCULATIONS

RC Circuit

1. From the known value of the frequency f, calculate the value of the angular frequency $\omega = 2\pi f$. Record the value of ω in Calculations Table 1.
2. Calculate the quantity $\sqrt{V_C^2 + V_R^2}$ for each case and record the values in Calculations Table 1. Calculate the percentage error in each of these values compared to the measured values of the generator voltage and record them in Calculations Table 1.
3. Calculate the values of C from the measured values of V_R and V_C for each value of R. Record each value of C in Calculations Table 1. Calculate the mean $\overline{C}$ and the standard error α_C for the four values of C and record them in Calculations Table 1.
4. Calculate the percentage error in the value of $\overline{C}$ compared to the known value of the capacitance C_k.

LCR Circuit

1. From the known value of f for each case, calculate the angular frequency $\omega = 2\pi f$ and record the values in Calculations Table 2.
2. Calculate the four values of V_L and V_r and record them in Calculations Table 2.
3. Calculate and record in Calculations Table 2 the four values of $V_L - V_C$ and $V_R + V_r$.
4. For each of the four cases calculate and record in Calculations Table 2 the value of the quantity $\sqrt{(V_L - V_C)^2 + (V_R + V_r)^2}$.
5. Calculate the percentage error in the quantity $\sqrt{(V_L - V_C)^2 + (V_R + V_r)^2}$ compared to the measured value of the generator voltage V. Record the values of those percentage errors in Calculations Table 2.

Name ______ Section ______ Date ______

37 LABORATORY 37 *Alternating Current* RC *and* LCR *Circuits*

PRE-LABORATORY ASSIGNMENT

1. In a series *RC* circuit such as the one in Figure 37-1, the following phase relationship exists between the generator voltage V, the capacitor voltage V_C, and the resistor voltage V_R. (a) V and V_R are in phase, V_C lags V_R by 90° (b) V_C and V_R are at angle ϕ, V leads V_C by 90° (c) V_C lags V_R by 90°, V lags V_R by angle ϕ (d) V_R lags V_C by ϕ, V_C leads V by 90°.

2. If a series *RC* circuit has $V_R = 12.6$ V and $V_C = 10.7$ V, the generator voltage must be (a) 12.6 V (b) 23.3 V (c) 1.9 V (d) 16.5 V. Show your work.

3. A series *RC* circuit has $\omega = 2000$ rad/s and $R = 300\ \Omega$. The voltage V_C is measured to be 4.76 V, and V_R is measured to be 6.78 V. What is the value of C? Show your work. (a) 2.37 μF (b) 5.00 μF (c) 1.17 μF (d) 4.67 μF.

4. A series *RC* circuit of $R = 500\ \Omega$ and $C = 3.00\ \mu$F is measured to have $V_R = 8.07$ V and $V_C = 6.68$ V. What is the current I, and what is the value of ω the angular frequency? Show your work.

   $I =$ ______ A $\qquad \omega =$ ______ rad/s

5. A series *LCR* circuit consists of an inductor of inductance L and resistance r, a capacitor C, a resistor R, and a generator of voltage V. Mark as true or false the following statements concerning the relative phase of V, V_L, V_r, V_C, and V_R.

______ 1. V_r and V_R are in phase.

______ 2. V_L leads V_C by 90°.

______ 3. V is at angle ϕ relative to V_R.

______ 4. $V_L - V_C$ is in phase with V_r.

______ 5. V_C lags V_R by 90°.

6. Measurements on the circuit described in Question 5 give $V_L = 10.76$ V, $V_C = 5.68$ V, $V_R = 6.32$ V, and $V_r = 3.75$ V. What is the generator voltage V? Show your work.

V = ______________ V

7. An inductor with $L = 150$ mH and $r = 200\,\Omega$ is in series with a capacitor, a resistor, and a generator of $\omega = 1000$ rad/s. The voltage across the inductor V_{ind} is measured to be 10.87 V, V_R is measured to be 4.65 V, and V_C is measured to be 5.96 V. What is the generator voltage V? (Hint—This is the measurement to be performed in this laboratory for *LCR* circuits. Use the appropriate equations to find V_L and V_r, and then use them and the values of V_R and V_C to calculate V.)

Name　　　　Section　　　　Date

Lab Partners

37 LABORATORY 37 *Alternating Current RC and LCR Circuits*

LABORATORY REPORT

Data Table 1

$f =$ Hz		$C_k =$ μF		
R (Ω)				
V_C (V)				
V_R (V)				
V (V)				

Calculations Table 1

$\omega =$ rad/s				
$\sqrt{V_C^2 + V_R^2}$				
% Error				
C (μF)				
$\overline{C} =$ μF	$\alpha_C =$ μF	% Error $\overline{C} =$		

Data Table 2

$r =$ Ω		$L =$ H		$C =$ μF
f (Hz)				
R (Ω)				
V (V)				
V_{ind} (V)				
V_C (V)				
V_R (V)				

Calculations Table 2

ω (rad/s)				
V_L (V)				
V_r (V)				
$(V_L - V_C)$ (V)				
$(V_R + V_r)$ (V)				
$\sqrt{(V_L - V_C)^2 + (V_R - V_r)^2}$ (V)				
% Error compared to V				

SAMPLE CALCULATIONS

1. $\omega = 2\pi f =$
2. $V = \sqrt{V_C^2 + V_R^2} =$
3. $C = (1/\omega R)(V_R/V_C) =$
4. $V_L = (V_{ind})\left(\omega L/\sqrt{(\omega L)^2 + r^2}\right) =$
5. $V_r = (V_{ind})\left(r/\sqrt{(\omega L)^2 + r^2}\right) =$
6. $V = \sqrt{(V_L - V_C)^2 + (V_R + V_r)^2} =$

QUESTIONS

1. Comment on the agreement between the measured generator voltage V and the quantity $\sqrt{V_C^2 + V_R^2}$ for the RC circuit data.

2. Do your results for Question 1 confirm that the capacitor has no resistance? State specifically how the data either do or do not confirm this expectation.

3. State carefully your evaluation of the precision of your measurements of the value of the capacitor in the RC circuit. State the evidence for your opinion.

4. Considering the given value of C_k as the true value, comment on the accuracy of your measurements of the capacitance.

5. Comment on the agreement between the measured generator voltage V and the quantity $\sqrt{(V_L - V_C)^2 + (V_R + V_r)^2}$ in the LCR circuit.

6. Do your results confirm the phasor diagram of Figure 37-2 as a correct model for the addition of the voltages in an *LCR* circuit? State why they do or do not confirm this model.

Oscilloscope Measurements

OBJECTIVES

- ❑ Investigate the fundamental principles and practical operation of the oscilloscope using signals from a function generator.
- ❑ Measure sine and other waveform signals of varying voltage and frequency.
- ❑ Compare voltage measurements with the oscilloscope to voltage measurements using an alternating current voltmeter.

EQUIPMENT LIST

- Oscilloscope (typical direct current to 20 Mhz), alternating current voltmeter (high frequency capability)
- Function generator (sine wave plus additional wave form such as a square wave or triangular wave), appropriate connecting wires (BNC to banana plug)

THEORY

The fundamental working part of an **oscilloscope** is a device called a **cathode-ray tube** (CRT). Its components include a heated filament to emit a beam of electrons, a series of electrodes to accelerate, focus, and control the intensity of the emitted electrons, two pairs of deflection plates that deflect the electron beam when there is a voltage between the plates (one pair for deflection in the horizontal direction and one pair for deflection in the vertical direction), and a **fluorescent screen** that emits a visible spot of light at the point where the beam of electrons strikes the screen. Together the heated filament and series of electrodes are called an electron gun. The electron gun and deflecting plates are arranged linearly inside an evacuated glass tube, and the fluorescent screen coats the glass tube at the opposite end of the tube from the electron gun as shown in Figure 38-1.

When there is no voltage between either pair of deflection plates, the electron beam will travel straight down the evacuated tube and strike the center of the fluorescent screen. When a constant voltage is applied between either the horizontal or vertical deflection plates, the beam will be displaced by a constant amount on the fluorescent screen in either the horizontal (x) or vertical (y) direction. The direction of the displacement depends upon the sign of the voltage, and the magnitude of the displacement is proportional to the voltage. If a time-varying voltage is applied to either set of deflecting plates, the displacement of the beam will vary with time as the applied voltage varies with time, and the electron beam spot will move on the screen as a function of time. When the beam strikes the screen the phosphor glow persists for approximately 0.1 s.

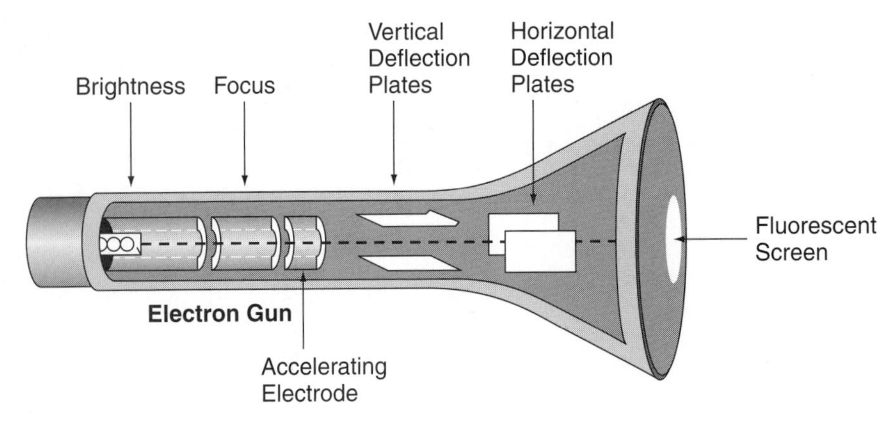

Figure 38-1 Cathode-ray tube.

We can deflect the electron beam in the horizontal (x) direction to represent a time scale by applying a time-varying sawtooth voltage waveform as shown in Figure 38-2.

When a voltage of that waveform and of the appropriate maximum voltage is applied to the horizontal plates, the beam spot will sweep across the fluorescent screen once each time the voltage linearly increases from its minimum up to its maximum. At the end of the sweep of the beam across the screen, the beam returns to the left of the screen. The time this takes will equal the period T of the sawtooth waveform. Because this waveform sweeps the beam across the screen, it is commonly called the **sweep generator.**

If the period T of the sweep generator is 1 s, the beam will clearly be recognizable as a spot that moves at constant speed across the tube face. If the period is as short as 0.1 s, the beam is no longer recognizable as a spot, but instead appears to be a somewhat pulsating line. This is because of the persistence of the phosphor, which causes the trace to still be glowing from one pass of the beam when another pass of the beam begins. For periods T of 0.01 s or less, the beam is moving across the screen so often that the persistence of the phosphor makes the trace appear as a steady line.

The oscilloscope is designed so that a series of specific sweep generator periods can be applied to the horizontal plates by selecting the position of a multiposition switch. The width of oscilloscope screen is fixed, usually 10 cm. Each different choice of period T represents a specific time per length of scale division in the horizontal direction. Typically these are chosen to decrease in a series of scales that are in the ratio 2:1:0.5. For a typical student-type oscilloscope, the time scales would be 19 settings ranging from 0.2 s/cm to 0.2 ms/cm. Because the screen is 10 cm wide, there is a factor of 10 between the period T and the time scale. If the period of the sweep generator is 10 ms, the time scale is 1 ms/cm. Time $t = 0$ is assumed to occur at the left of the screen, and time is assumed to increase to the right.

In the vertical direction the screen is typically smaller, usually about 8 cm total. The vertical input is calibrated directly in volts. The input voltage scale is also variable by the choice of a multiposition switch that selects the appropriate amplification of the input voltage over some chosen voltage range. The typical range of possible voltage scales is from 5 V/cm to 5 mV/cm. This choice of voltage scales allows a range of input voltages to be displayed with deflections on the oscilloscope screen that are large enough to be easily visible. For the choices stated, the maximum voltage that can be displayed on the screen is 20 V. The voltage can be either positive or negative polarity, so the vertical scale has its zero in the center of the screen to display both positive and negative voltages.

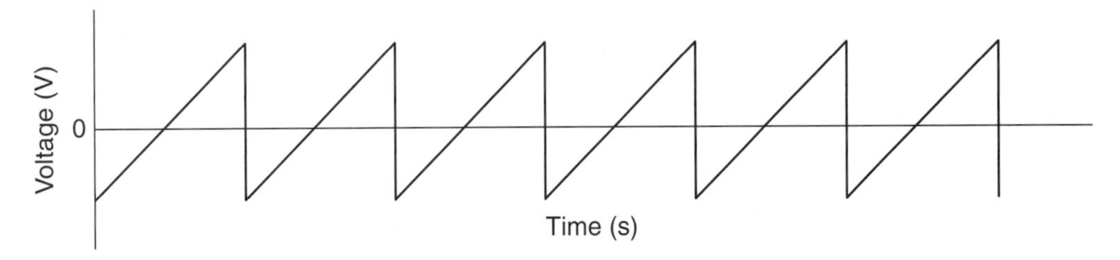

Figure 38-2 Sawtooth voltage waveform.

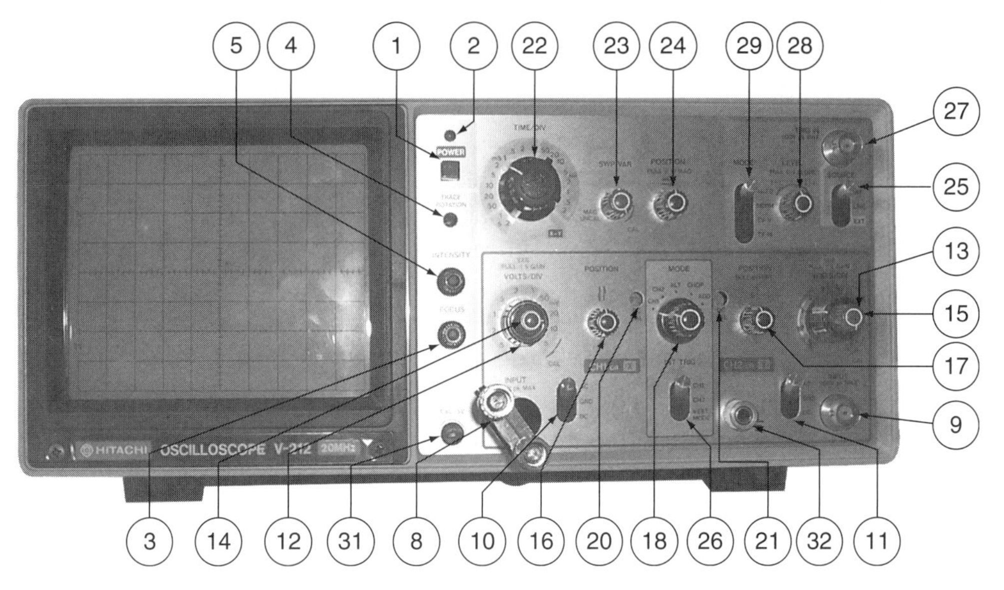

Figure 38-3 The Hitachi model V-212 oscilloscope.

The most common use of the oscilloscope is to use the time scale provided by the sweep generator to display the time variation of a voltage signal that is applied to the vertical plates. Usually this is some specific waveform that is repeated with a fixed frequency. For example, if a simple sine wave voltage is applied to the vertical plates, a display of the voltage versus time will be directly displayed on the oscilloscope screen as a sine wave trace of the beam with a maximum amplitude proportional to the maximum voltage of the signal, and with a period on the time scale of the oscilloscope that is equal to the period of the signal. If the voltage waveform applied to the vertical plates is a more complex waveform, the resulting trace on the screen will represent the shape of that complex waveform.

The discussion so far has ignored one important point, which involves the means to coordinate the starting time of the sweep generator with the starting point of the voltage signal that is to be displayed. We accomplish this by using some waveform as a "trigger" to start the sweep generator. The triggering waveform can be the same signal that is input to the vertical plates for analysis, a secondary external signal, or the 60 Hz line voltage. When the signal itself is used as the trigger for the sweep generator, the signal is observed on the oscilloscope as a steady display that is constant in time because the sweep generator is initiated at the same point on the repetitive vertical signal for each pass of the sweep generator. On most oscilloscopes this is referred to as internal triggering. That is the mode we will use in this laboratory.

EXPERIMENTAL PROCEDURE

The procedure refers to the Hitachi model V-212 in Figure 38-3. It is a typical student oscilloscope. If using another oscilloscope, refer to the instruction manual for the corresponding settings and controls.

In several of the instructions below, you are asked to draw what is on the oscilloscope display on the grids provided. In each of those cases, assume that the VOLTS/DIV and TIME/DIV are properly calibrated, and fill in the blank given for the values of VOLTS/DIV and TIME/DIV for the exercise associated with each set of grids. On the vertical scale 0 V is labeled. Label the full-scale voltage both positive and negative. The time scale is labeled with 0 s. Label the value of the full-scale time on the horizontal axis. Do this for each grid.

1. Turn on the power to the oscilloscope and let it come to thermal equilibrium for at least 10 minutes. Set the oscilloscope mode setting to CH1, the trigger source to INT, the trigger level to zero (center of range), trigger SLOPE to + (level knob pushed in), trigger MODE to AUTO, the INT TRIG to CH1, and CH1 to AC.

2. Set the TIME/DIV control to 1 ms/DIV, the SWP VAR control rotated fully clockwise to the CAL position, the VOLTS/DIV control to 1 V/DIV, and the VAR (PULL $\times$ 5 GAIN) control rotated fully clockwise to the CAL position.
3. Turn on the power to the function generator and let it come to thermal equilibrium for at least 10 minutes. Select a sine wave voltage, set the frequency $f = 100$ Hz, and connect the output of the function generator to the CH1 INPUT of the oscilloscope. Adjust the amplitude control of the function generator to zero. Adjust the VERTICAL POSITION control of the oscilloscope until the flat trace is exactly on the center line of the vertical display.
4. (a) Adjust the amplitude control of the function generator until the display on the oscilloscope is full-scale positive on the positive part of the cycle and full-scale negative on the negative part of the cycle. In the laboratory report section, carefully draw on the grid labeled 1A what is displayed on the screen. (b) Leaving all other parameters fixed, set the VOLT/DIV control to 2 V/DIV, and draw on the grid labeled 1B what is now displayed on the screen. (c) Leaving all other parameters fixed, set the VOLT/DIV control to 5 V/DIV, and draw on the grid labeled 1C what is now displayed on the screen.
5. (a) Leaving all other parameters fixed, set the VOLT/DIV control to 1 V/DIV, and select $f = 200$ Hz from the function generator. Draw on the grid labeled 2A what is now displayed on the screen. (b) Leaving all other parameters fixed, select $f = 400$ Hz from the function generator, and draw on the grid labeled 2B what is now displayed on the screen. (c) Leaving all other parameters fixed, select $f = 600$ Hz from the function generator, and draw on the grid labeled 2C what is now displayed on the screen.
6. (a) Leaving all other parameters fixed, set the VOLT/DIV control to 1 V/DIV, the TIME/DIV control to 2 ms/DIV, and select $f = 100$ Hz from the function generator. Note that the trigger slope control is still set at (+). Draw on the grid labeled 3A what is now displayed on the screen. (b) Leaving all other parameters fixed, pull out the trigger level control that sets the trigger slope to (−). Draw on the grid labeled 3B what is now displayed on the screen.
7. (a) Leaving all other parameters fixed, push in the trigger level control that sets the trigger slope to (+), and the trigger level is still set at zero. Draw on the grid label 4A what is now displayed on the screen. (b) Leaving all other parameters fixed, slowly turn the trigger level control clockwise, increasing the trigger level. Increase it only so long as the display remains triggered. At the maximum level that the display is triggered, draw on the grid labeled 4B what is displayed on the screen. (c) Leaving all other parameters fixed, slowly turn the trigger level control counterclockwise, decreasing the trigger level. Decrease it only so long as the display remains triggered. At the minimum level that the display is triggered, draw on the grid labeled 4C what is displayed on the screen.
8. Push the trigger level control in for (+) slope and turn the level back to zero. Set the TIME/DIV to 2 ms/DIV and set the function generator to a sine wave of $f = 100$ Hz. Use the alternating current voltmeter to set the output of the function generator to 1.00 V as read on the voltmeter. Input this sine wave to the oscil-loscope and measure the peak voltage of the sine wave. To measure the peak voltage of the sine wave, you are free to adjust the VOLT/DIV control to give the most accurate measurement possible. Generally this means adjusting the scale for as large a deflection as possible. Record the peak voltage of the sine wave as read from the oscilloscope in Data Table 1. Complete all the measurements in Data Table 1 from 1.00 V to 5.00 V. For each voltage, set the output from the generator using the voltmeter, and then read the voltage from the oscilloscope, each time choosing the VOLT/DIV that will allow the most accurate reading from the oscilloscope.
9. Set the function generator to output a triangular wave with $f = 1000$ Hz, and the TIME/DIV on the oscilloscope to 1 ms/DIV. Use the alternating current voltmeter to set the output of the function generator to 1.00 V as read on the voltmeter. Input this triangular wave to the oscilloscope and measure the peak voltage of the wave. Proceed as instructed for the sine wave above, this time measuring the voltages between 1.00 V and 5.00 V as read on the voltmeter. Record the results in Data Table 2.

10. The goal of this laboratory is to introduce students to the oscilloscope. Now simply experiment for yourself with the features of the oscilloscope. Input as many different frequencies and waveforms as time allows and attempt to learn everything you can about the operation of the oscilloscope by simply trying different settings of all of the oscilloscope controls.

CALCULATIONS

1. Perform a linear least squares fit to the data in Data Table 1 with the peak voltage read on the oscilloscope as the horizontal axis and the voltage as read on the voltmeter as the vertical axis. Determine the slope, the intercept, and the correlation coefficient. Record those values in Calculations Table 1.
2. Perform a linear least squares fit to the data in Data Table 2 with the peak voltage read on the oscilloscope as the horizontal axis and the voltage as read on the voltmeter as the vertical axis. Determine the slope, the intercept, and the correlation coefficient. Record those values in Calculations Table 2.

Name | Section | Date

38 LABORATORY 38 *Oscilloscope Measurements*

PRE-LABORATORY ASSIGNMENT

1. Describe the components that make up the electron gun in a cathode-ray tube.

2. Describe the voltage waveform that produces a linear time scale when applied to the horizontal plates of a cathode-ray tube.

3. When the electron beam strikes the fluorescent screen, the phosphor glow that results has persistence. Approximately how long does the glow persist?

4. A function generator outputs a sine wave of $f = 200$ Hz. It is input to an oscilloscope set at 1 ms/DIV. How many complete cycles of the sine wave are displayed on the oscilloscope? (Hint—The period of the sine wave T is related to the frequency f of the wave by $T = 1/f$, and there are 10 divisions on the time display of the oscilloscope.)

5. A typical student oscilloscope on its least sensitive calibrated scale can display a voltage up to a maximum of approximately (a) 1 V (b) 5 V (c) 20 V (d) 200 V.

6. A typical student oscilloscope on its most sensitive calibrated scale can display a voltage down to a minimum of approximately (a) 1 mV (b) 5 mV (c) 20 V (d) 200 mV.

7. A sawtooth wave with a period of 100 ms is applied to an oscilloscope with a screen 10 cm wide. What time is represented by 1 cm on the screen?

Name ____________________ Section __________ Date __________

Lab Partners ____________________

38 LABORATORY 38 *Oscilloscope Measurements*

LABORATORY REPORT

1A. TIME/DIV = __________

1A. VOLTS/DIV = __________

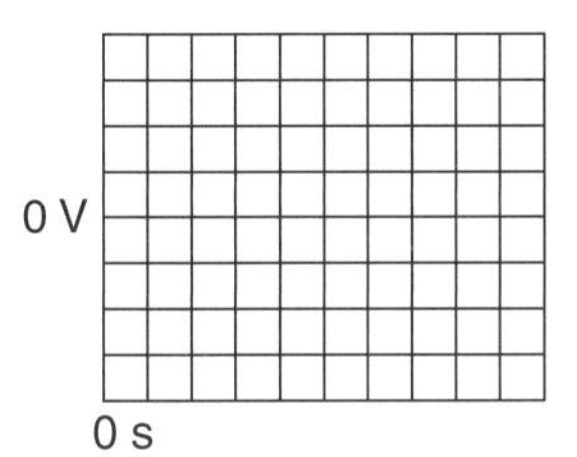

1B. TIME/DIV = __________

1B. VOLTS/DIV = __________

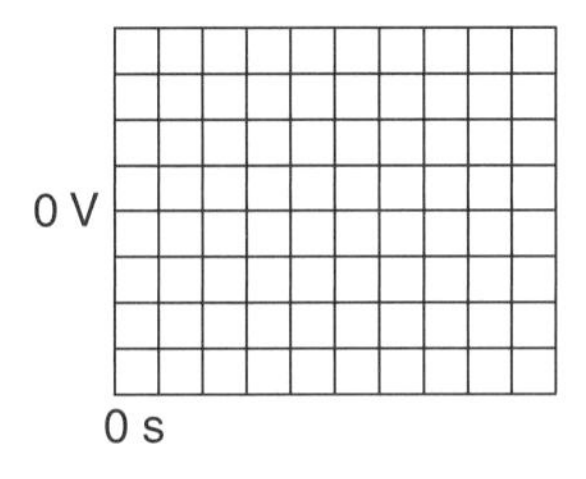

1C. TIME/DIV = __________

1C. VOLTS/DIV = __________

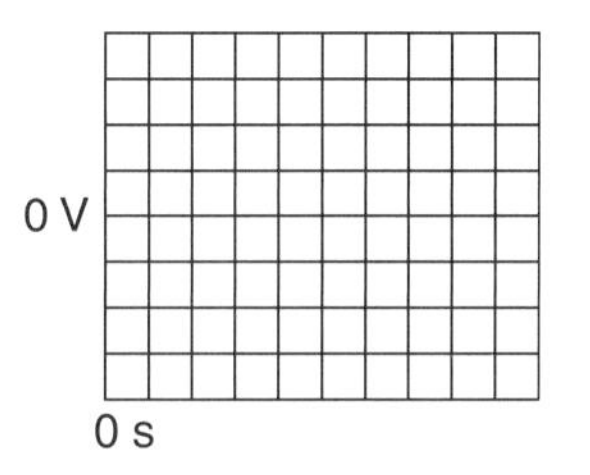

2A. TIME/DIV = __________

2A. VOLTS/DIV = __________

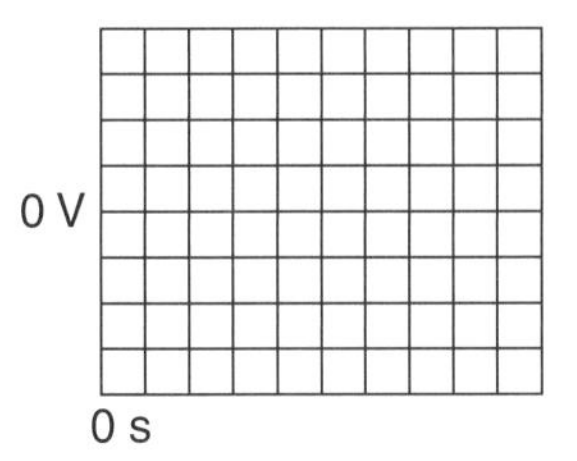

2B. TIME/DIV = __________

2B. VOLTS/DIV = __________

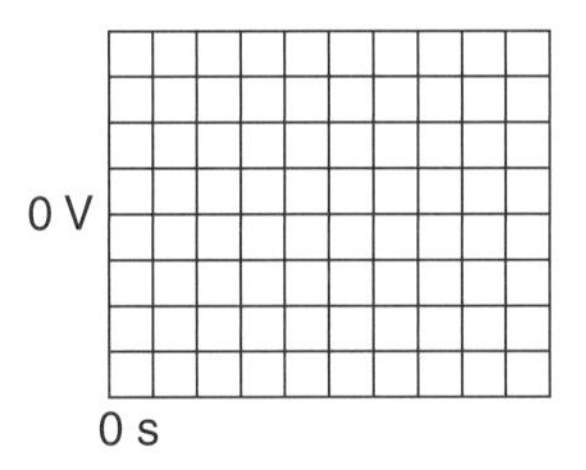

2C. TIME/DIV = __________

2C. VOLTS/DIV = __________

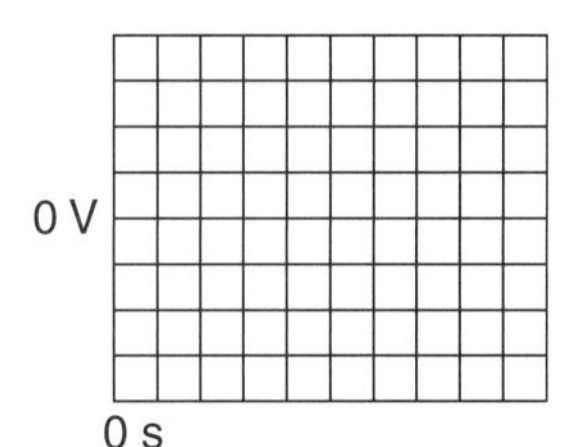

3A. TIME/DIV = __________

3A. VOLTS/DIV = __________

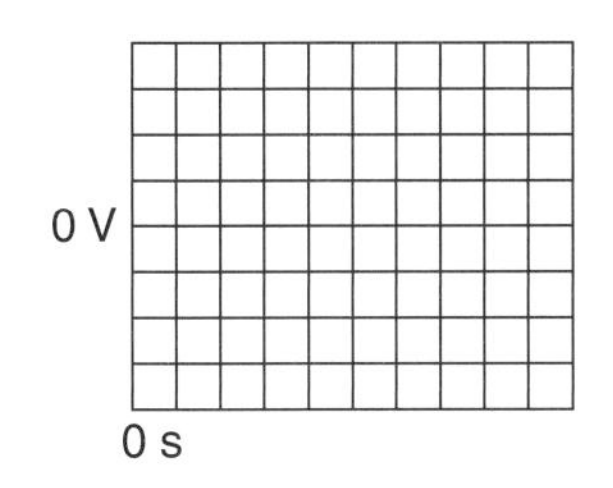

3B. TIME/DIV = __________

3B. VOLTS/DIV = __________

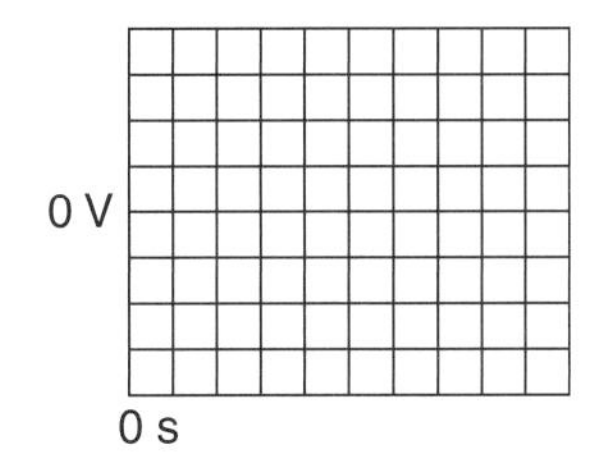

4A. TIME/DIV = __________
4A. VOLTS/DIV = __________

4B. TIME/DIV = __________
4B. VOLTS/DIV = __________

4C. TIME/DIV = __________
4C. VOLTS/DIV = __________

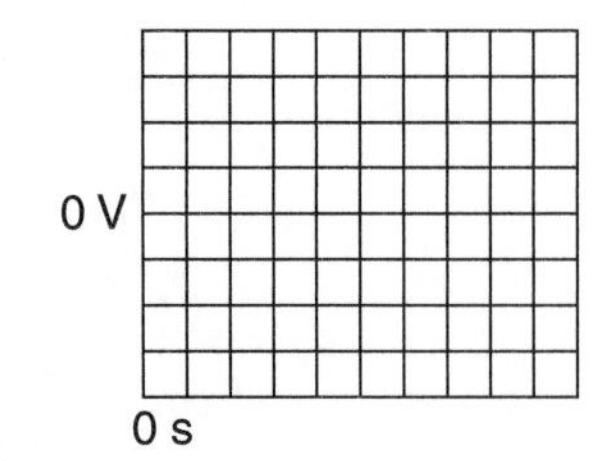

Data Table 1

Voltmeter (V)	*Oscilloscope* (V)
1.00	
2.00	
3.00	
4.00	
5.00	

Data Table 2

Voltmeter (V)	*Oscilloscope* (V)
1.00	
2.00	
3.00	
4.00	
5.00	

Calculations Table 1

Intercept =
Slope =
$r =$

Calculations Table 2

Intercept =
Slope =
$r =$

SAMPLE CALCULATIONS

None

QUESTIONS

1. In the grid labeled 2A, how many complete cycles are sketched in your figure? From your sketch, what is the period of the wave? Using this period, calculate the frequency of the wave for this sketch. Is it in agreement with the frequency used for this part of the experiment?

2. In your own words, explain why these two sketches in 3A and 3B appear as they do. They both have the trigger level zero, but one has a positive trigger slope and the other has a negative trigger slope.

3. Explain the appearance of sketches 4A, 4B, and 4C. They all have a positive trigger slope, but the trigger level of 4A is zero, the trigger level of 4B is positive, and the trigger level of 4C is negative.

4. For a sine wave, an alternating current voltmeter measures a root-mean-square value that is 0.707 of the peak value of the sine wave. Therefore the peak value measured on the oscilloscope should be 1/.707, or 1.414 times the voltmeter readings. The slope of the data in Data Table 1 that you calculated and recorded in Calculations Table 1 should be approximately 1.414. Calculate the percentage error between your slope for these data and 1.414.

5. For a triangular wave, an alternating current voltmeter measures a root-mean-square value that is 0.576 of the peak value of the triangular wave. Therefore the peak value measured on the oscilloscope should be 1/.576, or 1.736 times the voltmeter readings. The slope of the data in Data Table 2 that you calculated and recorded in Calculations Table 2 should be approximately 1.736. Calculate the percentage error between your slope for these data and 1.736.

6. An oscilloscope is set on a TIME/DIV setting of 50 ms. There are 10 divisions on the time scale. A sine wave on the oscilloscope display has exactly three full cycles of the sine wave that fit on the 10 divisions. What is the frequency of the wave?

Joule Heating of a Resistor

OBJECTIVES

- ❑ Investigate the dependence of the rise in temperature on the electrical energy input when an electric coil is immersed in water in a calorimeter.
- ❑ Determine an experimental value for the quantity commonly known as the **mechanical equivalent of heat** (*MEH*) and compare that value to the known value.

EQUIPMENT LIST

- Immersion heater coil to fit standard calorimeter, calorimeter, thermometer
- Direct current power supply (5A at 6 V), ammeter (0–5A), voltmeter (0–10 V)
- Laboratory timer, laboratory balance, calibrated masses

THEORY

When a resistor of resistance R has a current I at voltage V the **power** absorbed in the resistor is

$$P = I^2R = V^2/R = VI \qquad \text{(Eq. 1)}$$

Power is energy per unit time, and if the power P is constant, the **energy** U delivered in time t is given by

$$U = Pt \qquad \text{(Eq. 2)}$$

Substituting Equation 1 into Equation 2 gives the following expression for the electrical energy

$$U = VIt \qquad \text{(Eq. 3)}$$

When a resistor absorbs electrical energy, it dissipates this energy in the form of **heat** Q. If the resistor is placed in a **calorimeter,** the amount of heat produced can be measured when it is absorbed in the calorimeter. Consider the experimental arrangement shown in Figure 39-1 in which a resistor coil (also called an immersion heater) is immersed in the water in a calorimeter. The heat Q produced in the resistor is absorbed by the water, the calorimeter cup, and the resistor coil itself. This heat Q produces a rise in temperature ΔT. The heat Q is related to ΔT by

$$Q = (m_w c_w + m_c c_c + m_r c_r)\,\Delta T \qquad \text{(Eq. 4)}$$

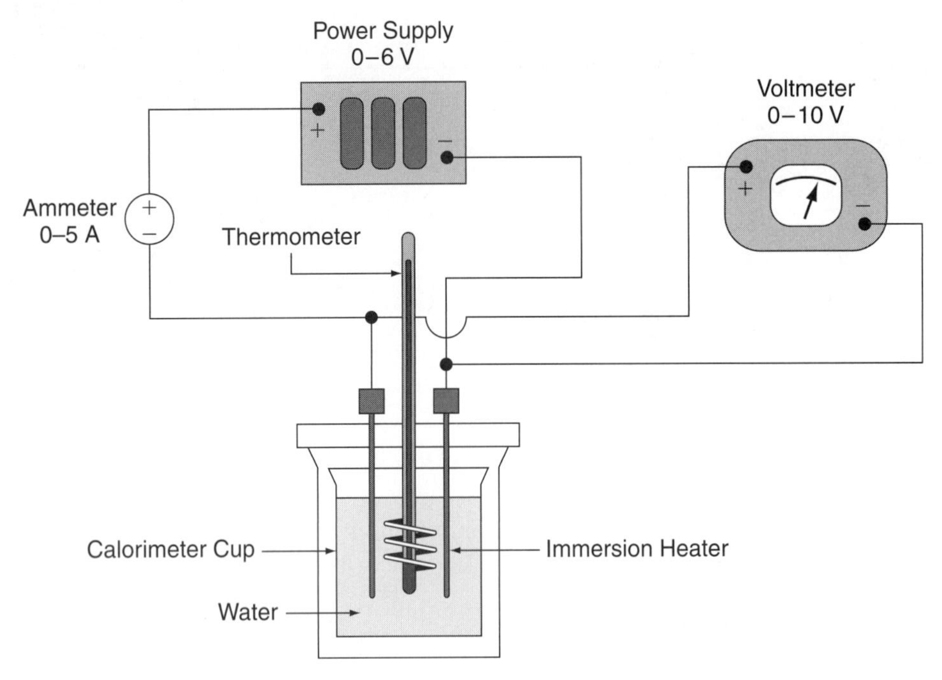

Figure 39-1 Experimental arrangement and circuit diagram for the calorimetric measurement of the heat produced in an immersion heater by an electrical current.

where each m and c are the masses and specific heats of the water, the calorimeter, and the resistor. Let mc stand for the sum of the product of mass and specific heat for the three objects that absorb the heat. In those terms the heat Q is given by the following:

$$Q = mc\,\Delta T \qquad \text{(Eq. 5)}$$

The electrical energy absorbed in the resistor is completely converted to heat. The equality of those two energies is expressed as

$$U(\text{J}) = MEH(\text{J/cal})\,Q(\text{cal}) \qquad \text{(Eq. 6)}$$

where MEH represents the conversion from **joules** to **calories.** Using the expression for U and Q from Equations 3 and 5 in Equation 6 leads to

$$VIt(\text{J}) = MEH(\text{J/cal})\,mc\,\Delta T(\text{cal}) \qquad \text{(Eq. 7)}$$

If a fixed current and voltage are applied to the resistor in a calorimeter, and the temperature rise ΔT is measured as a function of the time t, Equation 7 predicts that a graph with VIt(J) as the vertical axis and $mc\,\Delta T$ (cal) as the horizontal axis should produce a straight line with MEH as the slope.

EXPERIMENTAL PROCEDURE

1. Determine the mass m_c of the calorimeter cup and record it in the Data Table. Obtain from your instructor the specific heat of the calorimeter cup c_c, the mass of the resistor coil m_r, and its specific heat c_r, and record them in the Data Table.
2. Place enough water in the calorimeter cup to completely immerse the resistor coil. The water temperature should be a few degrees below room temperature. Be sure that the coil is completely covered by the water, but do not use any more water than is necessary. Determine the mass of the

water plus the calorimeter cup and record it in the Data Table. Determine the mass of the water by subtraction and record it as m_w in the Data Table.

3. Place the immersion heater in the calorimeter cup and construct the circuit shown in Figure 39-1. Check again that the immersion heater is below the water level. If it is not, add some more water and determine the mass of the water again.
4. Turn on the power supply and adjust the current between 4.0 and 5.0 A. Do this quickly and then turn off the supply with the output level still adjusted to the setting that produced the desired current. Do not allow the supply to stay on long enough to heat the water appreciably. Stir the system several minutes to allow it to come to equilibrium.
5. Determine the initial temperature T_i and record it in the Data Table. Estimate the thermometer readings to the nearest 0.1 C° for all temperature measurements.
6. With the power supply still set to the output level required to produce 4.0 A, turn on the power supply and simultaneously start the laboratory timer. Record the initial values of the current I and the voltage V in the Data Table. Let the timer run continuously and stir the system often. Measure and record the temperature T, the current I, and the voltage V every 60 s for 8 minutes. Record all data in the Data Table.

CALCULATIONS

1. Calculate the quantity mc where $mc = m_w c_w + m_c c_c + m_r c_r$ and record it in the Calculations Table.
2. Calculate the temperature rise ΔT above the initial temperature T_i from $\Delta T = T - T_i$ for each of the measured values of T and record the results in the Calculations Table.
3. Calculate the quantity $mc\ \Delta T$ for each case and record the results in the Calculations Table.
4. For each measurement of the voltage V and current I, calculate the product VI and record the results in the Calculations Table.
5. Calculate the mean $\overline{VI}$ and standard error α_{VI} for the values of VI and record the results in the Calculations Table.
6. Calculate the quantity $\overline{VI}\ t$ for each time t and record the results in the Calculations Table.
7. Perform a linear least squares fit to the data with $\overline{VI}\ t$ as the vertical axis and $mc\ \Delta T$ as the horizontal axis. Determine the slope MEH_{exp}, the intercept A, and the correlation coefficient r and record them in the Calculations Table.
8. Calculate the percentage error in the value of MEH_{exp} compared to the known value of $MEH = 4.186$ J/cal.

GRAPHS

1. Graph the data with $\overline{VI}\ t$ as the vertical axis and $mc\ \Delta T$ as the horizontal axis. Also show on the graph the straight line obtained from the linear least squares fit.

Name Section Date

39 LABORATORY 39 *Joule Heating of a Resistor*

PRE-LABORATORY ASSIGNMENT

1. What is the equation for the power P dissipated by a resistor of resistance R, current I, and voltage V?

2. What is the constant ratio between electrical energy (in joules) when it is converted completely to heat (in calories)? This is commonly referred to as the mechanical equivalent of heat (*MEH*).

3. A resistor has a current of 3.75 A when its voltage is 6.75 V. What is the resistance of the resistor? What power does it dissipate? Show your work.

4. A resistor has a resistance of 1.50 Ω and a voltage of 6.00 V across it. What is the current in the resistor? What power does it dissipate? Show your work.

5. If the resistor in Question 4 is immersed in water, how much energy does it deliver to the water in 350 s? Express your answer in joules and in calories. Show your work.

6. A resistor has a voltage of 6.65 V and a current of 4.45 A. It is placed in a calorimeter containing 200 g of water at 24.0°C. The calorimeter is aluminum (specific heat = 0.220 cal/g–C°), and its mass is 60.0 g. The heat capacity of the resistor itself is negligible. What is the temperature of the system 500 s later if all the electrical energy goes into heating the water and the calorimeter? Show your work.

Name
Section
Date

Lab Partners

39 LABORATORY 39 *Joule Heating of a Resistor*

LABORATORY REPORT

Data Table

Mass calorimeter + water	g	$T_i =$	°C
Mass calorimeter	g	c of calorimeter =	cal/g·C°
Mass water	g	c of water	cal/g·C°
Mass resistor coil	g	c of resistor coil	cal/g·C°

t (s)	V (V)	I (A)	T (°C)
0			
60			
120			
180			
240			
300			
360			
420			
480			

Calculations Table

ΔT (C°)	$mc\Delta T$ (cal)	VI (W)	$\overline{VI}t$ (J)
$mc = m_w c_w + m_c c_c + m_r c_r =$		cal/C°	
$\overline{VI} =$ W		$\alpha_{VI} =$ W	
$MEH_{exp} =$ J/cal	$A =$ J		$r =$
Percentage Error in $MEH_{exp} =$ %			

SAMPLE CALCULATIONS

1. $m_w = (m_c + m_w) - (m_c) =$
2. $mc = m_w c_w + m_c c_c + m_r c_r =$
3. $\Delta T = T - T_i =$
4. $mc\ \Delta T =$
5. $\overline{VI}\ t =$
6. % Error $MEH_{exp} =$

QUESTIONS

1. When the electrical power input is approximately constant, the temperature rise of the system should be proportional to the elapsed time. Do your data confirm this expectation? State the evidence for your answer.

2. What is the accuracy of your experimental value for *MEH*? State your evidence.

3. How long from the original starting time would it have taken to achieve a temperature of 50.0°C with the experimental arrangement you used? Show your work.

4. Assuming that one used the same heating coil and that its resistance did not change, how much would the power be increased if the voltage were increased by 50%? Show your work.

5. Suppose that the same mass of some liquid other than water were used in the calorimeter. If the liquid had a specific heat of 0.25 cal/g·C° would that tend to improve the results, make them worse, or have no effect on the results? Explain clearly the reasoning behind your answer.

Reflection and Refraction with the Ray Box

OBJECTIVES

- ❑ Investigate for reflection from a plane surface, the dependence of the angle of reflection on the angle of incidence.
- ❑ Investigate refraction of rays from air into a transparent plastic medium.
- ❑ Determine the index of refraction of a plastic prism from direct measurement of incident and refracted angles of a light ray.
- ❑ Investigate the focal properties of spherical reflecting and refracting surfaces.

EQUIPMENT LIST

- Ray box, 60.0° prism, plano-convex lens, circular metal reflecting surfaces
- Converging lens, diverging lens, protractor, straightedge, compass
- Sharp hard-lead pencil, black tape, several sheets of white paper

THEORY

Reflection

The reflection of light from a plane surface is described by the **law of reflection,** which states that the angle of incidence θ_i is equal to the angle of reflection θ_r.

The angles are measured with respect to a line perpendicular to the surface. Reflection from a plane mirror or a plane piece of glass are examples of the law of reflection.

In Figure 40-1(a) several incident rays and reflected rays are shown for a plane surface. The angle of incidence θ_i is seen to be equal to the angle of reflection θ_r.

Refraction

In general, light rays incident on a plane interface will be partially reflected and partially transmitted into the second medium. The transmitted ray undergoes a change in direction because the speed of light is different for different media. The ray is said to be **refracted.** This is illustrated in Figure 40-1(b). The angle of incidence is θ_1, and the angle of refraction is θ_2.

The speed of light in a vacuum is c ($\approx 3.00 \times 10^8$ m/s), the maximum possible speed of light. For any material the speed of light is v where $v \leq c$. A quantity called **the index of refraction** n for any medium is

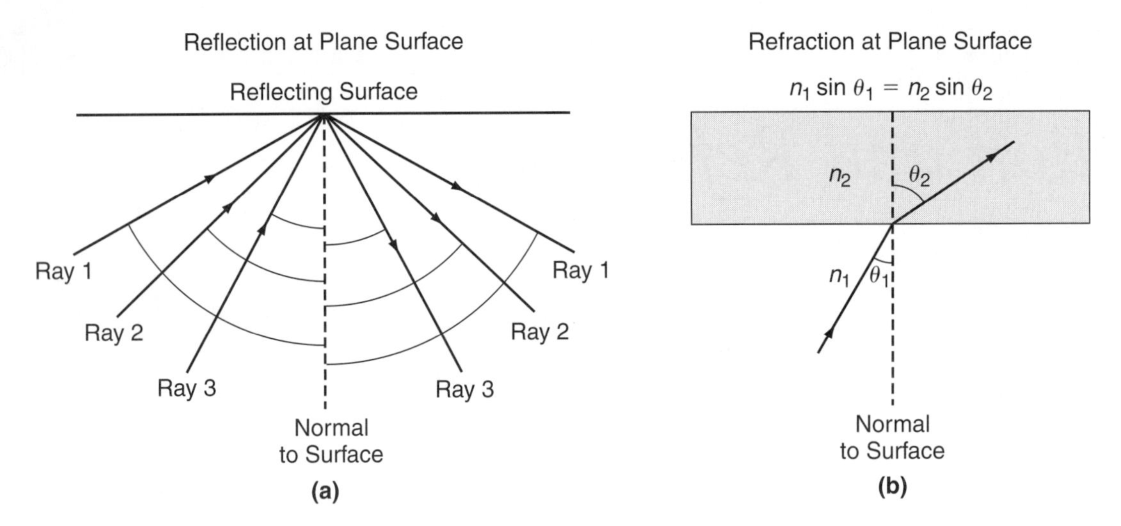

Figure 40-1 Illustration of reflection and refraction of light rays at a plane surface.

defined by $n = c/v$. Because $v \leq c$, the only allowed values of n are $n \geq 1$. The relationship (**Snell's law**) between the angle of incidence θ_1 and the refracted angle θ_2 is

$$n_1 \sin \theta_1 = n_2 \sin \theta_2 \qquad \text{(Eq. 1)}$$

When $n_1 > n_2$, Equation 1 implies that $\theta_1 < \theta_2$. This states that a ray going from a medium of a given index of refraction to one of a smaller index of refraction is bent away from the normal. If $n_1 < n_2$ then $\theta_1 > \theta_2$, and a ray going into a medium of larger index of refraction is bent toward the normal.

Focal Properties of Reflection and Refraction

Descriptions of the focal properties of reflection from spherical mirrors are shown in Figure 40-2. When reflection takes place from a concave spherical surface, incident parallel rays are converging and come to an approximate focus point. If R is the radius of curvature of the spherical surface, the focal point is a distance f (called the focal length) from the vertex of the spherical mirror where $f = R/2$. Incident parallel rays on a convex spherical mirror are diverging, but they appear to have come from a point. The distance from the vertex of the mirror to that point is called the focal length, and its magnitude is given by $f = R/2$. The focal length is positive for a concave converging mirror and negative for a convex diverging mirror.

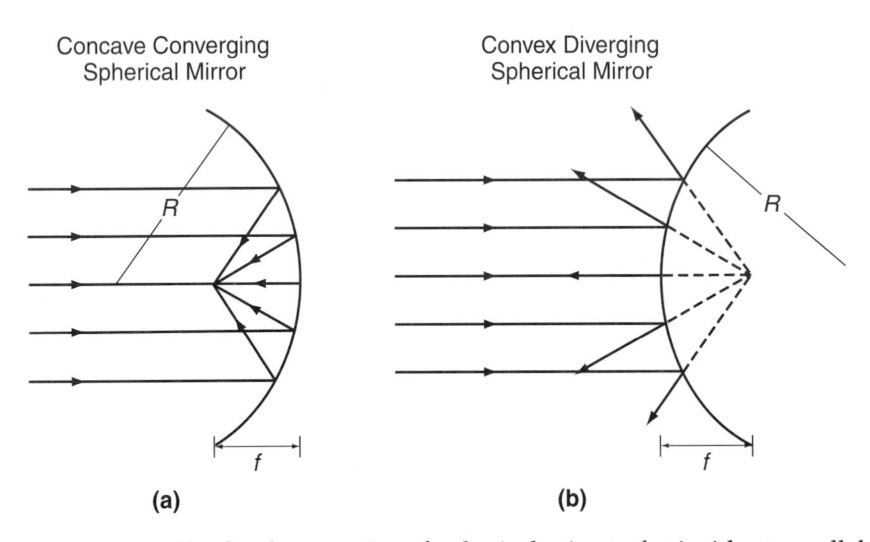

Figure 40-2 The focal properties of spherical mirrors for incident parallel rays.

EXPERIMENTAL PROCEDURE

Reflection

1. Use black tape to cover all of the slits from the ray box—except the central slit—to produce a single ray to examine reflection and refraction from a plane surface.
2. Place the 60.0° prism on a piece of white paper in the position shown in Figure 40-3(a). Draw a straight line along the face of the prism and place a small dot in the center of the line as shown.
3. Place the ray box about 0.15 m away from the prism and adjust the single ray to strike the plane surface at the position of the dot at an angle of incidence estimated to be about 60°. With a straightedge, draw a straight line in the direction of the incident ray and one in the direction of the reflected ray. This will produce the lines shown in Figure 40-3(b). Repeat this process two more times, once for an incident ray of about 45° and once for an incident ray of about 30° to produce the lines shown in Figure 40-3(c).
4. At the point of the dot construct a perpendicular to the face of the prism. Extend all six of the lines showing the ray directions until they intersect at the point of the dot to produce the lines shown in Figure 40-3(d). Use a protractor to measure the incident angles and reflected angles for each of the rays. Record all these angles (to the nearest 0.1°) in the Data Table.

Refraction

1. Place the prism on the paper as shown in Figure 40-4(a). Draw straight lines on the paper along two adjacent faces of the prism as shown in part (b) of the figure.
2. Place the ray box about 0.15 m away from the prism. Adjust the direction of the ray box so that the incident ray strikes one face of the prism at an angle of about 50° to a line drawn normally (90 degrees) to the prism face. Use a straightedge to draw a line in the direction of the incident ray and one in the direction of the refracted ray as shown in Figure 40-4(b) and Figure 40-5.

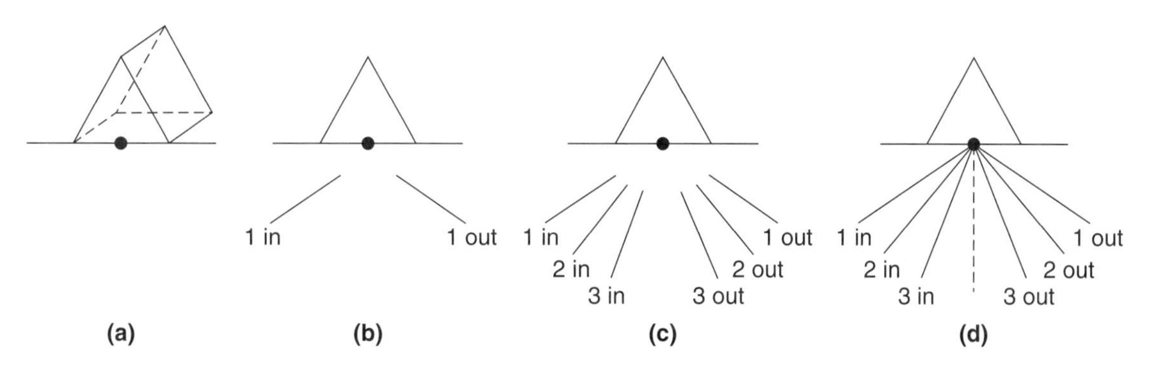

Figure 40-3 Tracing incident and reflected rays from a plane surface.

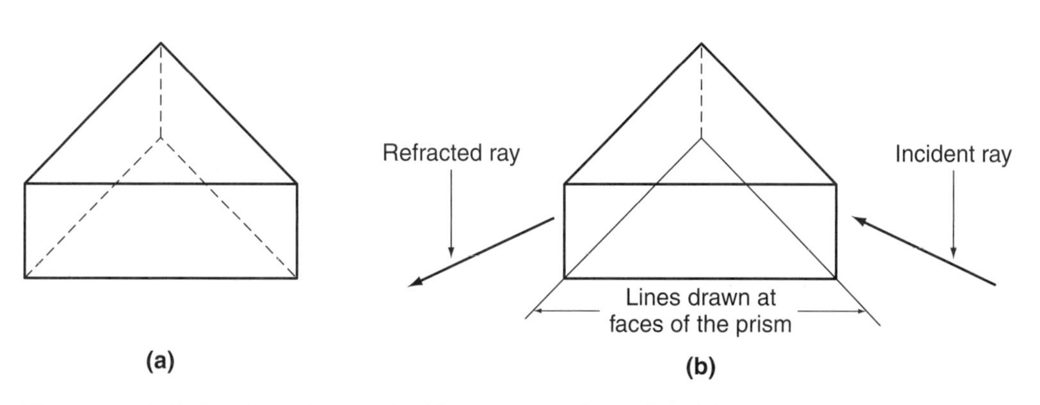

Figure 40-4 Refraction of a ray incident on one face of a 60° prism.

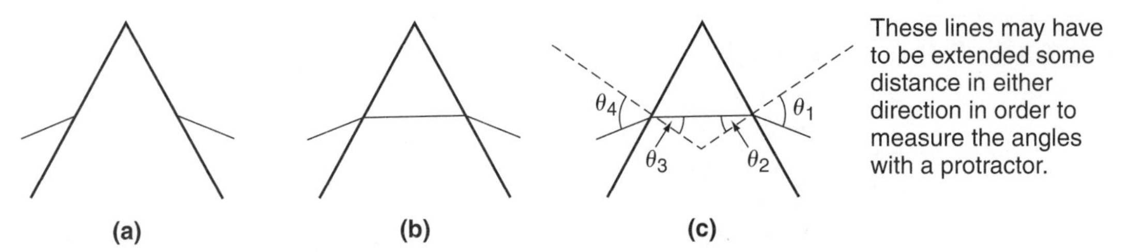

Figure 40-5 Step-by-step process to trace the rays and determine the four angles.

3. Using a separate sheet of paper for each ray, repeat Steps 1 and 2 for two other rays, one incident at an angle of about 60° with respect to the normal and the other incident at about 70° with respect to the normal.
4. Construct the lines tracing the path of each of the incident rays through the prism in the order of the steps shown in Figure 40-5. This will produce a figure from which the angles θ_1, θ_2, θ_3, and θ_4 can be determined with a protractor. Measure these angles for each of the three cases and record the values of all four angles (to the nearest 0.1°) in the Data Table.

Focal Properties of Reflection and Refraction

1. Remove the black tape from the ray box slits. Place the plastic plano-convex lens next to the slits to produce five parallel rays from the ray box. The lens may have to be rotated just slightly to produce the best set of parallel rays.
2. Place the circular metal reflector on a piece of white paper and trace its outline on the paper. Place the ray box about 0.15 m away on the concave side of the reflector. Align the five parallel rays with the center of the reflector to produce a pattern like the one in Figure 40-2(a). Make a tracing of this pattern, and from it, measure the focal length of the concave reflector. Record it in the Data Table as f_{con}.
3. Turn the reflector around and repeat Step 2 on another piece of paper with the reflector now acting as a convex mirror. Trace the pattern, which should look like that of Figure 40-2(b). Extend the reflected rays back to the point from which they appear to come. Measure the focal length and record it in the Data Table as f_{div}.
4. Using a compass, construct a circular arc that is the same radius of curvature as the reflector. Record in the Data Table the radius of that constructed circle as R, the radius of curvature of the reflector.
5. Place the plastic converging and diverging lenses on separate pieces of paper and trace the ray pattern produced by the parallel beam of rays from the ray box. Patterns like those of Figure 40-3 should be observed. Measure the focal length of the converging lens and record it in the Data Table as f_{con}. Measure the diverging lens focal length and record it in the Data Table as f_{div}.

CALCULATIONS

Reflection

1. Calculate the difference $|\theta_i - \theta_r|$ between the measured values of the incident and reflected angles for each of the three rays and record them in the Calculations Table.

Refraction

1. According to Snell's law, at the first surface $(1)\sin\theta_1 = n\sin\theta_2$. The value of $n = 1$ has been used for air, and n is the index of refraction of the prism. At the second surface, the equation is $n\sin\theta_3 = (1)\sin\theta_4$. Solving these two equations for n gives

$$n = \frac{\sin\theta_1}{\sin\theta_2} \quad \text{and} \quad n = \frac{\sin\theta_4}{\sin\theta_3} \qquad \text{(Eq. 2)}$$

2. Use these two equations to calculate two values of n for each of the incident rays. These are not independent measurements because the errors made in drawing the rays to determine the angle tend to produce two values of n with compensating errors. Take the average of the two values calculated by Equation 2 as a single measurement and record in the Calculations Table the average value of n for each ray.
3. Calculate the mean $\bar{n}$ and standard error α_n for the three measurements of n and record them in the Calculations Table.

Focal Properties of Reflection and Refraction

1. Calculate the percentage difference between the value of f_{con} and the value of f_{div} for the reflector.
2. According to theory, the value of the focal length for the reflector should be equal to $R/2$. Calculate the percentage difference between the measured value of $R/2$ and the focal lengths f_{con} and f_{div} for the reflector.

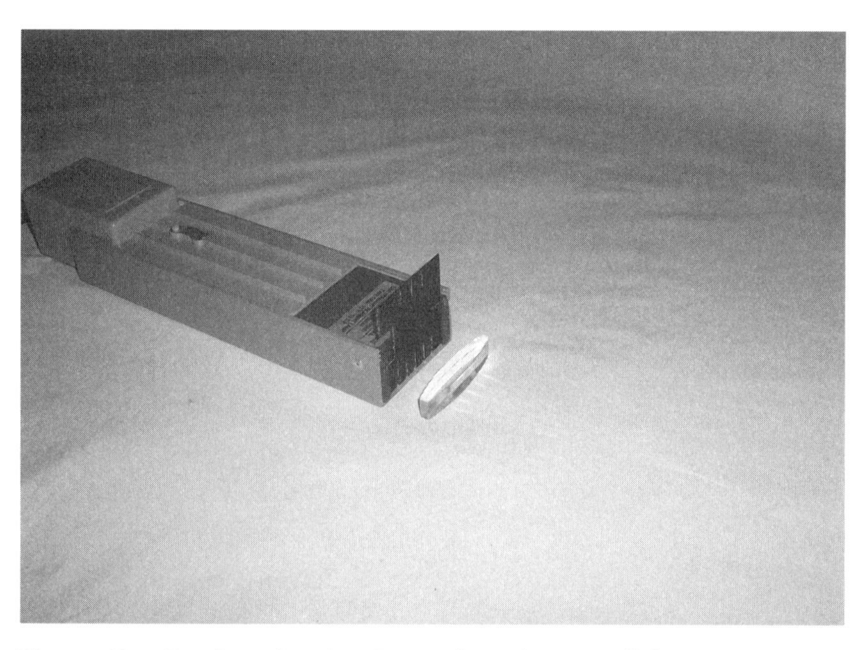

Figure 40-6 Ray box showing focus of incident parallel rays.

Name Section Date

40 LABORATORY 40 *Reflection and Refraction with the Ray Box*

PRE-LABORATORY ASSIGNMENT

1. Define the index of refraction.

2. State the law of reflection. Use a diagram to define the angles involved.

3. State Snell's law. Define terms and angles using a diagram.

4. A light ray is incident on a plane interface between two media. The ray makes an incident angle with the normal of 25.0° in a medium of $n = 1.25$. What is the angle that the refracted ray makes with the normal if the second medium has $n = 1.55$? Show your work.

5. A 60.0° prism has an index of refraction of 1.45 as shown below. A ray is incident as shown at an angle of 60.0° to the normal of one of the prism faces. Trace the ray on through the prism and find the angles θ_2, θ_3, and θ_4 as defined in the laboratory instructions. Show your work.

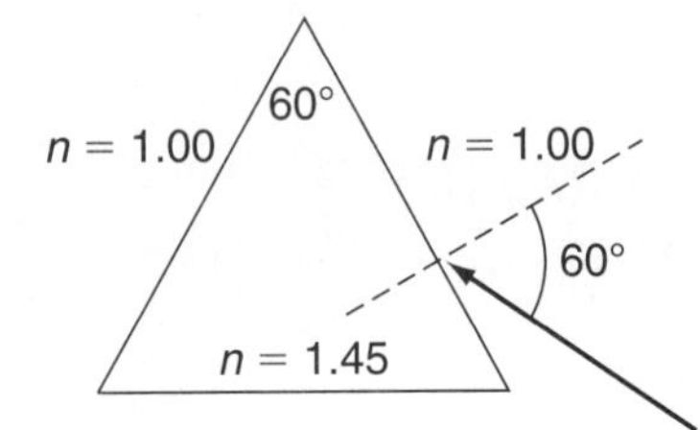

Name | Section | Date

Lab Partners

40 LABORATORY 40 *Reflection and Refraction with the Ray Box*

LABORATORY REPORT

Data Table

	Ray	θ_i	θ_r
Reflection			

Calculations Table

Angle Difference

	Ray	θ_1	θ_2	θ_3	θ_4
Refraction					

n	$\overline{n}$	α_n

Mirrors		f (cm)	R (cm)
	Concave f_{con}		
	Convex f_{div}		

% *Diff of f*	% *Diff of R/2*

Lenses		f (cm)
	Converging Lens	
	Diverging Lens	

SAMPLE CALCULATIONS

1. Angle Difference $= |\theta_i - \theta_r| =$

2. $n = \frac{1}{2}\left(\frac{\sin\theta_1}{\sin\theta_2} + \frac{\sin\theta_4}{\sin\theta_3}\right) =$

3. % Diff between f_{con} and $f_{div} =$

4. % Diff between f and $R/2 =$

QUESTIONS

1. Are your data consistent with the law of reflection? State your answer as quantitatively as possible.

2. State as quantitatively as possible the precision of your value for n, the index of refraction of the prism.

3. State how your data for the prism are evidence for the validity of Snell's law.

4. How well do your data for the focal lengths of the concave and convex mirror agree with the expectation that $f = R/2$? State your answer as quantitatively as possible.

5. Using the value of n determined for the prism, find the speed of light in the prism.

6. Inside the prism the wavelength of the light must change as well as the speed. Is a given wavelength longer or shorter inside the prism? Consider specifically light with a wavelength of 500 nm in air. What is the wavelength of this light inside the prism?

Focal Length of Lenses

OBJECTIVES

- ❑ Investigate the properties of converging and diverging lenses.
- ❑ Determine the focal length of converging lenses both by a real image of a distant object and by finite object and image distances.
- ❑ Determine the focal length of a diverging lens by using it in combination with a converging lens to form a real image.

EQUIPMENT LIST

- Optical bench, holders for lenses, a screen to form images, meter stick, tape
- Lamp with object on face (illuminated object), three lenses ($f \approx +20, +10, -30$ cm)

THEORY

When a beam of light rays parallel to the central axis of a lens is incident upon a **converging lens,** the rays are brought together at a point called the **focal point** of the lens. The distance from the center of the lens to the focal point is called f the **focal length** of the lens, and it is a positive quantity for a converging lens. When a parallel beam of light rays is incident upon a **diverging lens** the rays diverge as they leave the lens; however, if the paths of the outgoing rays are traced backward, the rays appear to have emerged from a point called the focal point of the lens. The distance from the center of the lens to the focal point is called the focal length f of the lens, and it is a negative quantity for a diverging lens. In Figure 41-1 two common types of lenses are pictured. In general, a lens is converging or diverging depending upon the curvature of its surfaces. In Figure 41-1 the radii of curvature of the surfaces of the two lenses are denoted as R_1 and R_2. The relationship that determines the focal length f in terms of the radii of curvature and the index of refraction n of the glass of the lens is called the **lens makers equation.** It is

$$\frac{1}{f} = (n-1)\left(\frac{1}{R_1} - \frac{1}{R_2}\right) \qquad \text{(Eq. 1)}$$

For the converging lens shown in Figure 41-1(a) the radius R_1 is positive and the radius R_2 is negative, but for the diverging lens of part (b), the radius R_1 is negative and the radius R_2 is positive. The signs of these radii are determined according to a sign convention that is described in all elementary textbooks.

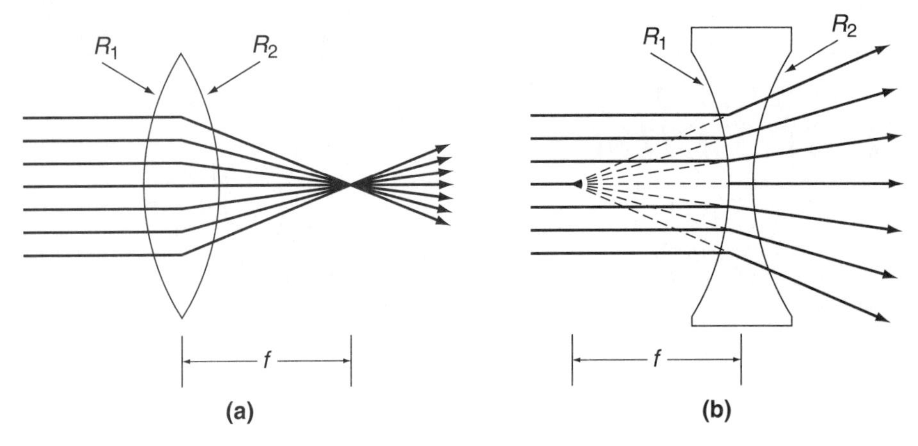

Figure 41-1 Ray diagram for converging and diverging lenses showing the definition of the focal length for both the converging case and the diverging case.

As an example, consider a double convex lens like the one shown in part Figure 41-1(a) made from glass of index of refraction 1.60 with radii of curvature R_1 and R_2 of *magnitude* 20.0 and 30.0 cm, respectively. According to the sign convention given above, that would mean $R_1 = +20.0$ cm and $R_2 = -30.0$ cm. Putting those values into Equation 1 gives a value for the focal length f of $+20.0$ cm.

Essentially, Equation 1 indicates that a lens that is thicker in the middle than at the edges is converging, and a lens that is thinner in the middle than at the edges is diverging. A lens can be classified as converging or diverging merely by taking it between one's fingers to see if it is thicker at the center of the lens than it is at the edge of the lens.

Lenses are used to form images of objects. There are two possible kinds of images. The first type, called a **real image,** is one that can be focused on a screen. For a real image, light actually passes through the points at which the image is formed. The second type of image is called a **virtual image**; light does not actually pass through the points at which the image is formed, and the image cannot be focused on a screen. Diverging lenses can form only virtual images, but converging lenses can form either real images or virtual images. If an object is farther from a converging lens than its focal length, a real image is formed. If the object is closer to a converging lens than the focal length, the image formed is a virtual one. Whenever a virtual image is formed, ultimately it will serve as the object for some other lens system to form a real image. Often the other lens system is the human eye, and the real image is formed on the retina of the eye.

In the process of image formation, the distance from an object to the lens is called the **object distance** p, and the distance of the image from the lens is called the **image distance** q. The relationship between the object distance p, the image distance q, and the focal length of the lens f is

$$\frac{1}{p} + \frac{1}{q} = \frac{1}{f} \qquad \text{(Eq. 2)}$$

Equation 2 is valid both for converging (positive f) and for diverging (negative f) lenses. Normally the object distance is considered positive. In that case a positive value for the image distance means that the image is on the opposite side of the lens from the object, and the image is real. A negative value for the image distance means that the image is on the same side of the lens as the object, and that the image is virtual.

If a lens is used to form an image of a very distant object, then the object distance p is very large. For that case, the term $1/p$ in Equation 2 is negligible compared to the other terms $1/q$ and $1/f$ in that equation. For the case of a very distant object, Equation (2) becomes

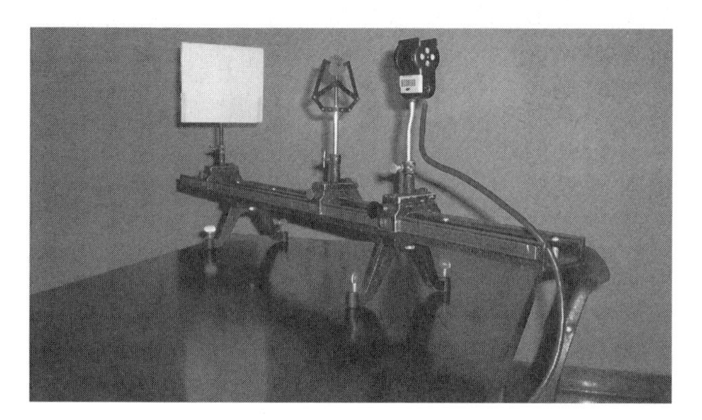

Figure 41-2 Optical bench with object, lens, and screen on which a real image is formed.

$$\frac{1}{q} = \frac{1}{f} \quad \text{(Eq. 3)}$$

For this case, the image distance is equal to the focal length. This provides a quick and accurate way to determine the focal length of a converging lens, but it is only applicable to a converging lens because the image must be focused on a screen. A diverging lens cannot form a real image, and this technique will not work directly for a diverging lens.

If two lenses with focal lengths of f_1 and f_2 are placed in contact, the combination of the two in contact acts as a single lens of effective focal length f_e. The effective focal length of the two lenses in contact f_e is related to the individual focal lengths of the lenses f_1 and f_2 by

$$\frac{1}{f_e} = \frac{1}{f_1} + \frac{1}{f_2} \quad \text{(Eq. 4)}$$

Equation 4 is valid for any combination of converging and diverging lenses. If the individual lenses f_1 and f_2 are converging, then the effective focal length f_e will also be converging. If one of the lenses is converging and the other is diverging, then the effective focal length can be either converging or diverging depending upon the values of f_1 and f_2. If the converging lens has a smaller magnitude than the diverging lens, then the effective focal length will be converging. We can use this fact to determine the focal length of an unknown diverging lens if it is used in combination with a converging lens with a focal length short enough to produce a converging combination.

EXPERIMENTAL PROCEDURE

Focal Length of a Single Lens

1. Place one of the three lenses in a lens holder on the optical bench and place the screen in its holder on the optical bench. Place the optical bench in front of a window in the laboratory and point the bench toward some distant object. Adjust the distance from the lens to the screen until a sharp, real image of the distant object is formed on the screen. You will be able to form such an image for only two of the three lenses. This experimental arrangement satisfies the conditions of Equation 3. The measured image distance is equal to the focal length of the lens. Record these measured image distances in Data Table 1 as the focal length of the two lenses for which the method works. Call the lens with the longest focal length A, the one with the shortest focal length B, and the one for which no image can be formed C.
2. Place lens B in the lens holder on the optical bench and use the lamp with the object painted on its face as an object. For various distances p of the object from the lens, move the screen until a sharp real image is formed on the screen. For each value of p measure the image distance q from the screen to the

lens. Make sure that the lens, the object, and the screen are at the center of their respective holders. Try values for p of 20, 30, 40, and 50 cm, determining the value of q for each case. If these values of p do not work for your lens, try other values until you find four values that differ by at least 5 cm. Record the values for p and q in Data Table 2.

Focal Length of Lenses in Combination

1. Place lens A and lens B in contact using masking tape to hold the edges of the two lenses parallel. Measure the focal length of the combination f_{AB} both by the very distant object method and by the finite object method. For the finite object method, just use one value of the object distance p and determine the image distance q. Record the results for both methods in Data and Calculations Table 3.
2. Place lens B and lens C in contact, using masking tape to hold the edges of the two lenses parallel. Repeat the measurements described in Step 1 above for these lenses in combination. Record the results in Data and Calculations Table 4.

CALCULATIONS

Focal Length of a Single Lens

1. Using Equation 2, calculate the values of the focal length f for each of the four pairs of objects and image distances p and q. Record them in Calculations Table 2.
2. Calculate the mean $\bar{f}$ and the standard error α_f for the four values for the focal length f and record them in Calculations Table 2.
3. The mean $\bar{f}$ represents the measurement of the focal length of lens B using finite object distances. Compute the percentage difference between $\bar{f}$ and the value determined using essentially infinite object distance in Data Table 1. Record the percentage difference between the two measurements in Calculations Table 2.

Focal Length of Lenses in Combination

1. From the data for lenses A and B, calculate the value of f_{AB} from the values of p and q. Record that value of f_{AB} in Calculations Table 3. Also record in that table the value of f_{AB} determined by the very distant object method.
2. Calculate the average of the two values for f_{AB} determined above. This average value of f_{AB} is the experimental value for the combination of these two lenses.
3. Using Equation 4, calculate a theoretical value expected for the combination of lenses A and B. Use the values determined in Data Table 1 by the distant object method for the values of f_A and f_B in the calculation. Record this value as $(f_{AB})_{theo}$ in Data and Calculations Table 3.
4. Calculate the percentage difference between the experimental value and the theoretical value for f_{AB}. Record it in Data and Calculations Table 3.
5. From the data for lenses B and C, calculate the value of f_{BC} from the values of p and q. Record that value of f_{BC} in Data and Calculations Table 4. Also record in that table the value of f_{BC} determined by the very distant object method.
6. Calculate the average of the two values for f_{BC} determined above. This average value is the experimental value for the combination of these two lenses.
7. Using the average value of f_{BC} determined in Step 6 and the value of f_B from Data Table 1 for the focal length of B, calculate the value of f_C, the focal length of lens C using Equation 4. Record the value of f_C in Data and Calculations Table 4.

Name _______________ Section _______________ Date _______________

41 LABORATORY 41 *Focal Length of Lenses*

PRE-LABORATORY ASSIGNMENT

1. Mark the following statements about lenses as true or false.

_____a. Incident parallel light rays converge if the lens's focal length is negative.

_____b. If the path of converging light rays is traced backward, the rays appear to come from a point called the focal point.

_____c. A double convex lens has a negative focal length.

_____d. The focal length of a lens is always positive.

2. A double convex lens is made from glass with an index of refraction of $n = 1.50$. The *magnitudes* of its radii of curvature R_1 and R_2 are 10.0 cm and 15.0 cm, respectively. What is the focal length of the lens? Show your work.

$f =$ ___________ cm

3. What is a real image? What is a virtual image?

4. For a diverging lens, state what kinds of images can be formed and the conditions under which those images can be formed.

5. For a converging lens, state what kinds of images can be formed and the conditions under which those images can be formed.

6. A lens has a focal length of $f = +10.0\,\text{cm}$. If an object is placed 30.0 cm from the lens, where is the image formed? Is the image real or virtual? Show your work.

7. An object is 16.0 cm from a lens. A real image is formed 24.0 cm from the lens. What is the focal length of the lens? Show your work.

8. One lens has a focal length of +15.0 cm. A second lens of focal length +20.0 cm is placed in contact with the first lens. What is the equivalent focal length of the combination of lenses? Show your work.

9. Two lenses are in contact. One of the lenses has a focal length of +10.0 cm when used alone. When the two are in combination, an object 20.0 cm away from the lenses forms a real image 40.0 cm away from the lenses. What is the focal length of the second lens? Show your work.

Name ____________________ Section __________ Date __________

Lab Partners ____________________

41

LABORATORY 41 *Focal Length of Lenses*

LABORATORY REPORT

Data Table 1

Lens	Image Distance (cm)	Focal Length (cm)
A		$f_A =$
B		$f_B =$

Data Table 2

p (cm)	q (cm)

Calculations Table 2

f_B (cm)	$\bar{f}_B$ (cm)	α_f (cm)	% Diff

Data and Calculations Table 3 (Lenses A & B)

q $(p = \infty)$	p	q	f_{AB} $p = \infty$	f_{AB} p & q	$\bar{f}_{AB}$	f_{AB} theo	% Diff

Data and Calculations Table 4 (Lenses B & C)

q $(p = \infty)$	p	q	f_{BC} $p = \infty$	f_{BC} p & q	$\bar{f}_{BC}$	f_C

SAMPLE CALCULATIONS

1. $f = \dfrac{pq}{p+q} =$
2. $\bar{f}_{AB} = (f_{AB^1} + f_{AB^2})/2 =$
3. $(f_{AB})_{theo} = (f_A)\,(f_B)/(f_A + f_B) =$
4. $f_C = (f_B)(f_{BC})/(f_B - f_{BC}) =$

QUESTIONS

1. Why is it not possible to form a real image with lens C alone?

2. Take lens C between your thumb and index finger. Is it thinner or thicker at the center of the lens than at the edge? Take lens B between your thumb and index finger. Is it thinner or thicker at the center of the lens than at the edge? From this information alone, what can you conclude about lenses C and B?

3. Consider the percentage difference between the two measurements of the focal length of lens B. Express α_f as a percentage of $\bar{f}$. Is the percentage difference between the two measurements less than the percentage standard error?

4. Compare the agreement between the experimental and theoretical values of f_{AB} the focal length of lenses A and B combined. Do these data suggest that Equation 4 is a valid model for the equivalent focal length of two lenses in contact?

5. If lens A and lens C were used in contact, could they produce a real image? State clearly the basis for your answer. You will need to do a calculation.

Diffraction Grating Measurement of the Wavelength of Light

OBJECTIVES

- ❑ Investigate the difference between continuous and discrete spectra.
- ❑ Investigate the characteristic spectra of individual gaseous elements.
- ❑ Determine the average line spacing for a diffraction grating assuming mercury wavelengths as known, and use it to determine the wavelengths of helium.

EQUIPMENT LIST

- Optical bench, diffraction grating (600 lines/mm replica grating)
- Spectrum tube power supply, mercury and helium discharge tubes
- Meter stick and slit arrangement, incandescent lightbulb (15 watt)

THEORY

When light is separated into its component **wavelengths,** the resulting array of colors is called a spectrum. If a light source produces all the colors of visible light, it is called a **continuous spectrum.** Generally, such sources of light are produced by heated solid metal filaments. An ordinary incandescent lightbulb with a tungsten filament produces a continuous spectrum.

Some light sources produce discrete wavelengths of light, and the spectrum appears as mostly dark with a few discrete lines of color at the wavelengths emitted by the source. Such light is produced by hot discharges of gas of a single chemical element, and the wavelengths of light emitted are characteristic of the electronic structure of that element. The spectrum is called a **discrete spectrum** or a line spectrum. The term **line spectrum** is used because the images produced usually are images of a narrow slit that is illuminated by the light source.

We can use several methods to separate light into its component wavelengths and produce a spectrum. This laboratory will use a **diffraction grating** to produce spectra from an incandescent lightbulb and from gas discharge tubes of mercury and helium. A transmission diffraction grating is a piece of transparent material ruled with a large number of equally spaced parallel lines. The distance between the lines is called the grating spacing d, and it is usually only a few times as large as a typical wavelength of visible light. The range of visible light wavelengths is from approximately 4×10^{-7} m to 7×10^{-7} m. It is customary to express the wavelength of light in units of nm (10^{-9} m). In those units the range of visible light is from 400 nm to 700 nm. A typical grating spacing d is in the range 1000 nm to 2000 nm.

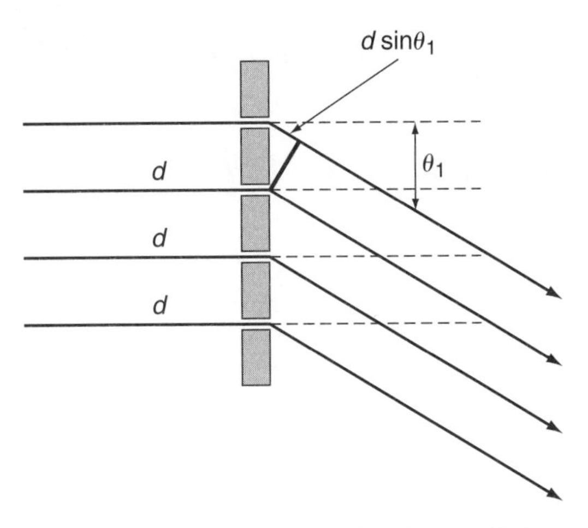

Figure 42-1 Ray diagram for the conditions of the first-order diffraction image.

The wavelengths of light determine the color of the light seen by the human eye. Starting from short wavelengths and going to long wavelengths, the order of colors is violet, blue, green, yellow, orange, and red. The actual range of the visible spectrum is somewhat different for individuals, and there may be a distinct difference in the ability of two laboratory partners to see the wavelengths at either end of the spectrum. It is often very difficult for some people to see the very short wavelengths.

Light rays that strike the transparent portion of the grating between the ruled lines will pass through the grating at all angles with respect to their original path. When deviated rays from adjacent rulings on the grating are in phase, an image of the source will be formed. This condition is satisfied when the adjacent rays differ in path length by an integral number of wavelengths of the light. Thus for a given wavelength λ a series of images will appear at angles θ_m that satisfy the equation

$$m\lambda = d \sin \theta_m \qquad \text{(Eq. 1)}$$

with m an integer. The first value of θ is θ_1 when $m = 1$, the second is θ_2 when $m = 2$. Figure 42-1 shows that the limit to the values of θ will be at $\theta = 90°$. This is referred to as the number of orders that can be seen and is determined by d and λ. Although it will be possible to see both first-order ($m = 1$) and second-order ($m = 2$) for the experimental arrangement used in this laboratory, measurements will be made only on the first-order images.

The experimental arrangement is shown in Figure 42-2. The discharge tube light source is viewed through the grating as shown. The distance L from the grating to the slit is chosen at a convenient fixed

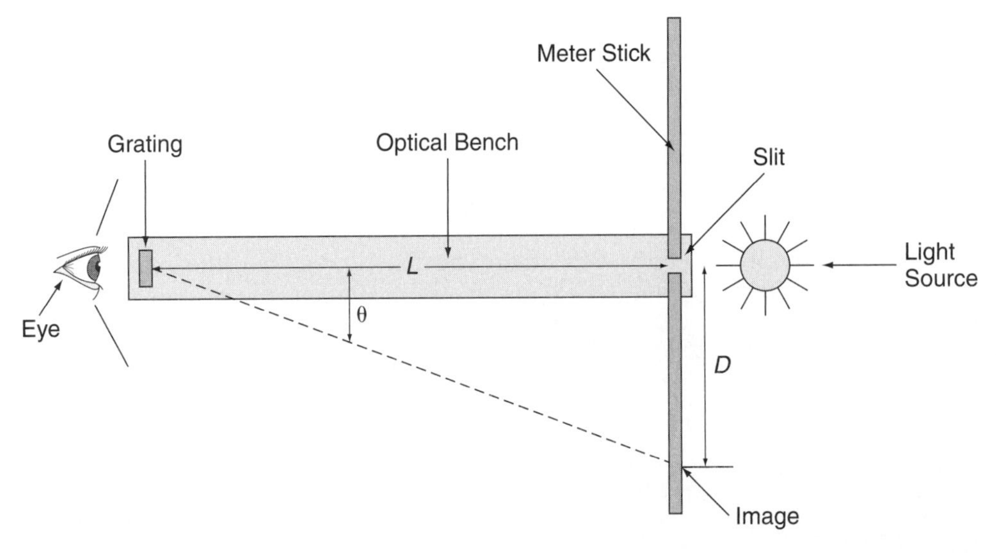

Figure 42-2 Arrangement of the diffraction grating, slit, light source, and optical bench.

value. The different wavelengths of the source will produce a first-order image at different angles and thus at different distances D from the slit as defined by Figure 42-2. We determine the angle θ corresponding to each wavelength by measuring D with L known.

If λ and m are known, we can determine d from Equation 1. First, a mercury light source will be used and its wavelengths given. A series of measurements will accurately determine the value of the grating spacing d. In the second part of the laboratory using this value of d for the grating, we will determine the wavelengths of a helium source.

EXPERIMENTAL PROCEDURE

1. Set up the experimental arrangement shown in Figure 42-2. Place the slit near one end of the optical bench just above the meter stick, which is held by the same holder that holds the slit. The meter stick should be perpendicular to the axis of the optical bench and should be level. The zero of the meter stick should be to the left with the markings increasing to the right. The slit should be at the 50.00 cm mark on the meter stick just above the meter stick so that its images can be located easily relative to a mark on the meter stick. Place the grating some distance L away from the slit with the plane of the grating perpendicular to the axis of the optical bench. Record the value (to the nearest 1 mm) of L in Data Table 1 and take all the data at this same value of L.
2. *Use extreme caution with the discharge tube power supply. It produces 5000 V and sufficient current to make it potentially lethal. Do not touch the supply electrodes while the supply is turned on.* With the power supply turned off and unplugged, place the mercury discharge tube into the electrode receptacles. Place the supply behind the slit with the discharge tube as close to the slit as possible. It may be necessary to place a block under the power supply to adjust the height of the discharge tube. The narrow portion of the discharge tube (which is the most intense) must be at the height of the slit.
3. *Now turn on the power supply. Do not accidentally touch the power supply electrodes while making the following adjustments.* While one partner looks through the grating directly at the slit, the other partner should make very fine adjustments in the position of the power supply to place the bright narrow portion of the discharge tube in alignment with the slit. Proper alignment is achieved when the slit is as bright as possible as seen by the person looking through the grating directly at the slit. It is extremely critical that the light source is positioned so that the slit is as bright as possible. The slightest movement of the light source relative to the slit after this adjustment has been made may severely alter the brightness of the images seen.
4. Look through the grating to the right and left of the slit. Just above the meter stick there should appear a series of images of the slit in various colors. It may be necessary to rotate the grating in its holder to place the images in the horizontal. The images may originally appear at any angle to the horizontal up to the extreme case of 90°, in which case they would be in the vertical. Rotate the grating until the images are horizontal and just above the meter stick.
5. In Data and Calculations Table 1 are listed seven wavelengths of mercury that should be prominent. They are listed in the order of increasing wavelength. They should appear in this order with the smallest wavelength at the smallest angle. Try to match the images that you see with the wavelengths given. It may be difficult to identify all seven of the lines. In particular, many people have difficulty seeing the violet lines clearly. While looking through the grating, locate the position of the first-order images that are to the right of the optical bench above the meter stick. One partner should locate the position of a given line by having the other partner move a small pointer (for example, a pencil point) along the meter stick until the pointer is in line with a given image. It may be helpful to use the small 15-watt light source to illuminate the meter stick to read the position once it has been located. Record (to the nearest 1 mm) the position P_R of each of the seven wavelengths in Data and Calculations Table 1.
6. Repeat the process for the images on the left of the optical bench that correspond to the seven wavelengths. Record (to the nearest 1 mm) the position P_L of each image in Data and Calculations Table 1.

7. Turn off the power supply and allow the discharge tube to cool. With the power supply turned off, remove the mercury discharge tube and replace it with the helium discharge tube. Position the power supply with the discharge tube aligned with the slit. Turn on the power supply and adjust the position of the supply for maximum brightness.
8. In Data and Calculations Table 2 are listed eight wavelengths of helium that should be visible. Again try to match the images that you see with the wavelengths given. Perform the same procedure as done above for mercury, measuring the positions P_R and P_L of each image on the right and the left. Record (to the nearest 1 mm) the data in Data and Calculations Table 2.
9. Place the 15-watt lightbulb behind the slit and observe the continuous spectrum. Locate the positions P_R and P_L of the following parts of the spectrum: (a) the shortest wavelength visible (b) the division between blue and green (c) the division between green and yellow (d) the division between yellow and orange (e) the division between orange and red (f) the longest wavelength visible. Record (to the nearest 1 mm) the position of these points in Data and Calculations Table 3.

CALCULATIONS

1. Calculate the distance D_R from the slit to each image on the right ($D_R = P_R - 50.0$) and calculate the distance D_L from the slit to each image on the left ($D_L = 50.0 - P_L$) for the mercury data in Data Table 1. Calculate the average distance $\overline{D} = (D_R + D_L)/2$ and calculate $\tan\theta = \overline{D}/L$, θ, and $\sin\theta$ for each image. Record (to three significant figures) those values in Data and Calculations Table 1.
2. Each of the measurements for mercury is an independent measurement for d the grating spacing. Use Equation 1 to calculate the seven values of d, and record them (to four significant figures) in Data and Calculations Table 1. Calculate the mean $\overline{d}$ and the standard error α_d for d and record them in Data and Calculations Table 1.
3. Calculate the values of D_R, D_L, and $\overline{D}$ for the helium data and use them to calculate $\tan\theta = \overline{D}/L$, θ, and $\sin\theta$ for each image. Use those values of $\sin\theta$ and $\overline{d}$ in Equation 1 to calculate the wavelengths of helium. Record all values (to three significant figures) in Data and Calculations Table 2.
4. From the data for the continuous spectrum, determine the wavelength that corresponds to the various points in the spectrum that were located. Calculate and record all the information called for in Data and Calculations Table 3.

Name Section Date

42

LABORATORY 42 *Diffraction Grating Measurement of the Wavelength of Light*

PRE-LABORATORY ASSIGNMENT

1. What is a continuous spectrum? What is a discrete spectrum?

2. What kind of light sources produce each type of spectrum?

3. The wavelengths produced by a hot gas of helium (a) form a discrete spectrum (b) form a line spectrum (c) are characteristic of the electronic structure of helium (d) all of the above are true.

4. A diffraction grating has a grating spacing of $d = 1500\,\text{nm}$. It is used with light of wavelength 500 nm. At what angle will the first-order diffraction image be seen? Show your work.

5. For a given wavelength λ and a diffraction grating of spacing d (a) an image is formed at only one angle (b) at least two orders are always seen (c) the number of orders seen can be any number and depends on d and λ (d) there can never be more than four orders seen.

6. The grating used in this laboratory (a) can produce only images in the horizontal (b) must be rotated in its holder until it produces the desired horizontal pattern (c) produces images only in the vertical direction (d) produces images only to the left of the slit.

7. A diffraction grating with $d = 2000$ nm is used with a mercury discharge tube. At what angle will the first-order blue-green wavelength of mercury appear? What other orders can be seen, and at what angle will they appear? Show your work.

8. The diffraction grating of Question 7 is used at a distance $L = 50.0$ cm from the slit. What is the distance D from the slit to the first-order image for the blue-green wavelength of mercury? Show your work.

9. What is the voltage and current of the spectrum tube power supply?

Name

Section

Date

Lab Partners

42

LABORATORY 42

Diffraction Grating Measurement of the Wavelength of Light

LABORATORY REPORT

Data and Calculations Table 1 (Mercury Spectrum)

Colors	λ (nm)	P_R (cm)	P_L (cm)	D_R (cm)	D_L (cm)	$\overline{D}$ (cm)	$\tan\theta$	θ	$\sin\theta$	d (nm)
Violet	404.7									
Violet	407.8									
Blue	435.8									
Blu-Gr	491.6									
Green	546.1									
Yellow	577.0									
Yellow	579.0									
$L =$			cm	$\overline{d} =$			nm	$\alpha_d =$		nm

Data and Calculations Table 2 (Helium Spectrum)

Colors	λ (nm)	P_R (cm)	P_L (cm)	D_R (cm)	D_L (cm)	$\overline{D}$ (cm)	*tan* θ	θ	*sin* θ	λ (nm)
Blue	438.8									
Blue	447.1									
Blue	471.3									
Blu-Gr	492.2									
Green	501.5									
Yellow	587.6									
Red	667.8									
Red	706.5									

Data and Calculations Table 3 (Continuous Spectrum)

Portion of Spectrum	P_R (cm)	P_L (cm)	$\overline{D}$ (cm)	*tan* θ	θ	*sin* θ	λ (nm)
Shortest Wavelength							
Division Blue and Green							
Division Green and Yellow							
Division Yellow and Orange							
Division Orange and Red							
Longest Wavelength							

SAMPLE CALCULATIONS

1. $D_R = P_R - 50.0 =$ and $D_L = 50.0 - P_L =$
2. $\overline{D} = (D_R + D_L)/2 =$
3. $\tan\theta = \overline{D}/L =$ and $\theta = \tan^{-1}(\theta) =$
4. $\sin\theta =$
5. $d = \lambda/\sin\theta =$
6. $\lambda = \overline{d}\sin\theta =$

QUESTIONS

1. Comment on the precision of your measurement of d.

2. List the accepted values of the eight wavelengths of helium below. Beside each one, show the percentage error in your measured values compared to these values including the sign of the error. Comment on the accuracy of your measurements.

3. If all the errors in Question 2 are of the same sign, it might be evidence of a systematic error. Based on this criterion, do your data show evidence of a systematic error? State your evidence for or against a systematic error.

4. If the grating had exactly 600 lines/mm, d would be 1667 nm. Use that value of d with the values of $\sin\theta$ in Table 2 to recalculate the wavelengths for helium. Are their percentage differences from the accepted values better or worse than in Question 2? Show your work.

5. Hydrogen has known emission lines of wavelength 656.3 nm and 434.1 nm. At what distance D away from the slit would each of these lines be observed in your experimental arrangement? Show your work.

Bohr Theory of Hydrogen—The Rydberg Constant

OBJECTIVES

- ❑ Investigate how well the visible light wavelengths of hydrogen predicted by the Bohr theory agree with experimental values.
- ❑ Determine an experimental value for the Rydberg constant from a fit of the measured values of hydrogen wavelengths to the form of the Balmer equation.

EQUIPMENT LIST

- Spectrometer, diffraction grating in holder (600 lines/mm or better)
- Hydrogen gas discharge tube, mercury discharge tube
- Power supply for the discharge tubes

THEORY

The spectrum from a hot gas of an element consists of **discrete wavelengths** that are characteristic of the element. In 1885, in an attempt to understand these spectra, Johann Balmer published an empirical relationship that described the visible spectrum of hydrogen. Although Balmer published the relationship in a somewhat different form, the modern equivalent is

$$\frac{1}{\lambda} = R_H\left(\frac{1}{2^2} - \frac{1}{n^2}\right) \quad n = 3, 4, 5, 6..... \qquad \text{(Eq. 1)}$$

where $R_H = 1.097 \times 10^7\ \text{m}^{-1}$ is a constant called the **Rydberg constant,** λ stands for the wavelength, and n is an integer that takes on successive values greater than 2. In 1913 Neils Bohr was able to derive the Balmer relationship by making a series of revolutionary postulates. The **Bohr theory** was historically of great importance in the developments that eventually led to modern quantum theory. In his attempts to explain the spectrum of hydrogen, Bohr was influenced by several recently developed theories. He incorporated concepts from the quantum theory of Max Planck, from the photon description of light by Albert Einstein, and from the nuclear theory of the atom suggested by Ernest Rutherford's α-particle scattering from gold. The central ideas of Bohr's theory are contained in a series of four postulates that are stated below.

(1) The electron moves in circular orbits of radius r_n around the nucleus under the influence of the Coulomb force between the negative electron and the positive nucleus.

(2) The electron of mass m can only have velocity v_n and orbits of radius r_n that satisfy the relationship $mv_n r_n = nh/2\pi$ where $h = 6.626 \times 10^{-34}$ J–s and $n = 1, 2, 3, 4, \ldots \infty$.

(3) In an allowed orbit the electron does not radiate energy. The atom is stable in these orbits, and this is called a stationary state. This postulate was a radical departure from classical physics. Classical electromagnetic theory predicts that an electron moving in a circle is accelerated and must radiate electromagnetic energy continuously.

(4) The atom radiates energy only when an electron makes a transition from one allowed orbit to another allowed orbit. If E_i and E_f stand for the energies of the initial and final stationary states, the energy radiated by the atom is in the form of a photon of energy $hf = E_i - E_f$ where f is the frequency of the photon.

With these postulates it is possible to derive an expression for the energy of the stationary states. They are given by

$$E_n = -\left(\frac{me^4}{8\varepsilon_0^2 h^2}\right)\frac{1}{n^2} \quad \text{with } n = 1, 2, 3, 4, \ldots . \infty \qquad \text{(Eq. 2)}$$

This expression for allowed energies can be used to obtain values for $1/\lambda$ predicted by the Bohr theory. The transitions that produce photons that correspond to the first four visible Balmer wavelengths are those from the states $n = 3$, 4, 5, 6, down to the $n = 2$ state. They are

$$\frac{1}{\lambda} = \frac{me^4}{8\varepsilon_0^2 ch^3}\left(\frac{1}{2^2} - \frac{1}{n^2}\right) \quad \text{with } n = 3, 4, 5, \text{ and } 6 \qquad \text{(Eq. 3)}$$

Bohr showed that the value of the constant $me^4/8\varepsilon_0^2 ch^3$ was in excellent agreement with the value of the Rydberg constant in Balmer's formula. This is striking confirmation of the validity of the Bohr theory of hydrogen.

The four wavelengths of the visible hydrogen spectrum that are easily seen and measured are in excellent agreement with the first four wavelengths predicted by Equation 3. This will be demonstrated by measuring the value of λ for those four wavelengths and performing a linear least squares fit with $1/\lambda$ as the vertical axis and the quantity $(1/2^2 - 1/n^2)$ as the horizontal axis with $n = 3$, 4, 5, and 6. The slope of the fit is an experimental value of the Rydberg constant R_H. The correlation coefficient of the least squares fit is a measure of the agreement of Bohr theory with the data.

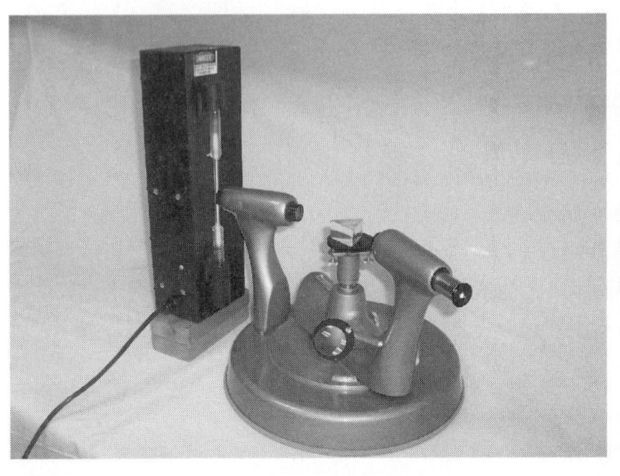

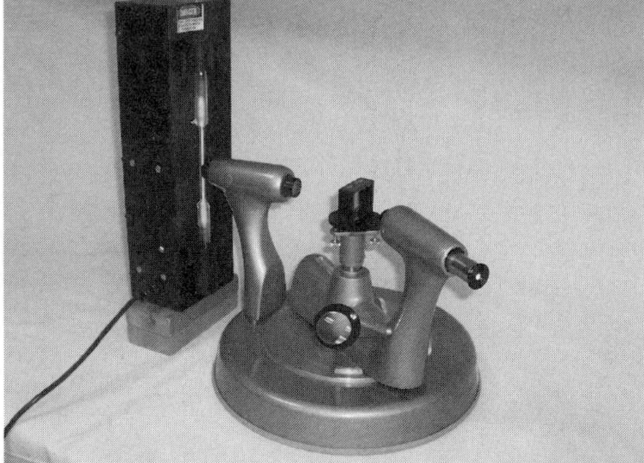

Figure 43-1 Spectrometer that can be used with a prism or with a diffraction grating.

The wavelengths will be measured with a diffraction grating spectrometer shown in Figure 43-1. Images of the slit for different wavelengths will appear in the first order at angles θ given by

$$\lambda = d \sin\theta \qquad \text{(Eq. 4)}$$

Initially, the grating spacing d will be assumed to be unknown, and the wavelengths of mercury will be considered as known. Measurements of the angles at which the mercury spectrum occur can then be used to determine d from Equation 4. Using that value of d, measurements of the angles at which the hydrogen wavelengths occur will allow the determination of those wavelengths.

EXPERIMENTAL PROCEDURE

1. Place the diffraction grating (in its holder) on the spectrometer table as shown in Figure 43-2. Place the mercury spectrum tube between the electrodes of the spectrum tube power supply. *DO NOT TOUCH THE HIGH VOLTAGE ELECTRODES WHILE THE POWER SUPPLY IS ON. IT PROVIDES A VOLTAGE OF 5000 V.*
2. Turn on the power supply and place the spectrum tube as close to the slit in the collimator tube as is possible. Rotate the telescope tube of the spectrometer until it is directly in line with the collimator tube. Adjust the slit in the collimator and the eyepiece of the telescope until a sharp image of the slit is obtained. The vertical crosshair of the telescope must also be in focus and in the center of the slit.
3. Rotate the telescope tube to the left or right until images of the spectral lines for mercury are located. The wavelengths of the mercury spectrum with the relative intensities in parentheses are: violet 4.047×10^{-7} m (1800), blue 4.358×10^{-7} m (4000), blue-green 4.916×10^{-7} m (80), green 5.461×10^{-7} m (1100), yellow 5.770×10^{-7} m (240), yellow 5.790×10^{-7} m (380). Rotate the telescope tube to the other side to be sure that all the lines can be located. This is just a preliminary check to be sure that all the lines are visible. It may not be possible to resolve the two yellow lines. If not, just assume one line at 5.780×10^{-7} m. *It is extremely important that the grating is never moved after it is originally positioned.*
4. If the spectrometer has a vernier scale capable of reading to one minute of arc, measure to the nearest one minute of arc the angle at which each wavelength of mercury occurs on both sides of 180°. Consult your instructor for directions in the use of the vernier scale. If the spectrometer does not have a vernier, estimate the angles with as much precision as possible. See Figure 43-2 for a description of θ_R and θ_L. Record the two angles for each of the wavelengths in Data and Calculations Table 1.
5. Without moving the diffraction grating, turn off the spectrum tube power supply. Carefully remove the mercury tube and replace it with the hydrogen tube. Turn on the supply and place the hydrogen tube as close to the slit as possible. *AGAIN, BE VERY CAREFUL NOT TO TOUCH THE HIGH*

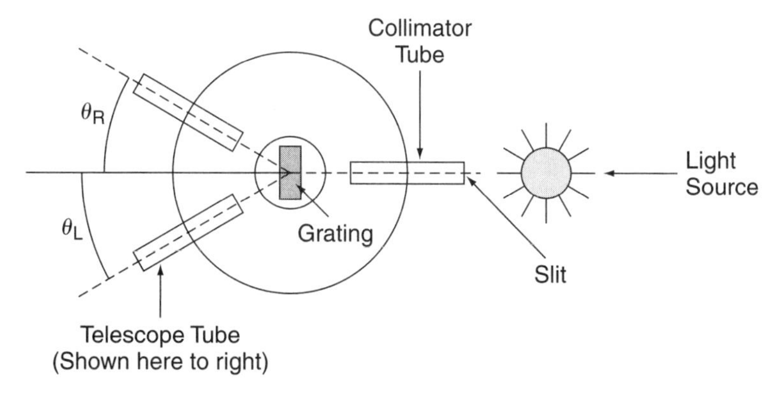

Figure 43-2 Experimental arrangement for the diffraction grating spectrometer.

VOLTAGE ELECTRODES WHILE MAKING THESE ADJUSTMENTS IN THE POSITION OF THE SUPPLY. Rotate the telescope tube back to 180° and carefully adjust the position of the hydrogen tube until a sharp image of the slit is seen directly through the grating. Everything should be in focus from the mercury measurements, and it should just be necessary to place the hydrogen tube in the correct position to give the brightest image. At all times be extremely careful not to move the grating.

6. Carefully measure with as much precision as possible the angle at which the first four wavelengths of the visible hydrogen spectrum occur on both sides of 180°. In order of increasing wavelength they are violet, blue, blue-green, and red. Record the two angles for each of the wavelengths in Data Table 2.

CALCULATIONS

1. For the mercury data in Data Table 1, calculate and record the diffraction angle θ for each wavelength from $\theta = |\theta_R - \theta_L|/2$. For each value of θ calculate a value for the grating spacing d using Equation 4. Calculate the mean of those values $\bar{d}$ and standard error $\alpha_{\bar{d}}$. Record all calculated values in Data and Calculations Table 1.
2. For the hydrogen data in Data Table 2 calculate the diffraction angle $\theta = |\theta_R - \theta_L|/2$ and record those values in Calculations Table 2. Use those values and the value of $\bar{d}$ to calculate values of the hydrogen wavelengths λ_{exp} and values for $1/\lambda_{exp}$ and record them in Calculations Table 2.
3. For each of the hydrogen wavelengths and its associated value of n, calculate the quantity $(1/4–1/n^2)$ and record the results in Calculations Table 2.
4. Perform a linear least squares fit with $1/\lambda_{exp}$ as the vertical axis and $(1/4–1/n^2)$ as the horizontal axis. Record the slope as $(R_H)_{exp}$, the intercept I, and the correlation coefficient r in Calculations Table 2.
5. Calculate the percentage errors for each of the experimental values of the hydrogen wavelengths and record them in Calculations Table 2.
6. Calculate the percentage error for the experimental value of the Rydberg constant $(R_H)_{exp}$ compared to the known value of $1.097 \times 10^7 \text{ m}^{-1}$. Record the results in Calculations Table 2.

GRAPHS

1. Make a graph with the experimental values $1/\lambda_{exp}$ as the vertical axis and $(1/4 - 1/n^2)$ as the horizontal axis. Also show on the graph the straight line obtained from the linear least squares fit to the data.

Name Section Date

43 LABORATORY 43 *Bohr Theory of Hydrogen—The Rydberg Constant*

PRE-LABORATORY ASSIGNMENT

1. State the Balmer formula for the wavelengths of the visible light spectrum of hydrogen.

2. Using the Balmer formula, calculate the first four wavelengths of the spectrum corresponding to $n = 3$, 4, 5, and 6. Show your work.

3. Describe the possible orbits of an electron in a hydrogen atom that are allowed by the Bohr theory.

4. What is a stationary state of the atom in Bohr theory?

5. State Bohr's postulate about the frequency f of light emitted when an electron makes a transition from a state of energy E_i to one of E_f.

6. In the Bohr theory, the Rydberg constant is equal to $me^4/8\varepsilon_0^2ch^3$. Using accepted values for the constants m, e, ε_0, c, and h, calculate the value of the Rydberg constant. Show your work.

7. A diffraction grating has a grating constant of $d = 1.500 \times 10^{-6}$ m. At what angle θ will the first-order image of light of wavelength 5.555×10^{-7} m appear? Show your work.

Name | Section | Date

Lab Partners

43 **LABORATORY 43** *Bohr Theory of Hydrogen—The Rydberg Constant*

LABORATORY REPORT

Data and Calculations Table 1 (Mercury Spectrum)

Colors	λ (10^{-7} m)	θ_R (degree)	θ_L (degree)	θ (degree)	d (10^{-7} m)	$\bar{d}$ (10^{-7} m)	α_d (10^{-7} m)
Violet	4.047						
Blue	4.358						
Blue-Green	4.916						
Green	5.461						
Yellow	5.770						
Yellow	5.790						

Data Table 2 (Hydrogen Spectrum)

Colors	λ (10^{-7} m)	n	θ_R (degree)	θ_L (degree)
Violet	4.102	6		
Blue	4.341	5		
Blue-Green	4.861	4		
Red	6.563	3		

Calculations Table 2 (Hydrogen Spectrum)

n	θ (degree)	λ_{exp} (m)	$1/\lambda_{exp}$ (m^{-1})	$1/4 - 1/n^2$	% *Error* λ
6					
5					
4					
3					
$(R_H)_{exp} =$		$I =$		$r =$	
Percentage Error in $(R_H)_{exp} =$					

SAMPLE CALCULATIONS

1. $\theta = |\theta_R - \theta_L|/2 =$
2. $d = \lambda/\sin\theta =$
3. $\lambda_{exp} = \bar{d}\ \sin\theta =$
4. % Error $\lambda_{exp} =$
5. $1/\lambda_{exp} =$
6. $1/4 - 1/n^2 =$
7. % Error $(R_H)_{exp} =$

QUESTIONS

1. Comment on the accuracy of your experimental value for the Rydberg constant R_H.

2. Comment on the accuracy of your experimental values for the wavelengths of hydrogen compared to the known values.

3. State the precision of your determination of the value of the grating spacing d. State clearly the basis for your answer.

4. Calculate the percentage error of your experimental values of the four wavelengths of hydrogen compared to the Balmer wavelengths you calculated in Question 2 of the Pre-Laboratory.

5. Using the Balmer formula (Equation 1), calculate the $n = 7$ wavelength for the hydrogen spectrum. Why was this wavelength not observed in the laboratory?

Simulated Radioactive Decay Using Dice "Nuclei"

OBJECTIVES

- ❑ Investigate the analogy between the decay of dice "nuclei" and radioactive nuclei.
- ❑ Demonstrate that both the number N and activity A of "nuclei" decrease exponentially.
- ❑ Determine experimental and theoretical values of the decay probability constant λ and the half-life for the dice "nuclei."

EQUIPMENT LIST

- 20-sided dice to simulate radioactive nuclei
- Three-cycle semilog graph paper

THEORY

One of the most noticeable differences between the classical physics known prior to 1900 and the modern physics since that time is the increased role that probability plays in modern physical theories. The exact behavior of many physical systems cannot be predicted in advance. Many situations involve a very large number of particles in which the behavior of any one particle is not predictable, but the average behavior of the collection of particles is quite predictable. One example is a sample of radioactive nuclei that emits alpha, beta, or gamma radiation. It is not possible to predict when any one radioactive nucleus will decay and emit a particle. However, because any reasonable sample of radioactive material contains such a large number of nuclei (at least 10^{12} nuclei), it is possible to predict the average rate of decay with high probability.

A basic concept of **radioactive decay** is that the probability of decay for each radioactive nucleus is constant. In other words, there are a predictable number of decays per second even though it is not possible to predict which nuclei among the sample will decay. A quantity called the **decay constant** λ characterizes this concept. It is the probability of decay per unit time for one radioactive nucleus. Because λ is constant, it is possible to predict the rate of decay for a radioactive sample. The value of the constant λ is different for each radioactive nuclide.

Consider a sample of N radioactive nuclei with a decay constant of λ. The rate of decay of these nuclei $\mathrm{d}N/\mathrm{d}t$ is related to λ and N by the equation

$$\frac{\mathrm{d}N}{\mathrm{d}t} = -\lambda N \qquad \text{(Eq. 1)}$$

The negative sign in the equation means that dN/dt is negative because the number of radioactive nuclei is decreasing. The number of radioactive nuclei at $t=0$ is designated as N_o. The question of interest is how many radioactive nuclei N are there at some later time t. The solution is found by rearranging Equation 1 and integrating it, subject to the condition that $N=N_o$ at $t=0$. The result of that procedure is

$$N = N_o e^{-\lambda t} \qquad \text{(Eq. 2)}$$

Equation 2 states that the number of nuclei N at some later time t decreases exponentially from the original number N_o that are originally present. The quantity λN is the number of decays per second, and it is called the **activity** A of the sample. It can be shown that it obeys the equation

$$A = A_o e^{-\lambda t} \qquad \text{(Eq. 3)}$$

Equations 2 and 3 state that both N and A decay exponentially with the same exponential factor. For measurements made on real radioactive nuclei, the activity A is usually all that can be measured.

The time it takes for N_o to be reduced to $N_o/2$, and the time it takes for A_o to be reduced to $A_o/2$, are the same. It is called the **half-life** $T_{1/2}$ of the decay. It is related to λ by

$$T_{1/2} = \frac{\ln(2)}{\lambda} = \frac{0.693}{\lambda} \qquad \text{(Eq. 4)}$$

Figure 44-1 shows two graphs of activity of a radioactive sample versus time, one semilog and the other linear. The laboratory to be performed does not involve the decay of real radioactive nuclei. Instead it is designed to illustrate the concepts described above by a simulated decay of dice "nuclei." In the laboratory, radioactive nuclei are simulated by a collection of 160 dice with 20 faces. Two of the 20 faces of each die are marked with a dot. The dice are shaken and thrown, and a dice "nucleus" has decayed if a marked face is up after the throw. In this simulation, the decay constant λ is equal to the probability of a marked face coming up, which is 2 out of 20. Thus the theoretical decay constant λ is 0.100. For the analogy, each throw of the dice is one unit of time.

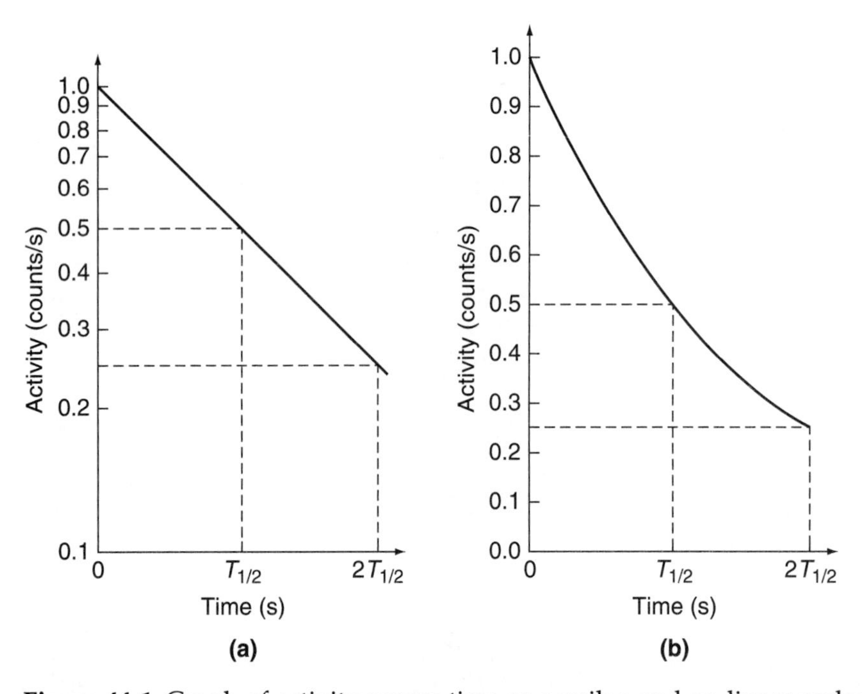

Figure 44-1 Graph of activity versus time on semilog and on linear scales.

A unique aspect of this simulation laboratory is that measurements can be taken on both the number remaining N and the number that decay. The number that decay is analogous to the activity A. For real radioactive nuclei, N cannot be measured directly but is inferred from measurements of A, the activity.

EXPERIMENTAL PROCEDURE

1. Place all 160 of the dice in the square plastic tray provided with the dice. Place the clear plastic cover over the dice and shake them vigorously. Remove all dice that come to rest with a marked face up. Remove only those that have a marked face that points directly upward. Record in Data and Calculations Table 1 the number of dice that decay (are removed) on the first throw of the dice. Also record the number of dice that remain after the ones that decay are removed.
2. Place the cover on the dice and shake the remaining dice vigorously. Remove the dice that come to rest with a marked face up on the second throw. Record in Data and Calculations Table 1 the number of dice removed on the second throw, and also record the number of dice that remain after the ones that decay are removed.
3. Continue this process of shaking the dice, removing the ones that have a marked face up, and recording the number of dice removed and the number of dice left for each throw. Continue this procedure for a total of 20 throws of the dice, or until all of the dice have been removed.
4. Each experimental group should record its data on the blackboard so that the class results can be plotted as a set of data with better statistics. Sum the total number of dice thrown originally for the entire class and sum the number removed at each throw of the dice. Record this class data in Data Table 2.

GRAPHS

1. On three-cycle semilog graph paper provided by the instructor, graph the results. Plot N the number of "nuclei" remaining on the log scale versus the number of throws on the linear scale, using $\times$ as a symbol. On the same piece of graph paper, plot the activity A (number removed each throw) on the log scale versus the number of throws on the linear scale, using $+$ for a symbol.
2. On a second piece of three-cycle semilog graph paper, graph the class data for N and A. With better statistics, these curves should be smoother than the individual data.
3. If the dice behaved exactly according to the theory described, all of the graphs described above would fall on a straight line on the semilog plot. The data for N will most likely show this trend better than the data for A.
4. For the individual data and the class data, draw a straight line that best fits the data. Do this for both N and A.

CALCULATIONS

1. For each of the 20 shakes of the dice, calculate the ratio of the number of dice removed after a given throw to the number shaken for that throw. In the simulation the number of dice removed is A, the activity, and the number shaken is N, the number of radioactive nuclei. Thus these ratios amount to an experimental value for λ. Note carefully that this ratio must be calculated with data from two different rows in Data and Calculations Table 1. For example, the number of dice *thrown* on the fourth throw is listed as the number of dice *remaining* after the third throw. Thus the ratio is calculated with the number removed on each row to the number remaining in the preceding row. Record these 20 values as λ_{exp} in Data and Calculations Table 1.

2. Calculate the mean of these 20 values for λ_{exp} and record it in Calculations Table 2 as $\overline{\lambda_{exp}}$.
3. The theoretical value of λ is 0.100. Record this value in Calculations Table 2 as λ_{theo}.
4. Calculate the percentage error in the value of $\overline{\lambda_{exp}}$ compared to λ_{theo}. Record this percentage error in Calculations Table 2.
5. Calculate the theoretical half-life from Equation 4 using the value of $\lambda = 0.100$. Record that value in Calculations Table 2 as $(T_{1/2})_{theo}$. For purposes of this calculation consider a fractional throw as possible.
6. Consider the straight line drawn through the data points of your individual data for N versus number of throws. The experimental half-life is the number of throws needed to go from any point on the line to one-half that value. Determine the number of throws needed to go from 120 to 60 on the straight line through your data. Record that number in Calculations Table 2 as $(T_{1/2})_{exp}$. For purposes of this determination, consider a fractional throw as possible.
7. Calculate the percentage error in the value $(T_{1/2})_{exp}$ compared to the value of $(T_{1/2})_{theo}$. Record that percentage error in Calculations Table 2.

Name Section Date

LABORATORY 44 *Simulated Radioactive Decay Using Dice "Nuclei"*

PRE-LABORATORY ASSIGNMENT

1. A typical sample of radioactive material would contain as a lower limit approximately how many nuclei? (a) 1000 (b) 10^6 (c) 10^{12} (d) 10^{23}

2. The theory of radioactive decay can predict when each of the radioactive nuclei in a sample will decay. (a) True (b) False

3. State the definition of the decay constant λ. What are its units?

4. For the simulation of a radioactive decay using 20-sided dice, what are the analogous quantities to the real quantities listed below?

   undecayed nucleus—

   decayed nucleus—

   time—

   decay constant—

5. What quantity can be measured for the simulation laboratory that cannot normally be directly measured in a true radioactive decay laboratory?

6. A radioactive decay process has a decay constant $\lambda = 1.50 \times 10^{-4}\,\text{sec}^{-1}$. There are 5.00×10^{12} radioactive nuclei in the sample at $t = 0$. How many radioactive nuclei are present in the sample one hour later? Show your work.

7. For the radioactive sample described in Question 6, what is the activity A (in decays per second) at $t = 0$? What is the activity one hour later? Show your work.

8. What is the half-life of the radioactive sample described in Question 6? Show your work.

Name Section Date

Lab Partners

44 LABORATORY 44 Simulated Radioactive Decay Using Dice "Nuclei"

LABORATORY REPORT

Data and Calculations Table 1

Throw	Removed (A)	Remaining (N)	λ_{exp}
0	0	160	
1			
2			
3			
4			
5			
6			
7			
8			
9			
10			
11			
12			
13			
14			
15			
16			

Data Table 2

Removed (A)	Remaining (N)
0	

(Continued)

(Continued)

Throw	*Removed (A)*	*Remaining (N)*	λ_{exp}
17			
18			
19			
20			

Removed (A)	*Remaining (N)*

Calculations Table 2

$\lambda_{theo} =$	$\overline{\lambda_{exp}} =$	% Error =
$(T_{1/2})_{theo} =$	$(T_{1/2})_{exp} =$	% Error =

SAMPLE CALCULATIONS

1. $\lambda_{exp} =$ (# removed)/(# thrown) =
2. $\overline{\lambda_{exp}} =$ (sum of 20 λ_{exp})/20 =
3. $(T_{1/2})_{theo} = (0.693)/(\lambda_{theo}) =$
4. $(T_{1/2})_{exp} = (0.693)/(\overline{\lambda_{exp}}) =$
5. % Error $\lambda =$
6. % Error $T_{1/2} =$

QUESTIONS

1. Do your data for *N* as a function of the number of throws give a reasonably straight line? Would a line with the same slope as you drew through the *N* versus number of throws fit reasonably well through the *A* versus number of throws plot?

2. Comment on the agreement between your experimental value for λ and the theoretical value for λ.

3. Comment on the agreement between your experimental value for the half-life and the theoretical value of the half-life.

4. Are the graphs for the class data smoother and more nearly a straight line than your individual data? Would you expect this to be true, and if so, why?

5. Calculate the half-life from the class data graph. Compare it to the theoretical value for the half-life. Would you expect it to show better agreement, and if so, why?

Geiger Counter Measurement of the Half-Life of ^{137}Ba

OBJECTIVES

- ❑ Measure the count rate versus voltage for a Geiger counter to determine its appropriate operating voltage.
- ❑ Measure the activity of ^{137}Ba as a function of time to determine its half-life.

EQUIPMENT LIST

- Geiger counter, scaler, timer, ^{90}Sr beta radiation source
- Minigenerator that produces ^{137}Ba from the decay of a parent nuclide of ^{137}Cs
- Disposable plachet (very thin small metal plate to contain a radioactive sample)
- Three-cycle semilog graph paper

THEORY

A basic concept of **radioactive decay** is that the probability of decay for each radioactive nucleus is constant. In other words, there are a predictable number of decays per second even though it is not possible to predict which nuclei among the sample will decay. A quantity called the **decay constant** λ characterizes this concept. It is the probability of decay per unit time for one radioactive nucleus. Because λ is constant, it is possible to predict the rate of decay for a radioactive sample. The value of the constant λ is different for each radioactive nuclide.

Consider a sample of N radioactive nuclei with a decay constant of λ. The rate of decay of these nuclei dN/dt is related to λ and N by the equation

$$\frac{dN}{dt} = -\lambda N \qquad \text{(Eq. 1)}$$

The negative sign in the equation means that dN/dt is negative because the number of radioactive nuclei is decreasing. The number of radioactive nuclei at $t = 0$ is designated as N_o. The question of interest is how many radioactive nuclei N are there at some later time t. The solution is found by rearranging Equation 1 and integrating it, subject to the condition that $N = N_o$ at $t = 0$. The result of that procedure is

$$N = N_o e^{-\lambda t} \qquad \text{(Eq. 2)}$$

Equation 2 states that the number of nuclei N at some later time t decreases exponentially from the original number N_o that are originally present. The quantity λN is the number of decays per second, and it is called the **activity** A of the sample. It can be shown that it obeys the equation

$$A = A_o e^{-\lambda t} \qquad \text{(Eq. 3)}$$

Equations 2 and 3 state that both N and A decay exponentially with the same exponential factor. The time it takes for N_o to be reduced to $N_o/2$, and the time it takes for A_o to be reduced to $A_o/2$, are the same. It is called the **half-life** $T_{1/2}$ of the decay. It is related to λ by

$$T_{1/2} = \frac{\ln(2)}{\lambda} = \frac{0.693}{\lambda} \qquad \text{(Eq. 4)}$$

To study these radioactive processes, we must detect the presence of these particles that are the product of the decay. We can build devices in many forms to accomplish the detection, but they all would have one feature in common. Every practical device that detects radiation allows the particles to interact with matter, and then uses that interaction as the basis for detection. The particular device we will use in this laboratory is called a **Geiger counter**. It consists of a tube in which the incident particle interacts, and a scaling circuit to count the pulses of electricity produced. A diagram of a Geiger tube is shown in Figure 45-1.

The Geiger tube is a small metal cylinder with a thin self-supporting wire along the axis of the cylinder. The wire is insulated from the cylinder. The cylindrical wall of the tube serves as the negative electrode (cathode), and the wire along the axis is the positive electrode (anode). At the entrance end of the tube there is a thin "window" formed by a very thin piece of fragile mica. Inside the counter is a special gas mixture that is ionized by any radiation that penetrates the "window."

In operation, a voltage is applied across the electrodes. The particular voltage for each tube must be determined experimentally. The applied voltage creates a large electric field in the tube, and the field is especially large in the region near the central wire. When radiation passes through the window and ionizes the gas, the large electric field causes an acceleration of the free electrons. These accelerated electrons cause additional ionizations that create an "avalanche" effect. The total number of ion-electron pairs created by a single incident particle is of the order of one million.

The electrons are more mobile and drift toward the positive central wire. When they arrive at the wire, their negative charge causes the voltage of the wire to be lowered, and this sudden drop in voltage creates a pulse that is counted by the electronic circuitry. Each pulse counted signifies the passage of a particle through the counter. The ions then recombine with electrons, leaving the gas neutral again and ready for the passage of another particle. The whole process takes a time of the order of 300 μs, and during that time period if another particle goes through the counter, it may not be counted. Thus one disadvantage of Geiger counters is this "dead time" during which counts may be missed. This is a negligible effect unless the count rate is very high.

The count rate of a Geiger counter is a function of the voltage applied across the electrodes. Therefore, the counter should be operated in a region where the rate at which the count rate changes with voltage is a

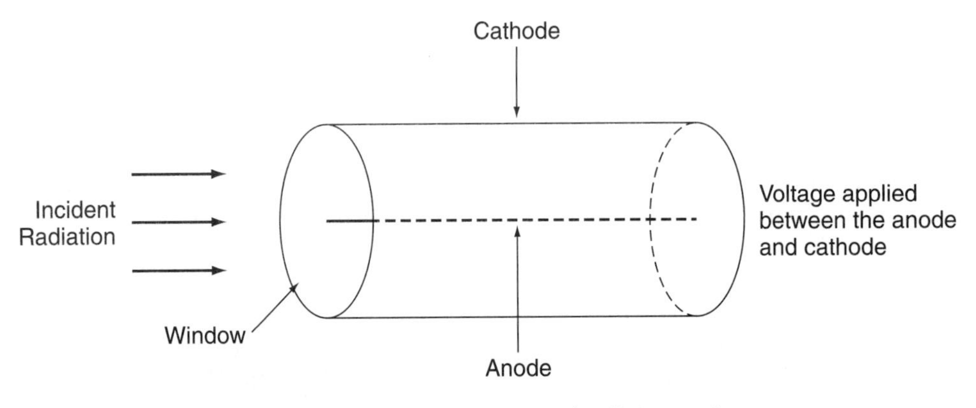

Figure 45-1 Diagram of the essential elements of a Geiger tube.

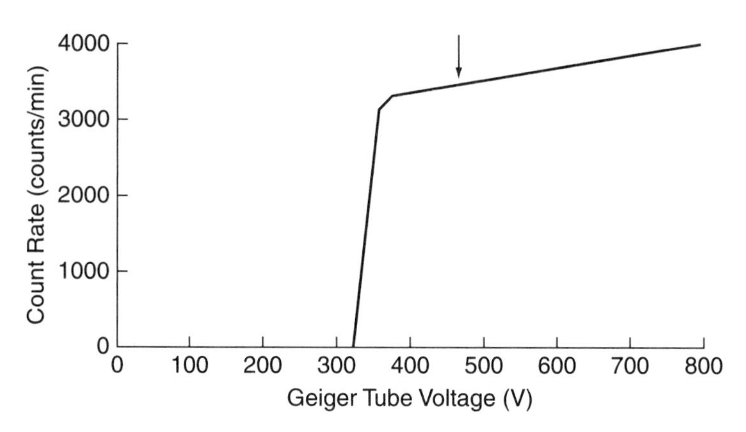

Figure 45-2 Typical count rate versus voltage for a Geiger tube.

minimum. This is accomplished experimentally by measuring the count rate of some fixed source of radiation as a function of the voltage applied to the tube. A graph of the count rate versus the voltage will be made from the data and the operating point will be chosen to be some voltage where the count rate versus voltage curve is as nearly flat as possible. This is referred to as the plateau region.

The counter needs a minimum voltage to produce pulses at all. Both this minimum voltage and the operating voltage are quite variable and depend upon the dimensions of the Geiger tube, and on the particular gas used in the tube. Therefore, the exact nature of the count rate versus voltage curve depends on the particular tube used. In general, the student-type Geiger tubes used in most undergraduate laboratories operate at about 500 V and do not have a very flat count rate versus voltage curve anywhere. Thus it is difficult to identify much of a plateau region. Typically, research-grade Geiger tubes operate in the neighborhood of 1000 V, and they have a much flatter count rate versus voltage curve. A typical count rate versus voltage curve for a student-type Geiger tube is shown in Figure 45-2.

EXPERIMENTAL PROCEDURE

Count Rate Versus Voltage to Determine Operating Voltage

1. All Geiger tubes have a maximum permissible operating voltage above which they are subject to breakdown. This may permanently damage the tube. Consult your instructor or the instruction manual for the specific maximum voltage of the particular Geiger tube used. *Do not ever exceed this maximum voltage.* Before plugging in the power cord of your instrument, make sure that the high voltage control is turned to the minimum setting.

2. Place the ^{90}Sr β source about 2 or 3 cm from the window of the Geiger tube. Once the source is positioned, take all the measurements for this procedure without changing the position of the source relative to the detector.

3. Turn on the power to the instrument, reset the scaler to zero, and start the counter in a continuous count mode with the high voltage still set to a minimum. Slowly increase the high voltage setting until counts begin to register on the scaler. Leave the voltage at this setting for which the Geiger tube just begins to count. This is called the threshold voltage.

4. Reset the scaler to zero, start the counter, and let it count for 1 minute. Repeat this procedure two more times for a total of three 1-minute counts at this voltage setting. Record the voltage and the three values of the count in Data Table 1.

5. Raise the high voltage setting by 50 V, repeat Step 4 at this new high voltage setting and record the high voltage and the counts for these three trials.

6. Continue this process up to the maximum permissible voltage of the Geiger tube. Be sure to obtain the proper maximum voltage for your tube either from your instructor or from the instruction manual for the instrument.

Half-life of ^{137}Ba

1. The graph of count rate versus voltage should be similar in shape to Figure 45-2. The operating point of the tube should be chosen to be just beyond the shoulder on the relatively flat portion of the graph. That region is shown by an arrow in Figure 45-2. If the graph of your data does not look like you expect, consult your instructor for help in picking the proper operating voltage for your Geiger counter. Make all the measurements in this procedure at the same voltage once it is determined.
2. With no radioactive source near the Geiger tube, reset the scaler to zero and count the background for 2 minutes. The activity will be determined by counting for 15.0 s intervals. Divide the number of counts obtained in 2 minutes by eight to obtain an average background count rate for 15.0 seconds. Record that background rate in Data Table 2.
3. The ^{137}Ba isotope must be prepared at the time of its use because it has such a short half-life. The Minigenerator contains ^{137}Cs, which decays with a 30-year half-life to produce ^{137}Ba as a daughter product. A saline solution of HCl is used to wash out the ^{137}Ba that has established equilibrium with the ^{137}Cs. The ^{137}Ba is created in an excited state and decays to the ground state of ^{137}Ba by gamma-ray emission with a half-life of less than 3 minutes.
4. Your instructor will prepare the sample when you are ready to count. *Very carefully review all of the remaining steps of the procedure so that you will be ready to begin counting the sample immediately after it is prepared.*
5. Set the scaler to the stop position and reset the scaler to zero.
6. As quickly as possible after you receive your sample, place the plachet containing the sample as close to the Geiger tube window as possible, to obtain the maximum counting rate.
7. As quickly as possible, start the scaler in the continuous count mode and start the external timer simultaneously. *Do not ever stop the timer. Let it run continuously for the rest of this procedure.*
8. When the timer reaches 15.0 seconds, stop the scaler. Record the counts obtained during the interval 0–15.0 seconds in the appropriate place in Data Table 2.
9. During the time interval when the timer reads between 15.0 and 30.0 s elapsed time, reset the scaler to zero and wait for the next counting period.
10. Start the scaler when the timer reads 30.0 s and stop the counter when the timer reads 45.0 s. Record the counts obtained in the interval 30.0–45.0 s in the appropriate place in Data Table 2.
11. Continue this process, letting the timer run continuously and alternately counting for a 15.0-s interval and resetting and waiting a 15.0-s interval until Data Table 2 is completed. If a mistake is made on one of the counting periods, just skip it and start at the next counting period.

CALCULATIONS

Count Rate Versus Voltage

1. Calculate the mean $\overline{C}$ and standard error α_C for the three trials of the count at each voltage. Calculate the square root of the mean count $\sqrt{\overline{C}}$ at each voltage. Record all calculated values in Calculations Table 1.
2. On linear graph paper plot the mean count rate $\overline{C}$ (counts/min) versus voltage to produce a graph like the one shown in Figure 45-2.

Half-life of ^{137}Ba

1. Each count period is 15.0 s long, during which time the activity changes continuously. For each of the counting periods, subtract the average background count for 15.0 s from the count in that time period. Divide that result for each period by 15.0 s and record it as the activity A in counts/s at the appropriate place in Calculations Table 2. This is actually the average activity during each 15.0-s

time interval, but it will be assumed that it approximates the instantaneous activity at the middle of each time interval.

2. Calculate the quantities (A_o/A) and $\ln(A_o/A)$ for each of the activities in Data Table 2 where A_o is taken to be the activity at $t = 0$.
3. The quantity $\ln(A_o/A)$ should vary linearly with t the time, and the slope should be equal to the disintegration constant λ. Perform a linear least squares fit of the data in Calculations Table 2 with $\ln(A_o/A)$ as the vertical axis and t the time as the horizontal. Record the value of the slope as the disintegration constant λ in Calculations Table 2.
4. Using Equation 4, calculate the experimental value of the half-life of ^{137}Ba from the experimental value of the disintegration constant λ.

GRAPHS

Half-life of ^{137}Ba

1. On three-cycle semilog graph paper, plot the activity in counts/second versus the time in seconds, using the log scale for the activity and the linear scale for time. Graph each activity at the time corresponding to the middle of the time interval as given in Calculations Table 2.

Name Section Date

45 LABORATORY 45 *Geiger Counter Measurement of the Half-Life of ^{137}Ba*

PRE-LABORATORY ASSIGNMENT

1. For the isotope of the element nickel $^{63}_{28}$Ni give the number of protons Z, the number of neutrons N, and the mass number A.

2. Show that Equation 3 in the laboratory can be expressed in the form $\ln(A_o/A) = \lambda t$. This is the form used to fit the data in the laboratory.

3. What is the basis for the detection of the particles from any radioactive decay?

4. The inside of a Geiger tube is filled with a (a) gas (b) liquid (c) solid (d) plasma.

5. The pulse that is counted in a Geiger tube is caused by a (a) rise in voltage from ions arriving at the anode (b) drop in the voltage from ions arriving at the anode (c) rise in voltage from electrons arriving at the anode (d) drop in voltage from electrons arriving at the anode.

6. In a Geiger tube the total time taken to create a pulse and then let the electrons recombine with the ions to again form a neutral state is about (a) 1 μs (b) 30 μs (c) 300 μs (d) 10,000 μs.

7. A typical operating voltage for a Geiger counter is about (a) 5 V (b) 50 V (c) 500 V (d) 5000 V.

8. In this laboratory the half-life of ^{137}Ba will be determined by measuring (a) the number of atoms left as a function of time (b) the activity that is constant (c) the activity that decreases as a function of time (d) the activity that increases as a function of time.

Name Section Date

Lab Partners

45 LABORATORY 45 *Geiger Counter Measurement of the Half-Life of* ^{137}Ba

LABORATORY REPORT

Data Table 1

Voltage	Count 1	Count 2	Count 3

Calculations Table 1

$\overline{C}$	α_C	$\sqrt{\overline{C}}$

Data Table 2

Background 2 minutes =

Count Period	*Counts*
0–15 s	
30–45 s	
60–75 s	
90–105 s	
120–135 s	
150–165 s	
180–195 s	
210–225 s	
240– 255 s	
270–285 s	
300–315 s	
330–345 s	
360–375 s	

Calculations Table 2

Background count for 15 s =

A (cts/s)	(A_o/A)	$ln(A_o/A)$	t (s)
			7.5
			37.5
			67.5
			97.5
			127.5
			157.5
			187.5
			217.5
			247.5
			277.5
			307.5
			337.5
			367.5
$\lambda =$ s^{-1}		$T_{1/2} =$ s	

SAMPLE CALCULATIONS

1. Background (15 s) = Total counts/8 =
2. A = (Counts – Background)/(15) =
3. (A_o/A) =
4. ln (A_o/A) =
5. $T_{1/2}$ =

QUESTIONS

1. For the count rate versus voltage data consider the standard error α_C and the square root of the mean count $\sqrt{\bar{C}}$. According to nuclear statistical theory, those quantities should be approximately equal. Calculate the percentage difference between them.

2. State the threshold voltage (voltage at which the counter begins to operate) for your Geiger tube.

3. On the semilog graph of the activity of ^{137}Ba, draw the best straight line that you can through the data points. From the straight line determine the half-life of the sample. Indicate exactly which points on the straight line are used to determine the half-life.

4. The accepted half-life of ^{137}Ba is 2.6 min. Calculate the percentage error of each of your determinations of this half-life. Show your work.

5. A physics professor purchased a ^{137}Cs source in September 1990, which had an activity of 2.00×10^5 disintegrations per second at that time. What is the activity of that source today? (The half-life of ^{137}Cs is 30.2 years.) Show your work.

Nuclear Counting Statistics

OBJECTIVES

- ❑ Investigate the counts from a radioactive source for 50 measurements under conditions in which the count rate should be approximately constant.
- ❑ Determine the standard deviation from the mean and standard error of the counts.
- ❑ Investigate how well the observed distribution of counts compares to that predicted by the normal distribution.
- ❑ Investigate whether or not $\sqrt{\overline{C}}$ approximates the standard deviation from the mean.

EQUIPMENT LIST

- Geiger counter (single unit containing Geiger tube, power supply, timer, and scaler)
- Long-lived radioactive source (such as ^{137}Cs or ^{60}Co)

THEORY

If all other sources of error are removed from a nuclear counting experiment, there remains an uncertainty due to the random nature of the nuclear decay process. It is assumed that there exists some **true mean** value of the count, which shall be designated as m. But we emphasize, do not assume that there is a true value for any individual count C_i. Although m is assumed to exist, it can never be known exactly. Instead, one can approach knowledge of the true mean m by a large number of observations. It can be shown that the best approximation to the true mean m is the mean $\overline{C}$, which is given by

$$\overline{C} = (1/n)\sum_{i=1}^{n} C_i \qquad \text{(Eq. 1)}$$

where C_i stands for the ith value of the count obtained in n trials. The **standard deviation from the mean** σ_{n-1} and the **standard error** α are defined in the usual manner as

$$\sigma_{n-1} = \sqrt{\sum_{i=1}^{n}(1/n-1)(\overline{C}-C_i)^2} \quad \text{and} \quad \alpha = \frac{\sigma_{n-1}}{\sqrt{n}} \qquad \text{(Eq. 2)}$$

These equations have been applied to essentially all of the measurements in this laboratory manual. In many of the cases where these ideas have been applied, they are somewhat questionable because the

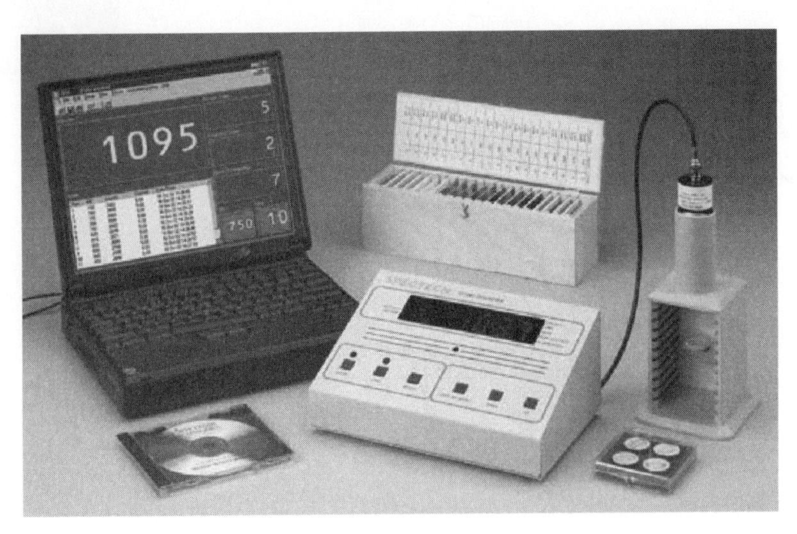

Figure 46-1 Geiger counter with timer-scaler and encapsulated radioactive sources. (Photo courtesy of Sargent-Welch Scientific Co.)

random errors are not necessarily the determining factor. For nuclear counting experiments, usually the random errors are the limiting factor, and these concepts generally do apply strictly to such measurements.

The way in which the measurements C_i are distributed around the mean $\overline{C}$ depends upon the statistical distribution. The binomial distribution is the fundamental law for the statistics of all random events including radioactive decay. Calculations are difficult with this distribution, and it is often approximated by another integral distribution called the Poisson distribution. For cases of m greater than 20, both the binomial and the Poisson distribution can be approximated by the **normal distribution.** It has the advantage that it deals with continuous variables, and thus calculations are much easier with the normal distribution. For most nuclear counting problems of interest, the normal distribution predicts the same results for nuclear counting that have been assumed for measurements in general. Approximately 68.3% of the measured values of C_i should fall within $\overline{C} \pm \sigma_{n-1}$, and approximately 95.5% of the measured values of C_i should fall within $\overline{C} \pm 2\,\sigma_{n-1}$.

There is one statistical idea valid for nuclear counting experiments that is not true for measurements in general. For any given single measurement of the count C in a nuclear counting experiment, an approximation to the standard deviation from the mean σ_{n-1} is given by

$$\sigma_{n-1} \approx \sqrt{C} \qquad \text{(Eq. 3)}$$

For a series of repeated trials of a given count, the most accurate determination is given by $\overline{C} \pm \alpha$. If only a single measurement of the count is made, the most accurate statement that can be made is given by $C \pm \sqrt{C}$.

In this laboratory, we will take a series of measurements of the same count to determine the distribution of the measurements about the mean. In addition, we will investigate the validity of Equation 3.

EXPERIMENTAL PROCEDURE

1. Consult your instructor for the operating voltage of the Geiger counter and set the Geiger counter to the proper operating voltage. Place a long-lived radioactive isotope on whichever counting shelf is necessary to produce between 500 and 700 counts in a 30 s counting period. For best results, the Geiger counter should have preset timing capabilities. If it does not and a laboratory timer is used, it would improve the timing precision if 60 s counting intervals are used. For whatever time is counted, between 500 and 700 counts should be recorded.

2. Repeat the count for a total of 50 trials. Make no changes whatsoever in the experimental arrangement for these 50 trials. Record each count in the Data Table. Do not make any background subtraction. Simply record the total count for each counting period.

CALCULATIONS

1. Calculate the mean count $\overline{C}$, the standard deviation from the mean σ_{n-1}, and the standard error α for the 50 trials of the count and record the results in the Calculations Table.
2. For each count C_i calculate $|C_i - \overline{C}|/\sigma_{n-1}$ and record the results in the Calculations Table.
3. Determine what percentage of the counts C_i are further from $\overline{C}$ than σ_{n-1} by counting the number of times a value of $|C_i - \overline{C}|/\sigma_{n-1} > 1$ occurs. Express this number divided by 50 as a percentage. Count the number of times that $|C_i - \overline{C}|/\sigma_{n-1} > 2$ occurs. Express this number divided by 50 as a percentage. Record these results in the Calculations Table.
4. Calculate $\sqrt{\overline{C}}$ and record its value in the Calculations Table.

GRAPHS

1. Construct a histogram of your data on linear graph paper. Consider the range of the data and arbitrarily divide the range into about 15 segments. For counts in the range used, this should give intervals of 8 or 10 counts. An example of some data is displayed in this manner in Figure 46-2. The mean of the data is 659 with $\sigma_{n-1} = 27$, and an interval of 10 has been chosen.

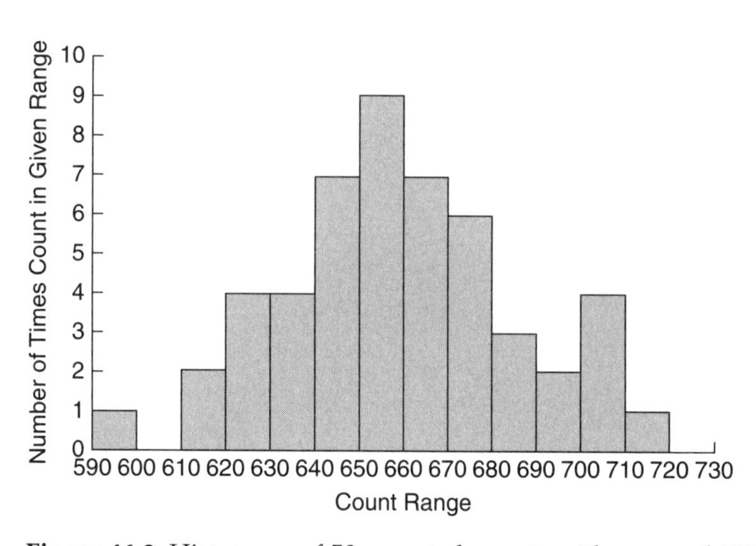

Figure 46-2 Histogram of 50 repeated counts with mean of 659.

Name | Section | Date

46

LABORATORY 46 *Nuclear Counting Statistics*

PRE-LABORATORY ASSIGNMENT

1. For nuclear counting experiments no true value of a given count is assumed. What quantity is assumed to have a true value?

2. What is the exact statistical distribution function that describes the statistics of nuclear counting experiments?

3. What statistical distribution function approximates nuclear counting statistics and is used because it deals with continuous variables? For what values of the true mean m is this distribution valid?

4. In a nuclear counting experiment a single measurement of C counts is obtained. What is the approximate value for σ_{n-1} for the count C?

5. According to the normal distribution function, when a given count is repeated 30 times, approximately how many of the results should fall in the range $\overline{C} \pm \sigma_{n-1}$? How many should fall in the range $\overline{C} \pm 2\sigma_{n-1}$?

6. A single count of a radioactive nucleus is made, and the result is 927 counts. What is the approximate value of σ_{n-1}?

7. A set of 10 repeated measurements of the count from a given radioactive sample was taken. The results were: 633, 666, 599, 651, 654, 690, 660, 659, 664, and 612. What is the mean count $\overline{C}$? What is the value of σ_{n-1}? What is the value of α? Which of the counts fall outside $\overline{C} \pm \sigma_{n-1}$? Is this approximately the number of cases expected? Show your work.

8. For the data in Question 7, is $\sqrt{\overline{C}}$ approximately equal to σ_{n-1}? Calculate the percentage difference between the two. Show your work.

Name | Section | Date

Lab Partners

46 LABORATORY 46 Nuclear Counting Statistics

LABORATORY REPORT

Data Table

i	C_i	i	C_i
1		18	
2		19	
3		20	
4		21	
5		22	
6		23	
7		24	
8		25	
9		26	
10		27	
11		28	
12		29	
13		30	
14		31	
15		32	
16		33	
17		34	

Calculations Table

i	$\|C_i - \overline{C}\|/\sigma_{n-1}$	i	$\|C_i - \overline{C}\|/\sigma_{n-1}$
1		18	
2		19	
3		20	
4		21	
5		22	
6		23	
7		24	
8		25	
9		26	
10		27	
11		28	
12		29	
13		30	
14		31	
15		32	
16		33	
17		34	

(Continued)

(Continued)

i	C_i
35	
36	
37	
38	
39	
40	
41	
42	

i	C_i
43	
44	
45	
46	
47	
48	
49	
50	

i	$\lvert C_i - \overline{C} \rvert / \sigma_{n-1}$	i	$\lvert C_i - \overline{C} \rvert / \sigma_{n-1}$
35		43	
36		44	
37		45	
38		46	
39		47	
40		48	
41		49	
42		50	
$\overline{C} =$		$\sigma_{n-1} =$	
$\alpha =$		$\sqrt{\overline{C}} =$	
% trial $> \sigma_{n-1}$ from mean $=$			
% trial $> 2\,\sigma_{n-1}$ from mean $=$			

SAMPLE CALCULATIONS

1. $\lvert C_i - \overline{C} \rvert / \sigma_{n-1} =$
2. $\sqrt{\overline{C}} =$
3. % trial $> \sigma_{n-1}$ from mean $=$
4. % trial $> 2\,\sigma_{n-1}$ from mean $=$

QUESTIONS

1. Consider the shape of the histogram of your data. Does it show the expected distribution relative to the mean of the data?

2. Compare the percentage of trials that have $|C_i - \overline{C}|/\sigma_{n-1} > 1$ with that predicted by the normal distribution. Compare the percentage of trials that have $|C_i - \overline{C}|/\sigma_{n-1} > 2$ with that predicted by the normal distribution.

3. What is the most accurate statement that you can make about the count from the sample based upon the data that you have taken?

4. Calculate the percentage difference between $\sqrt{\overline{C}}$ and σ_{n-1}. Do the results confirm the expectations of Equation 3?

Absorption of Beta and Gamma Rays

OBJECTIVES

- ❑ Investigate the difference in relative absorption by different types of materials for different kinds of radiation.
- ❑ Determine the absorption coefficient μ for gamma radiation in lead.

EQUIPMENT LIST

- Geiger counter (single unit with Geiger tube, power supply, timer, and scaler)
- Absorber set (lead and polyethylene), two-cycle semilog graph paper
- Gamma source (^{60}Co, for example), beta source (^{90}Sr, for example)

THEORY

Three different particles are emitted in the three types of natural radioactivity. In **α-decay** the emitted particle is a twice-ionized helium atom called an alpha particle. In **β-decay** the emitted particle is an electron or positron. For **γ-decay** the emitted particle is a high-energy photon. The different nature of these particles accounts for the differences in their relative absorption in matter. The most important characteristics that determine the interactions are the charge and the mass of the particles. Because they are charged and have large mass, alpha particles interact very strongly, lose energy in a series of interactions, and do not travel far in matter. For a typical α-decay the particle will not penetrate the thin window of a Geiger counter. Therefore we will use no α-decay sources in this laboratory.

Because the γ-ray photons have no mass or charge, they interact with matter in a fundamentally different way than charged particles interact. They interact less strongly with matter, and as a result are the most penetrating type of radiation. They require a greater thickness of material to be completely absorbed. Gamma rays interact with matter by the photoelectric effect, the Compton effect, and by positron-electron pair production. In each of these processes a photon is effectively completely removed from the beam in a single process, if it interacts at all. The consequence of this fact, that photons either do not interact at all or else are completely removed by an interaction, means that γ-ray intensity decreases exponentially with absorber thickness. Photons are the only particles from natural radioactivity to interact in this manner, and are the only ones to show an exact exponential decrease in matter.

It is found experimentally that when a beam of γ-rays of intensity I is incident on matter of thickness Δx, the change in intensity ΔI of the beam as it passes through the matter is proportional to the thickness Δx and to the incident intensity I. In equation form this is

$$\Delta I = -\mu I \, \Delta x \qquad \text{(Eq. 1)}$$

where μ is a constant of proportionality called the **absorption coefficient.** The constant μ has dimensions of inverse length and is commonly expressed in cm^{-1}. If the limit is taken so that the finite changes become differential, Equation 1 can then be integrated to give

$$I = I_o e^{-\mu x} \qquad \text{(Eq. 2)}$$

which is the characteristic exponential absorption described earlier. The term I_o is the intensity at thickness $x = 0$, and I is the intensity at thickness x. Equation 2 can be rewritten in the form

$$\ln(I_o/I) = \mu x \qquad \text{(Eq. 3)}$$

Equation 3 states that the quantity $\ln(I_o/I)$ should be proportional to x with μ as the constant of proportionality. This relationship will be used to determine μ. The relationship applies only to a source of pure γ-rays. The ^{60}Co source emits both γ-rays and β-rays, but the β-rays will be absorbed first, and the γ-rays that are left can be assumed to be a pure source of γ-rays.

The absorption of β-rays is completely different from the absorption of γ-rays. Beta rays are either electrons or positrons. In either case, they are charged particles with mass equal to the mass of the electron. As charged particles, they tend to lose their energy gradually in a series of collisions with the electrons in the atoms of the absorbing material. Generally each collision results in a relatively small energy loss, and a large number of collisions are necessary before all the energy of the incident β-ray is lost. As a consequence of the nature of this process, a beam of electrons, all of which have the same initial energy, has a definite range in a given type of absorber. Therefore, electrons do not exhibit absorption that is an exponential function of the thickness of the absorber.

The situation for the case of natural β-rays is complicated because not all of the betas have the same energy. Betas from a β-ray source have a spectrum of energy ranging from almost zero up to some maximum energy characteristic of the isotope. If an absorption experiment is performed for the spectrum of energies that is present for any β-ray emitter, often the intensity does in fact decrease in a nearly exponential manner. This is simply a fortuitous result of the combined effects of the initial energy

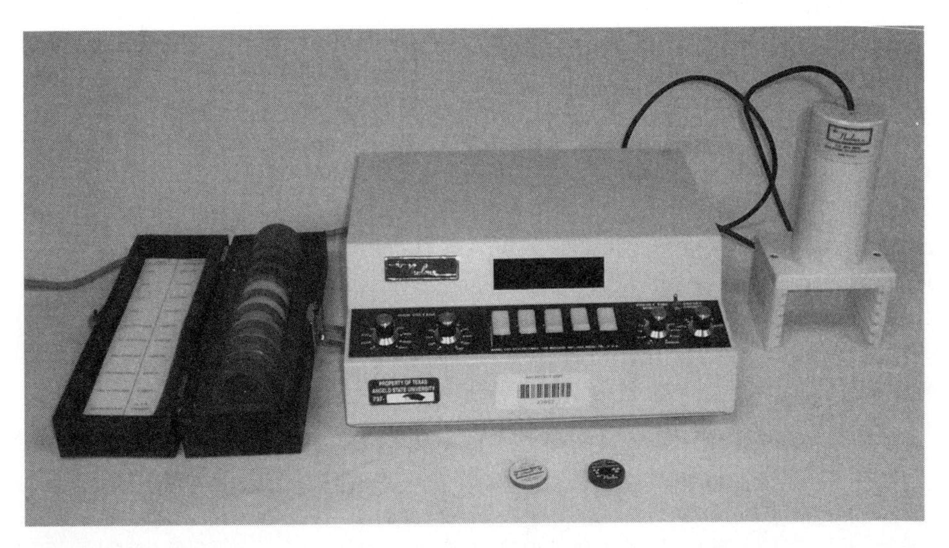

Figure 47-1 Geiger counter and set of standard absorbers described in procedure.

distribution, of back scattering into the detector, and of the true range-energy relationship. We will perform measurements on the absorption of β-rays from a ^{90}Sr source that we can assume to be a pure beta source.

EXPERIMENTAL PROCEDURE

Gamma Absorption

1. Consult your instructor for the operating voltage of the Geiger counter and set the Geiger counter to the proper operating voltage. Once it is set, leave the voltage unchanged for the rest of the laboratory. Place the ^{60}Co source in the third shelf position and place the empty absorber holder in the second shelf position. Be sure that the source is not moved during the rest of the procedure. For best results, let the Geiger tube stabilize for at least 10 min before proceeding.
2. Reset the scaler to zero and determine intensity I (the number of counts) from the source with no absorber in the holder. Count for a period of 60 s.
3. Using lead absorbers, determine the intensity I (counts) in a 60 s period for the following values of absorber thickness: 0.079, 0.159. 0.318, 0.635, and 0.953 cm. These choices assume the use of a standard set of absorbers 1/32, 1/16, 1/8, and 1/4 inch thick. Any known values of thickness that cover the stated range are satisfactory. Record in Data Table 1 the results of the count for each absorber and the thickness of the absorber.
4. Using polyethylene absorbers, determine the intensity I (counts) in a 60 s period for the following values of absorber thickness: 0.051, 0.076, 0.159, 0.318, 0.635, 0.953, and 1.27 cm. Again a standard set of absorbers is assumed, but any known values of thickness that cover the range are satisfactory. Record in Data Table 2 the count for each absorber and the thickness of the absorber.

Beta Absorption

1. Remove the ^{60}Co source and place a ^{90}Sr β-ray source in the third shelf position. With the empty absorber holder in the second shelf position, determine the number of counts in 60 s and record the results in Data Tables 3 and 4.
2. Using lead absorbers, determine the number of counts in 60 s for the following values of absorber thickness: 0.079, 0.159, 0.318, 0.635, and 0.953 cm.
3. Using the polyethylene absorbers, determine the number of counts in 60 s for the following values of thickness: 0.010, 0.020, 0.051, 0.076, 0.159, 0.318, and 0.635 cm.

CALCULATIONS

1. Although the ^{60}Co source has been referred to as a γ-ray source, it also has some beta activity. In the data taken with the lead absorber, the betas from ^{60}Co are completely absorbed by the first thickness of lead used. The rest of the data for count versus absorber thickness represents the absorption of γ-rays alone. Let the count for the 0.079 cm thick lead absorber represent the initial intensity I_o of γ-rays alone. Record that intensity in the Calculations Table as I_o. Calculate the increase in thickness that each absorber represents relative to the first absorber. Call this increase in thickness x_1 where $x_1 = x - 0.079$. For each of the absorbers beyond the first, record the values of I and x_1 in Calculations Table 1.
2. According to Equation 3, the quantity $\ln(I_o/I)$ should be proportional to x_1 for the assumption made in defining x_1 above. Perform a linear least squares fit with $\ln(I_o/I)$ as the vertical axis and x_1 as the horizontal axis. Record the slope as μ and record the correlation coefficient r in Calculations Table 1.

GRAPHS

1. On semilog graph paper, graph the intensity (counts) versus x for the ^{60}Co data from Data Tables 1 and 2. Use the linear scale for x. According to the theory, the part of the intensity due to γ-rays should be linear on this semilog plot. This means that the linear portion will occur only after the β-rays are absorbed. For the lead absorbers, this will occur after the first absorber. For the polyethylene absorbers, it will take several of the absorbers to remove the β-rays. A typical set of data for ^{60}Co using lead absorbers is shown in Figure 47-2. Note that the zero absorption is not on the straight line because of the β-ray contribution. The line is not a fit to the data, but simply a line drawn by hand.
2. On semilog graph paper, graph the intensity (counts) versus x for the ^{90}Sr data for betas on polyethylene from Data Table 4. Although it is only an approximation, these data may be somewhat linear.

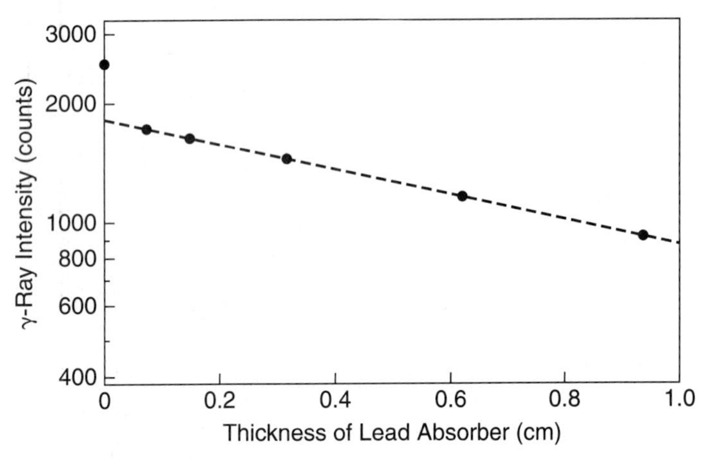

Figure 47-2 Typical data for ^{60}Co using lead absorbers.

Name Section Date

LABORATORY 47 *Absorption of Beta and Gamma Rays*

PRE-LABORATORY ASSIGNMENT

1. What are the names of the three types of natural radioactivity? Describe the nature of the particle produced in each.

2. Which kind of nuclear radiation undergoes a true exponential absorption as a function of absorber thickness? What property of its interaction with matter causes this to happen?

3. Which kind of natural radioactivity produces particles with a continuous spectrum of energy?

4. Which type of natural radioactivity produces particles that penetrate matter the least? What characteristics of these particles cause them to be stopped in less matter?

5. State the form of the equation for the absorption of γ-rays in matter. Define all the terms used in the equation.

6. What type of radiation is the most penetrating? What properties of the particles from this radiation cause them to penetrate matter so well?

7. A pure γ-ray source has a count rate of 5000 counts in one minute with no absorber between the source and the detector. An absorber of thickness 0.375 cm is placed between the source and the detector. The number of counts in the detector in 1 minute is now 3245. What is the absorption coefficient μ of the material? Show your work.

8. In the experimental arrangement of Question 7, if an additional 0.235 cm of the same material is placed between the detector and source, what is now the count in 1 minute? Show your work.

Name Section Date

Lab Partners

47 LABORATORY 47 *Absorption of Beta and Gamma Rays*

LABORATORY REPORT

Data Table 1 γ-rays on Lead

x (cm)	I (counts)
0	

Data Table 3 β-rays on Lead

x (cm)	I (counts)
0	

Data Table 2 γ-rays on Polyethylene

x (cm)	I (counts)
0	

Data Table 4 β-rays on Polyethylene

x (cm)	I (counts)
0	

Calculations Table 1

$I_o =$	counts		
x (cm)	I (counts)	x_1 (cm)	$ln(I_o/I)$
$\mu =$	(cm^{-1})	$r =$	

SAMPLE CALCULATIONS

1. $x_1 = x - 0.079 =$
2. $\ln(I_o/I) =$

QUESTIONS

1. For the semilog graph of the data of Table 1, at what thickness are all the betas absorbed? After the betas are absorbed, does the graph of absorption of gammas show only a linear behavior?

2. For the semilog graph of the data from Data Table 2, does the absorption of the betas take place over several absorbers? At what thickness are all of the betas absorbed? After the betas are absorbed, does the graph of absorption of gammas show only a linear behavior?

3. Comment on the data in Data Table 3 for the intensity of ^{90}Sr radiation versus the thickness of lead absorber. What is your conclusion about the absorption of betas in lead?

4. Comment on the semilog graph of the data from Data Table 4 for the intensity of ^{90}Sr radiation versus thickness of polyethylene absorber. Is the graph approximately linear? If it is not linear over the whole range, is it at least linear over some portion of the range?

5. For gammas of the approximate energy of the ^{60}Co gammas in lead, the approximate value of the absorption coefficient is $\mu = 0.65\ \text{cm}^{-1}$. Considering this as the accepted value, calculate the accuracy of your measurement of μ.

*Correlation Coefficients**

This table shows the probability of obtaining a given correlation coefficient r for two variables for which there is in fact no correlation. This probability is a strong function of the number of data points n. As an illustration of the table, consider the case of $n = 12$. There is a 10% probability of obtaining a value of $r \geq 0.497$, a 2% probability of obtaining a value of $r \geq 0.658$, and a 0.1% probability of obtaining a value of $r \geq 0.823$ for data for which no actual correlation exists. For many cases in the laboratory manual, you will take data that produce values of r greater than the 0.1% probability for the particular value of n. In those cases, one can conclude that the data are very strong evidence for a linear relationship between the variables.

Probability (%)

n	10	5	2	1	0.1
3	0.988	0.997	0.999	1.000	1.000
4	0.900	0.950	0.980	0.990	0.999
5	0.805	0.878	0.934	0.959	0.992
6	0.729	0.811	0.882	0.917	0.974
7	0.669	0.754	0.833	0.874	0.951
8	0.621	0.707	0.789	0.834	0.925
9	0.582	0.666	0.750	0.798	0.898
10	0.549	0.632	0.716	0.765	0.872
11	0.521	0.602	0.685	0.735	0.847
12	0.497	0.576	0.658	0.708	0.823
15	0.441	0.514	0.592	0.641	0.760
20	0.378	0.444	0.516	0.561	0.679

*This table is adapted from Table VI of Fisher and Yates, *Statistical Tables for Biological, Agricultural, and Medical Research*, published by Oliver & Boyd, Ltd., Edinburgh, by permission of the authors and publishers.

Properties of Materials

Table II A Density of Substances (kg/m³)

Substance	Density	Substance	Density
Aluminum	2.7×10^3	Cork	$0.22 - 0.26 \times 10^3$
Brass	8.4×10^3	Oak wood	$0.60 - 0.90 \times 10^3$
Copper	8.9×10^3	Maple wood	$0.62 - 0.75 \times 10^3$
Gold	19.3×10^3	Pine wood	$0.35 - 0.50 \times 10^3$
Iron	7.85×10^3	Alcohol, ethyl	0.79×10^3
Lead	11.3×10^3	Alcohol, methyl	0.81×10^3
Nickel	8.7×10^3	Mercury	13.6×10^3
Steel	7.8×10^3	Pure water	1.000×10^3
Zinc	7.1×10^3	Sea water	1.025×10^3

Table II B Specific Heats (Calories/gram – C°)

Substance	Specific Heat	Substance	Specific Heat
Aluminum	0.22	Mercury	0.033
Brass	0.092	Steel	0.12
Copper	0.093	Tin	0.054
Iron	0.11	Water	1.000
Lead	0.031	Zinc	0.093

Table II C Thermal Coefficients of Expansion $(C°)^{-1}$

Substance	α	Substance	α
Aluminum	24×10^{-6}	Brass and bronze	19×10^{-6}
Copper	17×10^{-6}	Lead	29×10^{-6}
Pyrex glass	3.2×10^{-6}	Ordinary glass	9×10^{-6}
Steel	11×10^{-6}	Concrete	12×10^{-6}
Gold	14×10^{-6}	Tin	27×10^{-6}

Table II D Resistivities and Temperature Coefficients

Substance	Resistivity(Ω–m)	Temperature Coefficient $(C°)^{-1}$
Aluminum	2.82×10^{-8}	3.9×10^{-3}
Copper	1.72×10^{-8}	3.9×10^{-3}
Silver	1.59×10^{-8}	3.8×10^{-3}
Gold	2.44×10^{-8}	3.4×10^{-3}
Nickel-silver	33×10^{-8}	0.4×10^{-3}
Tungsten	5.6×10^{-8}	4.5×10^{-3}
Iron	10×10^{-8}	5.0×10^{-3}
Lead	22×10^{-8}	3.9×10^{-3}
Carbon	3.5×10^{-5}	-0.5×10^{-3}

Some Physical Constants

Quantity	Symbol	Value[b]
Atomic mass unit	u	$1.660\ 538\ 73\ (13) \times 10^{-27}$ kg $931.494\ 013\ (37)$ MeV/c^2
Avogadro's number	N_A	$6.022\ 141\ 99\ (47) \times 10^{23}$ particles/mol
Bohr magneton	$\mu_B = \frac{eh}{2m_e}$	$9.274\ 008\ 99\ (37) \times 10^{-24}$ J/T
Bohr radius	$a_0 = \frac{h^2}{m_e e^2 k_e}$	$5.291\ 772\ 083\ (19) \times 10^{-11}$ m
Boltzmann's constant	$k_B = \frac{R}{N_A}$	$1.380\ 650\ 3\ (24) \times 10^{-23}$ J/K
Compton wavelength	$\lambda_C = \frac{h}{m_e c}$	$2.426\ 310\ 215\ (18) \times 10^{-12}$ m
Coulomb constant	$k_e = \frac{1}{4\pi\epsilon_0}$	$8.987\ 551\ 788\ldots \times 10^9$ N·m^2/C^2 (exact)
Deuteron mass	m_d	$3.343\ 583\ 09\ (26) \times 10^{-27}$ kg $2.013\ 553\ 212\ 71\ (35)$ u
Electron mass	m_e	$9.109\ 381\ 88\ (72) \times 10^{-31}$ kg $5.485\ 799\ 110\ (12) \times 10^{-4}$ u $0.510\ 998\ 902\ (21)$ MeV/c^2
Electron volt	eV	$1.602\ 176\ 462\ (63) \times 10^{-19}$ J
Elementary charge	e	$1.602\ 176\ 462\ (63) \times 10^{-19}$ C
Gas constant	R	$8.314\ 472\ (15)$ J/K·mol
Gravitational constant	G	$6.673\ (10) \times 10^{-11}$ N·m^2/kg^2

Quantity	Symbol	Value[b]
Josephson frequency–voltage ratio	$\frac{2e}{h}$	$4.835\,978\,98\,(19) \times 10^{14}$ Hz/V
Magnetic flux quantum	$\Phi_0 = \frac{h}{2e}$	$2.067\,833\,636\,(81) \times 10^{-15}\,\mathrm{T \cdot m^2}$
Neutron mass	m_n	$1.674\,927\,16\,(13) \times 10^{-27}$ kg $1.008\,664\,915\,78\,(55)$ u $939.565\,330\,(38)\,\mathrm{MeV}/c^2$
Nuclear magneton	$\mu_n = \frac{eh}{2m_p}$	$5.050\,783\,17\,(20) \times 10^{-27}$ J/T
Permeability of free space	μ_0	$4\pi \times 10^{-7}\,\mathrm{T \cdot m/A}$ (exact)
Permittivity of free space	$\epsilon_0 = \frac{1}{\mu_0 c^2}$	$8.854\,187\,817\ldots \times 10^{-12}\,\mathrm{C^2/N \cdot m^2}$ (exact)
Planck's constant	h $\hbar = \frac{h}{2\pi}$	$6.626\,068\,76\,(52) \times 10^{-34}\,\mathrm{J \cdot s}$ $1.054\,571\,596\,(82) \times 10^{-34}\,\mathrm{J \cdot s}$
Proton mass	m_p	$1.672\,621\,58\,(13) \times 10^{-27}$ kg $1.007\,276\,466\,88\,(13)$ u $938.271\,998\,(38)\,\mathrm{MeV}/c^2$
Rydberg constant	R_{H}	$1.097\,373\,156\,854\,9\,(83) \times 10^7\,\mathrm{m^{-1}}$
Speed of light in vaccum	c	$2.997\,924\,58 \times 10^8$ m/s (exact)

[a]These constant are the values recommended in 1998 by CODATA, based on a least-squares adjustment of data from different measurements. For a more complete list, see P. J. Mohr and B. N. Taylor, *Rev. Mod. Phys.* 72:351, 2000.

[b]The numbers in parentheses for the values above represent the uncertainties of the last two digits.

Solar System Data

Body	Mass (kg)	Mean Radius (m)	Period (s)	Distance from the Sun (m)
Mercury	3.18×10^{23}	2.43×10^{6}	7.60×10^{6}	5.79×10^{10}
Venus	4.88×10^{24}	6.06×10^{6}	1.94×10^{7}	1.08×10^{11}
Earth	5.98×10^{24}	6.37×10^{6}	3.156×10^{7}	1.496×10^{11}
Mars	6.42×10^{23}	3.37×10^{6}	5.94×10^{7}	2.28×10^{11}
Jupiter	1.90×10^{27}	6.99×10^{7}	3.74×10^{8}	7.78×10^{11}
Saturn	5.68×10^{26}	5.85×10^{7}	9.35×10^{8}	1.43×10^{12}
Uranus	8.68×10^{25}	2.33×10^{7}	2.64×10^{9}	2.87×10^{12}
Neptune	1.03×10^{26}	2.21×10^{7}	5.22×10^{9}	4.50×10^{12}
Pluto	$\approx 1.4 \times 10^{22}$	$\approx 1.5 \times 10^{6}$	7.82×10^{9}	5.91×10^{12}
Moon	7.36×10^{22}	1.74×10^{6}	—	—
Sun	1.991×10^{30}	6.96×10^{8}	—	—

Physical Data Often Used

Average Earth-Moon distance	3.84×10^{8} m
Average Earth-Sun distance	$1{,}496 \times 10^{11}$ m
Average radius of the Earth	6.37×10^{6} m
Density of air (20°C and 1 atm)	1.20 kt/m^3
Density of water (20°C and 1 atm)	1.00×10^{3} kg/m^3
Free-fall acceleration	9.80 m/s^2
Mass of the Earth	5.98×10^{24} kg
Mass of the Moon	7.36×10^{22} kg
Mass of the Sun	1.99×10^{30} kg
Standard atmospheric pressure	1.013×10^{5} Pa

Some Prefixes for Powers of Ten

Power	Prefix	Abbreviation	Power	Prefix	Abbreviation
10^{-24}	yocto	y	10^{1}	deka	da
10^{-21}	zepto	z	10^{2}	hecto	h
10^{-18}	atto	a	10^{3}	kilo	k
10^{-15}	femto	f	10^{6}	mega	M
10^{-12}	pico	p	10^{9}	giga	G
10^{-9}	nano	n	10^{12}	tera	T
10^{-6}	micro	μ	10^{15}	peta	P
10^{-3}	milli	m	10^{18}	exa	E
10^{-2}	centi	c	10^{21}	zetta	Z
10^{-1}	deci	d	10^{24}	yotta	Y

Standard Abbreviations and Symbols for Units

Symbol	Unit	Symbol	Unit
A	ampere	K	kelvin
u	atomic mass unit	kg	kilogram
atm	atmophere	kmol	kilomole
Btu	British thermal unit	L	liter
C	coulomb	lb	pound
°C	degree Celsius	ly	light-year
cal	calorie	m	meter
d	day	min	minute
eV	electron volt	mol	mole
°F	degree Fahrenheit	N	newton
F	farad	Pa	pascal
ft	foot	rad	radian
G	gauss	rev	revolution
g	gram	s	second
H	henry	T	tesla
h	hour	V	volt
hp	horsepower	W	watt
Hz	hertz	Wb	weber
in.	inch	yr	year
J	joule	Ω	ohm

Mathematical Symbols Used in the Text and Their Meaning

Symbol	Meaning
$=$	is equal to
$\equiv$	is defined as
$\neq$	is not equal to
$\propto$	is proportional to
$\sim$	is on the order of
$>$	is greater than
$<$	is less than
$\gg(\ll)$	is much greater (less) than
$\approx$	is approximately equal to
Δx	the change in x
$\sum_{i=1}^{N} x_i$	the sum of all quantities x_i from $x = 1$ to $i = N$
$\|x\|$	the magnitude of x (always a nonnegative quantity)
$\Delta x \to 0$	Δx approaches zero
$\frac{dx}{dt}$	the derivative of x with respect to t
$\frac{\partial x}{\partial t}$	the partial derivative of x with respect to t
$\int$	integral

Conversions

Length

1 in. = 2.54 cm (exact)
1 m = 39.37 in. = 3.281 ft
1 ft = 0.304 8 m
12 in. = 1 ft
3 ft = 1 yd
1 yd = 0.914 4 m
1 km = 0.621 mi
1 mi = 1.609 km
1 mi = 5 280 ft
1 μm = 10^{-6} m = 10^3 nm
1 ligh-tyear = 9.461×10^{15} m

Area

1 m^2 = $10^4 cm^2$ = 10.76 ft^2
1 ft^2 = 0.092 9 m^2 = 144 $in.^2$
1 $in.^2$ = 6.452 cm^2

Volume

1 m^3 = $10^6 cm^3$ = 6.102×10^4 $in.^3$
1 ft^3 = 1.728 $in.^3$ = 2.83×10^{-2} m^3
1 L = 1 000 cm^3 = 1.057 6 qt = 0.035 3 ft^3
1 ft^3 = 7.481 gal = 28.32 L = 2.832×10^{-2} m^3
1 gal = 3.786 L = 231 $in.^3$

Mass

100 kg = 1 t (metric ton)
1 slug = 14.59 kg
1 u = 1.66×10^{-27} kg = 931.5 MeV/c^2

Force

1 N = 0.224 8 lb
1 lb = 4.448 N

Velocity

1 mi/h = 1.47 ft/s = 0.447 m/s = 1.61 km/h
1 m/s = 100 cm/s = 3.281 ft/s
1 mi/min = 60 mi/h = 88 ft/s

Acceleration

1 m/s^2 = 3.28 ft/s^2 = 100 cm/s^2
1 ft/s^2 = 0.304 8 m/s^2 = 30.48 cm/s^2

Pressure

1 bar = 10^5 N/m^2 = 14.50 lb/$in.^2$
1 atm = 760 mm Hg = 76.0 cm Hg
1 atm = 14.7 lb/$in.^2$ = 1.013×10^5 N/m^2
1 Pa = 1 N/m^2 = 1.45×10^{-4} lb/$in.^2$

Time

1 yr = 365 days = 3.16×10^7s
1 day = 24 h = 1.44×10^3 min = 8.64×10^4 s

Energy

1 J = 0.738 ft · lb
1 cal = 4.186 J
1 Btu = 252 cal = 1.054×10^3 J
1 eV = 1.6×10^{-19} J
1 kWh = 3.60×10^6 J

Power

1 hp = 550 ft·lb/s = 0.746 kW
1 W = 1 J/s = 0.738 ft · lb/s
1 Btu/h = 0.293 W

Some Approximations Useful for Estimation Problems

1 m ≈ 1 yd
1 kg ≈ 2 lb
1 N ≈ $\frac{1}{4}$ lb
1 L ≈ $\frac{1}{4}$ gal
1 m/s ≈ 2 mi/h
1 yr ≈ $\pi \times 10^7$s
60 mi/h ≈ 100 ft/s
1 km ≈ $\frac{1}{2}$ mi

The Greek Alphabet

Alpha	A	α	Iota	I	ι	Rho	P	ρ
Beta	B	β	Kappa	K	κ	Sigma	Σ	σ
Gamma	Γ	γ	Lambda	Λ	λ	Tau	T	τ
Delta	Δ	δ	Mu	M	μ	Upsilon	Υ	υ
Epsilon	E	ϵ	Nu	N	ν	Phi	Φ	ϕ
Zeta	Z	ζ	Xi	Ξ	ξ	Chi	X	χ
Eta	H	η	Omicron	O	o	Psi	Ψ	ψ
Theta	Θ	θ	Pi	Π	π	Omega	Ω	ω